INTRODUCTION TO
PROFESSIONAL ENGINEERING
IN CANADA

Gordon C. Andrews

P.Eng., Professor Emeritus, University of Waterloo

J. Dwight Aplevich

P.Eng., Professor Emeritus, University of Waterloo ·

Roydon A. Fraser

P.Eng., Professor, University of Waterloo

Carolyn G. MacGregor

Associate Professor, University of Waterloo

PEARSON

Prentice
Hall

Toronto

Library and Archives Canada Cataloguing in Publication

Introduction to Professional Engineering in Canada / Gordon C. Andrews ... [et al.]. — 3rd ed.

Includes index.
ISBN 978-0-13-515360-4

1. Engineering—Vocational guidance—Canada.   I. Andrews, G. C. (Gordon Clifford), 1937-

TA157.I57 2009   620.0023   C2008-903410-4

ISBN-13: 978-0-13-515360-4
ISBN-10: 0-13-515360-3

Vice President, Editorial Director: Gary Bennett
Acquisitions Editor: Michelle Sartor
Marketing Manager: Michelle Bish
Developmental Editor: Mary Wong, Ben Zaporozan
Production Editor: Patricia Jones
Copy Editor: Laurel Sparrow
Production Coordinator: Sarah Lukaweski
Permissions Researcher: Amanda McCormick
Art Director: Julia Hall
Cover Designer: Anthony Leung
Cover Image: Getty Images

9 10 11 CP 15 14 13
Printed and bound in Canada.

The excerpts from Regulation 941 under the Professional Engineers Act in Sections 3.1.1 and 3.1.3 are copyright © Queen's Printer for Ontario, 1990. This is an unofficial version of Government of Ontario legal materials.

The terms 386 processor, 486 DX processor, Pentium, and Itanium in Figure 9.5 are trademarked terms by the Intel Corporation.

Acknowledgment for Figure 17.3 is given to the Air-Conditioning and Refrigeration Institute and the Association of Professional Engineers and Geoscientists of New Brunswick for the use of their registered trademark.

The iron ring and the words "iron ring" are protected by Canadian and U.S. trademarks and copyrights that belong exclusively to The Corporation of the Seven Wardens Inc./Société des Sept Gardiens Inc.

Preface

This book is intended to explain the elements of what every beginning engineering student should know about the engineering profession in Canada, emphasizing basic skills and knowledge that are well known to practising engineers and particularly useful to students. The book has evolved through several versions and has served engineering students for more than a decade and a half. The first edition was adopted by several Canadian institutions, and comments received have influenced later editions, which have been extensively revised to assist students as well as instructors. The four parts of the book are organized as follows:

Part I
The engineering profession, Chapters 1 to 5

Chapters 1, 2, and 3 emphasize that "engineering" is more than a simple set of university courses. It is an organized profession, with strict requirements for admission, a code of ethics, and professional regulations. Chapter 4 explains how engineering societies focus on activities outside the main mandate of the provincial regulatory organizations, and Chapter 5 gives advice for studying and writing examinations to help students, both in their immediate task of acquiring an engineering education and for lifelong learning.

Part II
Engineering communications, Chapters 6 to 9

In emphasizing the value of communicating effectively in print, Chapter 6 reviews technical document types, including some that are not typically discussed in English courses, and Chapter 7 reviews basic writing techniques. Chapter 8 contains a thorough discussion of the purpose and structure of engineering reports and the mechanics of writing them. Many students need this knowledge immediately because they may be required to write formal reports in their courses or during their first work-term or internship. Our philosophy has also been to attempt to show how to write a report that will stand up to challenge, such as might happen in a court of law. Chapter 9 describes elementary principles for the use and design of graphics and the essential rules for including graphics in engineering documents.

Part III
Engineering measurements, Chapters 10 to 14

Measurement is at the core of the applied physical sciences, and its correct interpretation is crucial to many engineering decisions. Part III introduces techniques for interpreting, manipulating, and presenting measured quantities. Chapter 10 contains a careful treatment of unit systems, with emphasis on SI units and rules for including them in written documents. Systematic and random measurement errors are treated in Chapter 11, together with correct use of significant digits. Chapter 12 introduces methods for estimating propagated error and provides motivation for the elementary application of statistics and probability to measurements in Chapters 13 and 14.

Part IV
Engineering practice,
Chapters 15 to 21

This part introduces several distinctive topics that make engineering a profession of practice, rather than of purely academic study. Chapter 15 is a basic introduction to the design process, emphasizing the iterative nature of design. Chapter 16 is a brief introduction to résumés, interviews, and the relationship of engineers to business. The rapidly changing domain of intellectual property is described in Chapter 17 in order to show how the results of the creative process can be protected and exploited. Chapter 18 briefly describes project planning and illustrates several common planning techniques using examples. Chapter 19 introduces safety in engineering design and practice and lists guidelines for eliminating workplace and other hazards. Chapter 20 introduces the complex problem of calculating and managing risks associated with large systems. Chapter 21 describes environmental issues and sustainability, with an emphasis on global warming, climate change, and the responsibilities of engineers.

How to use this book

The authors and others have used this book in several introductory engineering courses, supplemented with videos, exercises, assignments, and an introduction to software tools to suit the engineering discipline and course goals. The assignments usually include a formal engineering report on a topic requiring basic analysis or design. Conceptual knowledge is tested by several short examinations during the term. Portions of the material—particularly the report-writing and design chapters—are also used in higher-level courses.

The Quick Quiz questions at the end of each chapter are intended for students to answer in a few minutes in order to solidify their knowledge of each chapter immediately after reading it.

The order in which material from this book is presented can be varied considerably. We advise introducing the professional matters in Part I early, together with an introduction to Part II if student assignments include writing engineering reports. The chapters in Part III on measurements and errors can then be covered, ending the course with chapters on design, intellectual property, or safety from Part IV. However, a different order may suit other courses.

This book is intended for beginning engineering students. One of the authors (Gordon Andrews) has authored a more advanced textbook, *Canadian Professional Engineering & Geoscience: Practice & Ethics,* which is intended for senior courses, practising engineers, and graduates preparing to write provincial professional engineering practice exams (see reference [3] in Chapter 2).

Text enrichment site

To supplement the book, we have prepared a text enrichment site (www.pearsoned.ca/text/andrews) for instructors and students. Here you will find such items as links to professional organizations and electronic files of all the figures in the book.

Acknowledgments

The authors would like to thank all those who assisted, directly or indirectly, in this and previous versions of this book. We gratefully acknowledge the students, teaching assistants, and many colleagues who provided suggestions, comments, and criticism. Surin Kalra, Bob Pearce, and the late Alan Hale assisted immensely with early material and are gratefully remembered. Comments and suggestions from Ewart Brundrett, Don Fraser, Rob Gorbet, June Lowe, Roy Pick, John Shortreed, Carl Thompson, Neil Thomson,

Barry Wills, and others were very much appreciated. The director of first-year studies at the University of Waterloo created a positive environment for the development of this book, and we would like to thank Jim Ford, Gord Stubley, Bill Lennox, and Ajoy Opal, who held this position in recent years.

We are grateful to the following instructors who provided formal reviews for this edition: Michael Sjoerdsma (Simon Fraser University), Judith Dimitriu (Ryerson University), and Edward Kinley (University of New Brunswick).

We are indebted to people who helped us to get permission to use copyrighted material and to the copyright holders. In particular, we would like to thank Connie Mucklestone and Johnny Zuccone of PEO, Melissa Mertz of APEGNB, Lynne Vanin of MD Robotics, Paul Deslauriers of Waterloo FSAE, Nick Farinaccio of Chipworks, Richard Van Vleck of American Artifacts, and Mark Menzer of ARI.

We thank Connie Mucklestone of PEO for her thorough reading and detailed comments on Part I of the first edition of the book.

We also thank Anne Sharkey and co-workers at the Standards Council of Canada, who provided useful comments about our discussion of engineering standards.

Dennis Burningham, MSc, chemical engineer and UK industry consultant, provided his personal insight into sustainability, and many of his notes and comments are woven into Chapter 21. The authors would like to acknowledge this assistance and express thanks.

We owe thanks to the people at Pearson Education Canada who helped us to meet our publishing schedule: Mary Wong, Michelle Sartor, Laurel Sparrow, Sarah Lukaweski, Patricia Jones, and those who worked on earlier editions. Paul McInnis deserves special thanks for his perspicacious suggestions about the first edition.

Every effort has been made to avoid errors and to credit sources. The authors would be grateful for advice concerning any errors or omissions. In particular, we would like to know, from students, what you particularly liked or disliked in this book and, from instructors, what changes you would like to see in later editions.

Gordon C. Andrews
J. Dwight Aplevich
Roydon A. Fraser
Carolyn G. MacGregor

Contents

Part I	The Engineering Profession	1

Chapter 1	An Introduction to Engineering	3

Chapter 2	The Licensed Professional Engineer	19

Chapter 16	**The Engineer in Business**	245

Chapter 17	**Intellectual Property**	257

List of figures

List of tables

Part I

The Engineering Profession

Figure I.1   Design of the Canadarm and Canadarm2 required expertise in several engineering disciplines: materials, structures, robotics, electronics, control systems, computers, and software. The Canadarm is listed among the five most significant Canadian engineering achievements of the 20th century. This is an artist's rendering of the mobile servicing system, which includes the Canadarm2 and other elements. (Courtesy of MD Robotics)

By choosing engineering, you have taken the first step toward a challenging and rewarding profession. In future years you will share the great sense of personal achievement that is typical of engineering, as your ideas move from the design office or computer lab to the production line or construction site. This book is intended to acquaint future engineers with many basic engineering concepts. However, in passing along this basic information, we hope that the excitement and creativity of engineering also show through.

Part I introduces the reader to the engineering profession.

Chapter 1 **An introduction to engineering:** Most people know that engineers wear iron rings, but how is engineering defined, exactly? This chapter gives you a working definition of engineering and explains the difference between engineers and other technical specialists, such as research scientists, technicians, and technologists.

Chapter 2 **The licensed professional engineer:** An engineering education is not just a set of related university courses; it is the entry point to a legally recognized profession with strict requirements for admission, a code of ethics, and professional regulations. This chapter tells you how the engineering profession is organized and how you get into it.

Chapter 3 **Engineering ethics:** The public expects all professional people to be honest, reliable, and ethical. How does the public expectation apply to engineering? This chapter includes the code of ethics, how it applies to typical engineering practice, the proper use of the engineer's seal, and the significance of the iron ring.

Chapter 4 **Engineering societies:** Engineering societies help you by publishing technical papers, organizing conferences and engineering contests, and presenting short courses. Such assistance is very useful to engineering undergraduates, but it is even more important after you graduate. This chapter explains the role of engineering societies and describes several societies that may be of interest to you.

Chapter 5 **Advice on studying and exams:** You are investing a lot of time and money in your engineering education. Would you like to protect this investment? If you can master the fundamental advice in this chapter, you should be able to guarantee academic success and still have enough free time to enjoy many other interesting aspects of university life.

Chapter 1

An Introduction to Engineering

An engineering degree opens the door to a highly respected and highly structured profession, so graduation is a major achievement. Engineering students look forward to the happy day when they receive their engineering degrees, as shown in Figure 1.1, and their iron rings, as described in Section 3.7. Chapter 1 of this book introduces you to the engineering profession by exploring the following topics:

- a definition of engineering,
- the role of engineers in relation to other technical specialists,
- characteristics of some important engineering disciplines,
- engineering in Canada, including Canadian engineering accomplishments and the distribution of engineers by province,
- the talents and skills needed to become an engineer,
- some of the engineering challenges faced by society.

Figure 1.1 Two engineering graduates, diploma in hand and Ritual of the Calling of an Engineer in memory (as described in Section 3.7), celebrate their achievements. Graduation is the first important step to becoming a professional engineer.

The Further Study questions at the end of this chapter (Section 1.7) may help you to master the main ideas in the chapter, and also to confirm your decision to choose engineering as a career.

1.1 What is an engineer?

The term *engineer* comes to English from the Latin word *ingenium*, meaning talent, genius, cleverness, or native ability. Its first use was to describe those who had an ability to invent and operate weapons of war. Later, the word came to be associated with the design and construction of works, such as ships, roads, canals, and bridges, and the people skilled in these fields were non-military, or *civil* engineers. The meaning of the term *engineering* depends, to some degree, on the country. In England, people with practical skills have been called engineers since the time of the Industrial Revolution. In North America, more emphasis was placed on formal training, as a result of a recognized need for trained engineers during the wars of the 1700s and 1800s, and the modelling of early American engineering programs on French engineering schools [1].

In Canada, the title *Professional Engineer* is restricted by law for use by persons who have demonstrated their competence and have been licensed by a provincial licensing body, referred to in this book as a *provincial Association*. Exceptions are permitted for stationary and military engineers, who are subject to other regulations. The legal definition of engineering and the licensing of engineers are discussed in Chapter 2. Licensing is important, because the designations *engineer, engineering, professional engineer, P.Eng., consulting engineer* and their French equivalents are official marks held by Engineers Canada, and only licensed engineers may use these titles. Engineers Canada, known until 2007 as the Canadian Council of Professional Engineers (CCPE), acts on behalf of the provincial Associations of Professional Engineers and the Ordre des ingénieurs du Québec.

Although early civilizations produced significant engineering achievements, tools and techniques evolved especially rapidly in the 18th and 19th centuries during the Industrial Revolution. Advances in mathematics and science in this period permitted the prediction of strength, motion, flow, power, and other quantities with increasing accuracy. Recently, the development of computers, inexpensive electronic communication, and the Internet have placed huge amounts of information within easy reach, and the role of the engineer is rapidly evolving, together with the rest of modern society. Therefore, a modern definition of engineering must be broad enough to allow for change. The following definition is adequate for most informal discussions, although a more specific definition, as well as a legal definition, is given in the next chapter:

> An *engineer* is a person who uses science, mathematics, experience, and judgment to create, operate, manage, control, or maintain devices, mechanisms, processes, structures, or complex systems, and who does this in a rational and economic way subject to human, societal, and environmental constraints.

This definition emphasizes the rational nature and technological base of engineering, but it does not fully express the human context of the profession. An engineering career involves problem solving, designing, and building, which can give great pleasure to the engineer and others. There are many opportunities for friendship with team members, and engineering projects exercise communication, management, and leadership skills. Engineering decisions may involve societal or ethical questions. Personal experience and judgment are needed for many decisions, because complex projects sometimes affect society, and social effects are not always reducible to scientific principles or mathematical theorems.

Engineer and writer Henry Petroski emphasized the human side of engineering in *To Engineer is Human* [2]. He contends that in reaction to constantly changing requirements imposed by clients and society, it is human nature to extend design methods to their limits, where unpredicted failures sometimes occur, and to create new design methods as a result of those failures. Samuel Florman, also an engineer and prolific writer, suggests in *The Existential Pleasures of Engineering* [3] that since humankind first began to use tools, the impulse to change the world around us has been part of our nature. Thus, he argues, to be human is to engineer.

1.2   The role of the engineer

Engineering is usually a team activity. Because of the great complexity of many projects, engineering teams often include persons with widely different abilities, interests, and education, who cooperate by contributing their particular expertise to advance the project. Although engineers are only one component of this diverse group, they contribute a vital link between theory and practical application. A typical technical team might include scientists, engineers, technologists, technicians, and skilled workers, whose activities and skills may appear at first glance to overlap. The following paragraphs describe the tasks performed by different members of a typical engineering team. These are broad categorizations, so exceptions are common.

Research scientist   The typical research scientist works in a laboratory, on problems that expand the frontiers of knowledge, but which may not have practical applications for many years. A doctorate is usually the basic educational requirement, although a master's degree is sometimes acceptable. The research scientist typically supervises research assistants and will usually be a member of several learned societies in his or her particular field of interest, but will not usually be a member of a self-regulating profession. The *raison d'être* of pure science is the understanding of natural phenomena, and in a project team, the main responsibility of the scientist is to provide scientific analysis. Although many scientists apply their knowledge to practical tasks, science does not have the same legal structure or responsibilities as engineering.

Engineer   The engineer typically provides the key link between theory and practical applications. The engineer must have a combination of extensive theoretical knowledge, the ability to think creatively, the knack for obtaining practical results, and the ability to lead a team toward a common goal. The bachelor's degree is the basic educational requirement, although the master's degree or doctorate is useful and preferred by some employers. In Canada, all work that is legally defined as engineering must be performed or supervised by a licensed professional engineer, who is required, by provincial law, to be a member or licensee of the provincial Association of Professional Engineers (or in Québec, the Ordre des ingénieurs). Membership confers the right to use the title *Professional Engineer* (P.Eng. or in Québec, ing.).

Technologist   The technologist typically works under the direction of engineers in applying engineering principles and methods to complex technical problems. The basic educational requirement is graduation from a three-year technology program at a community college or equivalent, although occasionally a technologist may have a bachelor's degree, usually in science, mathematics, or related subjects. The technologist often supervises the work of others and is encouraged to have qualifications recognized by a technical society. In Ontario, for example, the Ontario Association of Certified Engineering Technicians and Technologists (OACETT) confers the title of *Certified Engineering Technologist* (C.E.T.). This is a voluntary organization, and the title is beneficial, but not legally essential, for working as a technologist. The fundamental technical difference between technologists and engineers is usually the greater theoretical depth of the engineering education and the greater hands-on experience implied by the technology diploma.

Technician The technician typically works under the supervision of an engineer or technologist in the practical aspects of engineering, such as making tests and maintaining equipment. The basic educational requirement is graduation from a two-year technician program at a community college or its equivalent. Associations such as OACETT may confer the title *Certified Engineering Technician* (C.Tech.) on qualified technicians, although the title is not essential to obtain work as a technician.

Skilled worker The skilled worker typically carries out the designs and plans of others. Such a person may have great expertise acquired through formal apprenticeship, years of experience, or both. Most trades (electrician, plumber, carpenter, welder, pattern maker, machinist, and others) have a trade organization and certification procedure.

Each of the above groups has a different task, and there are considerable differences in the skills, knowledge, and performance expected of each. In particular, a much higher level of accountability is expected from the professional engineer than from other members of the engineering team. The engineer is responsible for competent performance of the work that he or she supervises. In fact, engineers may be held legally accountable not only for their own acts but also for advice given to others. Judgment and experience are often as important as mathematics, science, and technical knowledge; liability insurance is becoming essential in the public practice of engineering. These professional aspects are discussed further in Chapter 2.

As you begin your engineering career, you should recognize that the categories described above are not rigid. Movement from one group to another is possible, but is not always easy.

1.3   Engineering disciplines

Most people can name a few branches of engineering: civil, electrical, mechanical, and chemical engineering, perhaps. However, the number of engineering disciplines is much larger than is commonly known. The Canadian Engineering Accreditation Board publishes an annual list of accredited Canadian engineering degree programs [4]. The list changes slightly from year to year as new programs are created or old ones dropped. Equating equivalent French and English names and then removing all duplicates reduces the full list of accredited programs to a smaller set of distinct specializations, illustrated in Figure 1.2. In 2008, there were 244 accredited programs and 60 specializations. The figure shows the names occurring more than once and the number of occurrences of each name on the list. The five most common branches make up more than half of the total number of accredited programs in Canada.

In the United States there are approximately eight times as many accredited engineering degree programs as in Canada, with twice as many program names, but the most common names are typically identical and in the same order as in Canada.

Choosing your program You must choose an engineering discipline that is right for you. Many students choose before they apply for university, but if you have not made a choice or have doubts, you should get further advice from your professors, guidance counsellors, or perhaps best of all, a friend who is a practising engineer. Libraries also have many references, such as

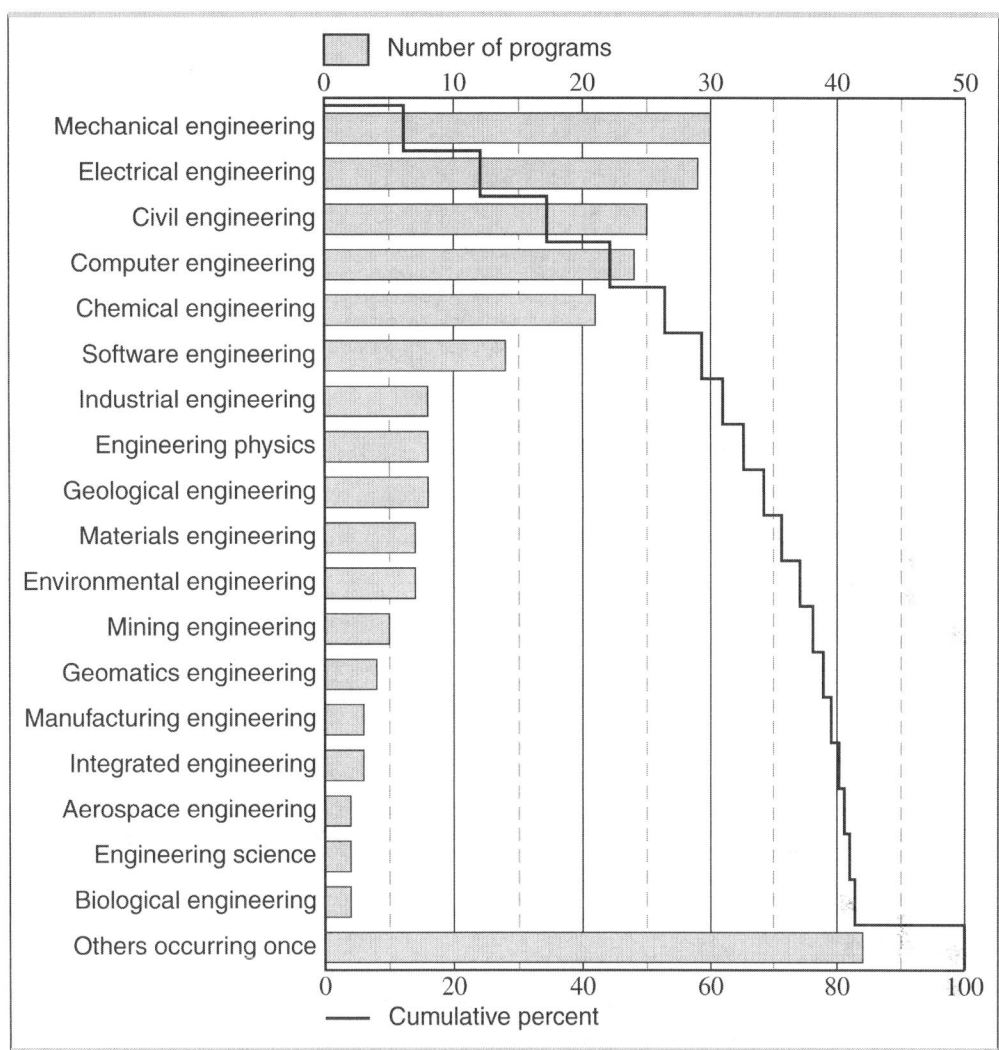

Figure 1.2   Names of accredited Canadian engineering programs in 2008. Of the 244 program names published in reference [4], those occurring more than once are listed on the left, with the bottom row showing that 42 names occur once. The most common program is mechanical engineering, which is found at 30 institutions. Although there are 60 distinctly named specializations, the cumulative percent line shows that the five most common disciplines make up more than 50 % of the total.

encyclopedias and university calendars showing the courses offered for each discipline. The Internet is also an important source of information and advice; see reference [5] for example.

Your choice of university program (that is, your engineering discipline) should be a conscious decision, because it will likely shape the rest of your life. It is wise to consider all relevant factors before you commit to a program. For example, typical factors to consider are the courses that really motivate or inspire you, the type of work you want to do, your preference for mathematical or applied work, your preference for the design office or field-work, or even where you want to work after graduation.

The first step is to gather information about engineering programs. Only then can you make an informed decision. To compare programs, an easy method is simply to write a list of the advantages and disadvantages for each program, based on the factors that are most important to you. You can then make a subjective, but informed, decision.

However, you might like to try the computational decision-making method in Section 15.5.1. To use this method, create a chart with a column for each program choice, and a row for each factor that is important in your decision. Assign weights to each factor, depending on its importance to you, then rank each program according to how well it satisfies each factor. The total benefit for each program is found by a simple calculation that is explained in detail in Section 15.5.1. The "best" program is one that receives the largest sum from the weights that you gave the factors. The computational method sometimes yields surprising results! A spreadsheet is useful, because you may want to recalculate with different factors and weights until you are confident that you have the right result. This repetitive process ensures that all relevant factors are systematically considered before a final decision is made.

To help you confirm your knowledge of engineering disciplines, the following paragraphs describe a few well-known specializations. The basic courses are similar in many disciplines, especially in the first year or two. Consequently, switching from one discipline to another is usually easiest then.

Chemical   Chemical engineers use knowledge of chemistry, physics, biology, and mathematics to design equipment and processes for the manufacture of chemicals and chemical products. The chemical process industries are skill- and capital-intensive, and require highly specialized designers to achieve competitive manufacturing. Some chemical engineers design equipment and processes for the extraction of minerals, petroleum, and other natural resources. Others design new processes for sustaining and reclaiming the natural environment.

Civil   Civil engineers design and supervise the construction of roads, highways, bridges, dams, airports, railways, harbours, buildings, water supply systems, and sewage systems. Many civil engineers specialize in one phase of their discipline, such as highway, sanitary, soil, structural, transportation, or hydraulic engineering. The largest number of licensed engineers in private practice is in civil engineering.

Computer   Computer engineers use a combination of electrical engineering and computer science subjects to design, develop, and apply computer systems. They must understand electronic circuits, computer hardware, computer interfacing and interconnection, software development, and algorithms. Most importantly, they must be able to select the optimal combination of hardware and software components required to satisfy specified performance criteria.

Electrical   Electrical engineers design and supervise the construction of systems to generate, transmit, control, and use electrical energy. Specialists in the power field typically design and develop heavy equipment (such as generators and motors), and transmission lines and distribution systems, including the complex equipment needed to control these powerful devices. Specialists in electronics and communications design and develop devices and systems for transmitting data, solid-state switching, microwave relays, computer logic circuits, and computer hardware.

Environmental   Environmental engineers respond to needs for improved air and water quality and efficient waste management. Environmental engineering is a rapidly growing discipline, concerned with site assessment and approval, air quality in buildings, monitoring and control devices, and a host of investigative, instrumentation, and other support activities. These environmental applications require a broad engineering knowledge, including topics from chemistry, physics, soils engineering, mechanical design, fluid mechanics, and meteorology.

Geological   Geological engineers use knowledge of the origin and behaviour of geological materials to design structures, such as foundations, roads, or tunnels built on or through these materials, and to develop exploration or extraction methods for petroleum and minerals. They may also work with civil engineers in the geotechnical design or construction of roads, airports, harbours, waste disposal systems, and other civil works. Geological engineers are also in increasing demand for ground-water and environmental impact studies and for new petroleum recovery projects such as tar-sands oil recovery.

Industrial   Industrial engineers use probability, statistics, and other mathematical subjects for the design of efficient manufacturing processes. They assume responsibility for quality control, plant design, and the allocation of material, financial, and human resources for efficient production. Automation, materials handling, environmental protection, robotics, human factors, and data processing are some of the specialized subjects required.

Mechanical   Mechanical engineers enjoy a wide scope of activity, including the design, development, manufacture, sale, and maintenance of machinery, ranging from appliances to aircraft. They may be involved with engines, turbines, boilers, pressure vessels, heat exchangers, or machine tools. They may specialize in fields such as machine design, heating, ventilating and refrigeration, thermal and nuclear power generation, manufacturing, quality control, or production scheduling. There are many mechanical engineers employed in related fields, such as mining, metallurgy, transportation, oil refining, and chemical processing.

Software   Software engineering is a new discipline that combines classical engineering project-management skills with the specialized tools and knowledge that are required to design, build, analyze, and maintain complex computer software. In addition to intimate knowledge of the program-design aspects of computer science, the software engineer requires a sufficiently broad understanding of the natural sciences in order to work with specialists in other disciplines and to design correct software to be embedded in machines.

Other programs   Some engineering programs have a relatively broad set of requirements, with the opportunity to choose specialized options or projects in later years. Systems design engineering, engineering science, and génie unifié are examples of such programs. Graduates may work in interdisciplinary areas requiring computer expertise and broad problem-solving and group management abilities. Some engineering disciplines are industry-related, for example, aerospace, agricultural, biosystems, building, forest, mining, and petroleum engineering. Finally, some engineering disciplines are directly related to specific scientific knowledge: engineering physics, management engineering, mathematics and engineering, materials engineering, mechatronics engineering, and nanotechnology engineering are examples.

Other careers   An engineering degree can be excellent preparation for a career in another discipline, such as law, medicine, or business, to name only a few possibilities. If this is your intention, emphasize breadth in your choice of elective courses, since your breadth of knowledge combined with your problem-solving skills will be key future assets.

1.4   Engineers across Canada

In 2006, the annual membership survey conducted by Engineers Canada showed that 155 485 licences were held by professional engineers practising in Canada, including holders of temporary engineering licences, but not including geologists, geophysicists, or geoscientists. The distribution of engineering licences, by province and territory, is shown in Table 1.1.

Most engineers are found in the industrialized areas of Ontario and Québec, with the next largest numbers in the resource-rich provinces of Alberta and British Columbia. Comparing the coasts, the West Coast (British Columbia) has many more engineers than the four Atlantic Provinces combined (Nova Scotia, New Brunswick, Prince Edward Island, and Newfoundland and Labrador)—in fact, about half again as many. Yukon and Prince Edward Island have the fewest of Canada's engineers. Nunavut is included in the total for the Northwest Territories.

Table 1.1   Distribution of the 155 485 engineering licences in Canada by province and territory in 2006 (data from reference [6]). Many engineers (about 12 %) are licensed in more than one province or territory.

Province or Territory	Association Acronym	Practising Engineers	Percentage of Total
Ontario	PEO	60 987	39.2
Québec	OIQ	35 368	22.7
Alberta	APEGGA	27 422	17.6
British Columbia	APEGBC	14 137	9.1
Saskatchewan	APEGS	3 559	2.3
Manitoba	APEGM	3 556	2.3
Nova Scotia	APENS	3 548	2.3
New Brunswick	APEGNB	3 391	2.2
Newfoundland and Labrador	PEG-NL	1 814	1.2
Northwest Territories and Nunavut	NAPEGG	894	0.6
Prince Edward Island	APEPEI	413	0.3
Yukon	APEY	396	0.3

| 1.5 | **Canadian engineering achievements** |

The history of Canadian engineering is full of personalities, struggles, and achievements—too full, in fact, to describe in detail in these few pages. This section lists some important accomplishments of Canadian engineers and concludes with sources of further information on this fascinating subject.

The centennial of engineering as an organized profession in Canada was celebrated in 1987. As part of the celebration, the Engineering Institute of Canada (EIC), Engineers Canada, and the Association of Consulting Engineers of Canada (ACEC) assembled a jury to identify the top engineering achievements of the previous century. The choices were based, among other criteria, on originality, ingenuity, creativity, the importance of the contribution to engineering, and whether the accomplishment initiated significant social and economic change. The jury received many nominations and reviewed a final list of 110 engineering projects, which was then reduced to the 10 most significant achievements.

Another formal review was held in 1999 to identify the five most significant Canadian engineering achievements of the 20th century. The review was sponsored by the organizers of the 1999 National Engineering Week and was held at the National Museum of Science and Technology in Ottawa. The selection criteria included special ingenuity, Canadian content, scope, and diversity.

The centennial list is given below, followed by the five achievements selected from the 20th century and some other well-known Canadian engineering accomplishments.

Transcontinental railway
The Canadian Pacific Railway linked Canada from coast to coast in 1885, a massive project for its time.

St. Lawrence Seaway
The seaway is a series of canals and waterways that opened the Great Lakes to ocean-going ships in 1959.

Synthetic rubber plant
The synthetic rubber plant of Polymer/Polysar (later Polysar, now Bayer Rubber) at Sarnia, Ontario, was built during World War II and was of great importance to the war effort.

Athabasca oil sands
The commercial oil sands development in Northern Alberta showed the feasibility of recovering oil from the Athabasca oil sands. Canada has more oil than Saudi Arabia, but the oil must be separated from the sand.

Very-high-voltage transmission
Hydro-Québec was the first electrical utility to develop transmission lines at a very high voltage (735 kV) for long-distance power transmission.

Nuclear power
The CANDU nuclear power system is an outstanding Canadian design that produces electric power using natural uranium fuel and heavy-water cooling, avoiding the need for expensive fuel enrichment.

Beaver aircraft
The De Havilland DHC 2 Beaver aircraft was designed and built in 1947 to open and explore Canada's North. The Beaver is a robust all-metal aircraft with excellent short takeoff and landing ability with heavy loads.

Alouette satellite With the Alouette I, Canada became the third nation with a satellite in orbit, after the Soviet Union and the United States. Alouette I was launched in 1962 for the purpose of studying the atmosphere. The satellite operated for 10 years and was one of the most successful satellites ever launched.

Snowmobile Joseph-Armand Bombardier invented the snowmobile in 1937. The revolutionary idea provided essential winter travel for snowbound Canadians, and developed into a huge new manufacturing industry.

Trans-Canada telephone network Completed in 1958 as the world's longest microwave network, a Canada-wide chain of microwave towers enabled the reliable transmission of rapidly increasing telecommunication traffic. The Anik satellite, launched in 1972, other geo-stationary satellites, and fibre-optic transmission have enhanced the network.

CPR Rogers Pass project The Rogers Pass rail link through the Selkirk range of the Rocky Mountains was notorious for heavy snow and steep grades for trains to negotiate. The $500 million, 34 km Rogers Pass project, completed in 1989, included 17 km of surface route, six bridges totalling 1.7 km, and two tunnels. The 14.7 km Mount Macdonald Tunnel, the pride of the project, is the longest railway tunnel in the western hemisphere.

Confederation Bridge The 12.9 km Confederation Bridge, linking New Brunswick and Prince Edward Island across the Northumberland Strait, is the longest bridge in the world crossing salt water and subject to winter ice hazards.

Canadarm The Canadarm remote manipulator system, one of the main Canadian contributions to the U.S. space shuttle program, was designed by Canadian engineers and built in Canada. The Canadarm deploys and retrieves satellites and other cargo from the space shuttle cargo bay. A more recent enhanced version, Canadarm2 (shown on page 1) is Canada's contribution to the International Space Station.

IMAX William Shaw, one of the early members of the IMAX Corporation, was the engineer responsible for developing an Australian patent (see Chapter 17) into the "rolling loop" film-transport mechanism in the IMAX projector, which produces stable high-resolution images on an immense screen.

Pacemaker John A. Hopps, working in 1949 with medical colleagues at the Banting and Best Institute in Toronto, discovered a method for restarting a heart that had stopped beating. This led Hopps to develop the first pacemaker at Canada's National Research Council in 1950. A poll of the Canadian public identified the Hopps pacemaker as the achievement that "made them most proud to be Canadian."

Rideau Canal The Rideau Canal is the oldest continuously operated canal system in North America. It was constructed after the War of 1812 to provide a navigable route, well away from the U.S. border, from Lake Ontario to the Ottawa River. Under the direction of military engineer Colonel John By, the canal was constructed entirely by hand, employing thousands of labourers to build the 45 locks along its 200 km length.

Avro Arrow The Avro Arrow was a Canadian-designed all-weather fighter–interceptor that first flew in March 1958. Its flight performance was decades ahead of other aircraft of the time, including Mach 2 speed at 15 000 m (50 000 ft) in normal flight. Prime Minister Diefenbaker abruptly cancelled development in September 1958, and many engineers emigrated to the United States to work in the American space program.

CN Tower The CN Tower was built in 1976 as the world's tallest free-standing structure, rising to a height of 553 m. A telecommunications centre, the tower is also a popular tourist attraction. In 1995, the American Society of Civil Engineers (ASCE) listed the CN Tower as one of the seven wonders of the modern world.

Toronto domed stadium The Toronto domed stadium features a retracting roof, supported by steel arches. It was opened in 1989 as Skydome, but has since been renamed.

Winnipeg floodway In 2008, the International Association of Macro Engineering Societies added the Winnipeg floodway to its list of major engineering achievements. The floodway diverts floodwaters around the city of Winnipeg, and its construction between 1963 and 1968 required moving more earth than was moved for the construction of the Suez canal.

1.5.1   Sources related to Canadian engineering history

Many excellent books have been written on Canadian engineering history; we particularly recommend references [7–10]. The following organizations all maintain web sites, and most produce printed material on Canadian engineering history.

The Engineering Institute of Canada (EIC), the Canadian Academy of Engineering, and the Canadian Society for Civil Engineering (CSCE) are discussed in Chapter 4. The EIC lists over 70 sources for Canadian engineering history on its web site and has published a series of historical and biographical working papers on engineering topics. The Academy web site lists historical publications of interest to Canadian engineers, and the CSCE web site contains links to civil engineering history and heritage.

A search in *The Canadian Encyclopedia* [11] will lead to an explanation and discussion of almost any topic from Canadian engineering history.

The archives of the Canadian Broadcasting Corporation contain historical reports under the headings "science & technology" and "disasters & tragedies."

The University of British Columbia and other institutions host web sites that describe well-known engineering failures and disasters, most of which are not Canadian.

1.6   Challenges and opportunities for engineering

You will encounter many personal challenges in your education and in your professional life, but you are also a member of a global society, living on a fragile planet. Crises aggravated by rising world population and by increasingly rapid movement of people, goods, money, and disease may affect you during your working life. However, climate change caused by global warming is likely the most important challenge to society that you will face during your engineering career. The problem is illustrated humorously in Figure 1.3, but the reality is more serious. Canada generates most of its electric power from fossil fuel plants, which emit greenhouse gases that cause global warming and climate change.

**Climate change—
A major engineering
challenge**

Any doubts about global warming and climate change were set to rest in 2007, when the authoritative Intergovernmental Panel on Climate Change (IPCC) issued its comprehensive reports [12]. As a prominent British engineer stated, after reviewing the first IPCC report: "There is now virtually no doubt that climate change is directly linked to mankind's profligate consumption of energy" [13]. The IPCC conclusions are discussed in greater detail in Chapter 21. The message to be drawn from the conclusions is that climate change in the next few decades will cause intense weather, floods, droughts, crop failures, drinking-water shortages, and mass migration that will seriously disrupt our lives unless we take immediate action to combat it.

Figure 1.3   This editorial cartoon, which depicts our dependence on electric power, appeared after the massive power failure in Ontario and the northeastern United States in August 2003. Sophisticated computers and basic life support alike depend on a heavily loaded power network. Electric power generation, which currently depends largely on fossil fuels, leads to global warming and climate change. Developing sustainable electrical power is a major challenge for engineers. (Reprinted with permission from *The Globe and Mail*.)

The challenge for engineers is to aggressively reduce our greenhouse gas emissions. This requires new, intelligent efforts to conserve energy and to avoid waste, such as the invention of innovative methods to reduce, reuse or recycle materials. Engineers must also think ahead, and adapt our current infrastructure for the inevitable effects of climate change. We must design our infrastructure more robustly, to withstand the intense windstorms, hurricanes, and floods that climate change will inflict in future decades. Engineers are, or should be, becoming better informed about the consequences of climate change—we must be part of the solution, not part of the problem. Climate change is a serious challenge, but several authors have suggested ways to fight it, adapt to it, or even profit from it [14, 15].

**The role of
engineers in society**

Engineers must take a greater role in political and social debate. Canadian engineers are rarely found in the upper levels of our government, and are seldom involved in policy-making, even when the issues have an engineering content. For example, global warming is not entirely a technical problem. Wasteful practices, accepted by society for centuries, must be changed. Fighting climate change will require some effort and sacrifice, and we must ensure that the effort is rewarded, and that the sacrifice is spread evenly. Constraints that are sufficient to contain global warming will not be easily accepted. Engineers must speak out in the political and ethical debate over climate change and similar issues, so that society's attitudes, laws, and way of life change, appropriately.

Engineering opportunities   More positively, Canada's engineering students live today in a world of unparalleled opportunity. The information revolution is still in its early stages, and it will continue to multiply our strength and intelligence, and help to create a world of diversity, freedom, and wealth that could not have been imagined by earlier generations.

The challenges to the engineer, and the opportunities, have never been greater.

1.7   Further study

1.   Choose the best answer for each of the following questions.

(a) In Canada, the title of *engineer* (implying that the holder is a professional engineer):

  i. is not restricted by law.

  ii. may be used only by people employed in technical jobs.

  iii. may be used only by graduates of accredited university engineering programs.

  iv. may be used only by people licensed under provincial or territorial laws.

(b) Engineering projects are usually a team activity, and the engineer typically

  i. provides the key link between the theory and the practical applications.

  ii. takes responsibility when engineering work affects public safety.

  iii. sets the "factor of safety" between the system capacity and the expected load.

  iv. All of the above are correct.

(c) Technologists and technicians typically work with engineers on projects. Which of the following statements is true about their certification process?

  i. Certification as a technician or technologist is voluntary.

  ii. Certification as a technician or technologist is compulsory.

  iii. Technicians and technologists do not have a certification process.

  iv. Technicians and technologists must be certified as engineers-in-training.

(d) How many distinct names of accredited engineering programs are found in Canadian universities?

  i. fewer than 30   ii. approximately 60   iii. more than 100   iv. more than 200

(e) Engineering decisions must be based exclusively on the objective use of logic without letting personal experience influence them.

  i. false   ii. true

(f) A professional engineer who designs hydroelectric power transmission lines and distribution systems, including the control systems for these devices, is most likely to have studied in which of the following engineering programs?

  i. civil   ii. electrical   iii. industrial   iv. geological   v. software
  vi. mechanical

(g) A professional engineer who improves the quality of rivers, lakes, and the atmosphere by monitoring air and water quality and by solid and liquid waste management and disposal is most likely to have studied in which of the following engineering programs?

   i. civil   ii. environmental   iii. industrial   iv. geological
   v. manufacturing   vi. mechanical

(h) A professional engineer who develops software for digital telephone switching systems and designs control systems for automated manufacturing production lines is most likely to have studied in which of the following engineering programs?

   i. manufacturing   ii. electrical   iii. industrial   iv. computer
   v. software   vi. environmental

(i) Of the many challenges of the future, the one described in Chapter 1 as probably the most important is

   i. the population explosion.   ii. environmental pollution.
   iii. the rapid spread of disease.   iv. climate change caused by global warming.
   v. congestion in crowded cities.

(j) According to the Engineers Canada membership survey conducted in 2006, which provinces or territories have the most and the fewest licensed engineers?

   i. Ontario has the most, and PEI and Yukon have the fewest.
   ii. Alberta has the most, and Yukon and Northwest Territories have the fewest.
   iii. Ontario has the most, and Yukon and Northwest Territories have the fewest.
   iv. Québec has the most, and PEI and Yukon have the fewest.

2.   Select one of the many Canadian engineering achievements listed in Section 1.5, and write a brief summary of its development and impact.

3.   If personal characteristics are important in selecting a career or, in particular, a branch of engineering, what interests, aptitudes, and experiences led you to select the particular branch of engineering in which you are currently enrolled? How do your characteristics match the characteristics that you think would be necessary in your chosen branch?

4.   "An engineer is someone who can do for ten shillings what any fool can do for a pound"—Nevil Shute in *Slide Rule: The Autobiography of an Engineer*. See if you can find an equally pithy definition that could be used to interest high-school students in the profession.

5.   What is the difference between a profession and a job? Your opinions may be influenced by Chapters 2 and 3, which describe the engineering profession.

6.   Should the professional person be more concerned about the welfare of the general public than the average person? Should persons in positions of great trust, whose actions could cause great harm to the general public, be required to obey a more strict code of ethics with respect to that trust than the average person? You may find some hints by glancing at Chapters 2 and 3.

7.   When you graduate, you will want to know how intense the competition is for career positions. The number of students enrolled in your discipline is a rough indicator of the future supply of job applicants. Search the Internet to find enrollment statistics for engineering disciplines in Canada.

8.   Question 7 considers the supply of engineering graduates in your year of graduation. The demand for graduates in each discipline is more difficult to predict, but is cyclical in some industries and subject to major trends, such as the computer and communications revolution, the biological gene-modification revolution, and the demographics of birth, death, and immigration. Make a list of at least five possible jobs you might wish to do, and consider how each of them might be affected by current economic trends. Your conclusions cannot be exact, but might be very useful as you seek work-term or internship experience before graduation.

1.8   References

[1] L. P. Grayson, *The Making of an Engineer*. New York: John Wiley & Sons, Inc., 1993.

[2] H. Petroski, *To Engineer is Human: The Role of Failure in Successful Design*. New York: St. Martin's Press, 1985.

[3] S. C. Florman, *The Existential Pleasures of Engineering*. New York: St. Martin's Press, 1976.

[4] Canadian Engineering Accreditation Board, *Accreditation Criteria and Procedures*. Ottawa: Engineers Canada, 2007. <http://www.engineerscanada.ca/e/files/report_ceab.pdf> (March 9, 2008). Pre-publication data at <http://www.engineerscanada.ca/e/acc_programs_2.cfm> (September 10, 2008).

[5] Industry Canada, *Professional Engineering in Canada: Building a Nation*. Industry Canada, 1998. <http://epe.lac-bac.gc.ca/100/205/301/ic/cdc/pec/index.html> (March 9, 2008).

[6] Engineers Canada, *2006 Membership Survey*. Ottawa: Engineers Canada, 2007. Advance data provided to G. Andrews by S. Colasante, June 26, 2007.

[7] N. Ball, *Mind, Heart, and Vision: Professional Engineering in Canada 1877 to 1987*. Ottawa: National Museum of Science & Technology, 1988.

[8] J. J. Brown, *Ideas in Exile: A History of Canadian Invention*. Toronto: McClelland and Stewart, 1967.

[9] J. J. Brown, *The Inventors: Great Ideas in Canadian Enterprise*. Toronto: McClelland and Stewart, 1967.

[10] J. R. Millard, *The Master Spirit of the Age: Canadian Engineers and the Politics of Professionalism, 1887–1922*. Toronto: University of Toronto Press, 1988.

[11] J. H. Marsh, ed., *The Canadian Encyclopedia*. Toronto: McClelland and Stewart, 1988. <http://www.thecanadianencyclopedia.com/> (March 9, 2008).

[12] S. Solomon, D. Qin, M. Manning, Z. Chen, M. Marquis, K. B. Averyt, M. Tignor, and H. L. Miller, eds., *Climate Change 2007: The Physical Science Basis. Contribution of Working Group I to the Fourth Assessment Report of the Intergovernmental Panel on Climate Change.* New York: Cambridge University Press, 2007. <http://www.ipcc.ch/ipccreports/ar4-wg1.htm> (March 9, 2008).

[13] I. Arbon, "Quoted on the BBC news," 2007. Energy, Environment & Sustainability Group, Institute of Mechanical Engineers, <http://news.bbc.co.uk/2/hi/science/nature/6324093.stm> (March 9, 2008).

[14] G. Monbiot, *Heat: How to Stop the Planet from Burning.* Toronto: Random House Anchor Canada, 2007.

[15] T. Homer-Dixon, *The Upside of Down.* New York: Knopf, 2006.

Chapter 2

The Licensed Professional Engineer

Your university engineering program is not just a set of related courses; it is the entry point into a challenging and respected profession with high admission standards, professional regulations, and a code of ethics. The history of the engineering profession in Canada is full of success stories. However, it also contains some tragic disasters, such as the collapse of the Québec Bridge (shown in Figure 2.1), which influenced the legal regulation of the profession. This chapter discusses the engineering profession in Canada, and how university graduates enter the profession. In particular, this chapter describes

- the characteristics of a profession,
- how the engineering profession is regulated by law, and
- the academic and experience requirements for entering the engineering profession.

Figure 2.1 The photo shows the wreckage of the Québec Bridge, which collapsed during construction on August 29, 1907, killing 75 workmen. The cause of the collapse was traced back to errors in the design, and the tragedy spurred discussion of the need for government regulation of the engineering profession. Provincial and territorial laws regulating engineering were passed during the decade following the tragedy.

2.1 Engineering is a profession

The following definition of a profession is attributed to S. C. Florman, an engineer and prolific writer:

> A profession is a self-selected, self-disciplined group of individuals who hold themselves out to the public as possessing a special skill derived from training and education and who are prepared to exercise that skill in the interests of others. [1]

Engineering satisfies this definition, as do the older organized professions, such as medicine and law. Engineers possess a high level of skill and knowledge, obtained from lengthy education and experience. Engineering is a creative vocation with a positive purpose. However, the typical working environment differs from that of other professionals. Most engineers are employees of companies and work in project teams; other professionals are often self-employed and work on a personal basis with clients. The difference is illustrated in the following quote, which emphasizes that the engineering profession is highly regarded, in spite of the employee status of most of its members:

> The hard fact of the matter is that people need physicians to save their lives, lawyers to save their property and ministers to save their souls. Individuals will probably never have an acute, personal need for an engineer. Thus, engineering as a profession will probably never receive the prestige of its sister professions. Although this may be an unhappy comparison, the engineer should take note that physicians and lawyers both feel that the prestige of their professions has never been lower, and they are mightily concerned; yet [...] engineers are considered to be sober, competent, dedicated, conservative practitioners, without such devastating problems as embezzlement or absconding members and without the constant references to malpractice and incompetence. [2]

2.2 Regulation of the engineering profession

Since all professions involve skill or knowledge that cannot be evaluated easily by the general public, governments usually impose some form of regulation or licensing on them. The purpose of regulation is to prevent unqualified persons from practising, to set standards of practice that protect the public, and to discipline unscrupulous practitioners.

The United States was the first country to regulate the modern practice of engineering. The state of Wyoming enacted a law in 1907 as a result of instances of gross incompetence during a major irrigation project [1]. In Canada, spurred by the Québec Bridge collapses of 1907 and 1916, provinces began to regulate engineering. For example, such a law was passed in New Brunswick in 1920. The Associations in several provinces now also regulate the profession of geoscience, beginning in Alberta in 1955. At present, all provinces and territories of Canada, and all of the United States, have licensing laws to regulate the engineering profession and the title of Professional Engineer.

There are three typical methods of government regulation: by direct control through government departments, by establishing independent agencies, and by permitting the professions themselves to be self-regulating bodies, which determine standards for admission, professional practice, and discipline of their members. In Canada, engineering is self-regulating, as are all of the major professions (such as medicine, law, and accounting). Each provincial and territorial government has passed a law that designates a licensing body, called an Association or (in Québec) an Ordre, that regulates the engineering profession.

In most of Canada, engineering and geoscience are viewed in law as branches of a single profession. Eight jurisdictions regulate engineers and geoscientists, together, in the same legislative acts (Alberta, British Columbia, Manitoba, New Brunswick, Newfoundland, Northwest Territories, Nunavut, and Saskatchewan). Three provinces regulate geoscientists as an independent self-governing profession, equivalent to the engineering profession (Ontario, Nova Scotia, and Québec). Prince Edward Island and Yukon regulate engineering, but do not regulate geoscience as a profession. In this book, discussion of the regulation of engineering typically includes the regulation of geoscience.

In the United States, agencies appointed by the state governments write the regulations and license engineers. In Britain, chartered institutes promote the practice of engineering for the public benefit, but the associated laws do not regulate the term *engineer*. The sign "engineer on duty" is found outside many garages.

| 2.2.1 | Case study of a critical event: The Québec Bridge tragedy |

Introduction   The Québec Bridge is the longest cantilevered span in the world, but the structure collapsed during construction, killing 75 workmen in the wreckage (shown in Figure 2.1). The story of the Québec Bridge is marked by grand ambition, regrettable negligence, and a tense drama in the final hours before the disastrous 1907 collapse. The redesigned bridge still stands today. The disaster is particularly relevant to engineers, because it stirred Canadians to demand regulation of the profession.

At the time, the 1907 tragedy was the worst design or construction accident that had ever occurred in North America, and remained so for over 70 years. (The collapse of a hotel walkway in Kansas City in 1981 killed 114 people and injured over 200.) The Québec Bridge collapse is discussed in detail in several recent texts [3, 4], but the inquiry reports and bridge description written shortly after the tragedy are still worth reading [5–7].

Description of the tragedy   The proponents of the Québec Bridge intended to build a cantilevered span of 550 m (1800 ft) between supports. This length was chosen mainly because of the location of the supporting bedrock. The span was a challenge because it exceeded the 521 m (1710 ft) spans on the Forth Bridge in Scotland, which were then the longest cantilevered spans in the world.

The Québec Bridge & Railway Company hired Theodore Cooper of New York as the consulting engineer responsible for designing the bridge and guaranteeing its strength.

Cooper was a senior engineer and had supervised many important projects. His assistant, Peter Szlapka, was a design engineer, educated in Germany, with 27 years of experience. Szlapka worked under Cooper's direction, drawing up the details of the bridge design. Cooper and Szlapka worked in New York; the beams, columns, and other bridge parts were fabricated at the Phoenix Company in Pennsylvania, and were shipped to the construction site in Québec, where hundreds of workmen were erecting the bridge, supervised by many site engineers and inspectors.

The construction scheme was to build each half of the bridge out from the opposite shores until the two halves met in mid-stream. The south side was started first, but as it reached out about 200 m, warning signs were observed—some of the compression members were bending. The site engineers notified Cooper by telegram that these members were deforming, and Cooper sought to visualize the problem and recommend a solution. The exchange of messages and attempts to find the reasons for the deformed members continued for about three weeks. By August 27, 1907, the situation was deemed "serious" by a senior site engineer, Norman McLure, who suspended erection of the bridge until Cooper could review the design and authorize work to continue. McLure then left for New York, where he convinced Cooper of the gravity of the problem. Cooper confirmed McLure's decision to suspend the construction and sent a telegram ordering "Add no more load to bridge till after due consideration of facts." Unfortunately, during McLure's absence and unknown to McLure or Cooper, chief site engineer Edward Hoare mistakenly ordered work to resume.

The telegraph message, which might have saved many lives, was relayed to the site at Québec. Testimonies disagree about whether or not the telegram was delivered, and if delivered, whether Hoare read it, but the workers were not recalled from the worksite. Ninety-two men were reportedly working on the structure when it collapsed at the end of the workday on August 29, 1907. A few reached safety in the midst of the falling girders and the ear-splitting fractures of beams and cables that reportedly were as loud as artillery explosions, but within less than a minute, 75 workmen were killed, crushed, or lay trapped and drowning in the cold water of the St. Lawrence River. The sound of the collapse was heard in Québec City, several kilometres away.

Fortunately, the falling wreckage missed the steamer Glenmont, which had passed under the structure only minutes before the collapse. The Glenmont attempted to pick up survivors, but none were found. The falling debris carried many trapped souls to death in the deep water.

The accident investigation A royal commission was immediately ordered to investigate the collapse and find its cause. The commission conducted a lengthy inquiry and wrote a very thorough summary of the design and the cause of the failure, and made many important recommendations, valuable to bridge designers around the world.

In assessing blame, the commission's most scathing criticism was levelled at design engineer Theodore Cooper and his assistant, Peter Szlapka. Although Cooper had personally sought and obtained ultimate design authority for the project, he visited the Québec site only when the supporting piers were being built and was never on-site during the actual bridge erection. Moreover, he visited the steel fabrication workshop in

Pennsylvania only three times during the project. In addition, the commission found the chief site engineer, Edward Hoare (who ordered work to resume on the fateful day), to be "not technically competent" to supervise the work. Communication problems were obviously also key factors in the tragedy, and better efforts should have been made to overcome the obstacles caused by the long distances between the design engineer, the fabricator, and the erection site.

In particular, the commission found serious errors and deficiencies in Cooper's design. Most importantly, Cooper's initial estimates of the dead-loads on the bridge were not recalculated as the design progressed. The design process normally starts with initial estimates of beam and column sizes, and the weights (dead-loads) of these estimated components are then used to calculate the stresses in the structure. If the stresses are too high, then the sizes are changed and the stresses are recalculated. This iterative process must be performed as many times as necessary, but Cooper and Szlapka failed to do so. Cooper's load and size estimates were low, leading to erroneously low dead-load estimates. The actual stresses were well above the limits for safe design. Moreover, Szlapka later criticized Cooper's curved compression chords, which buckled during the collapse, as being introduced for "artistic" reasons.

The new bridge design   Within a year after the tragedy, the federal government recognized that the Québec Bridge was an essential link in the railway routes that were being built across Canada, and authorized a new bridge construction project. The earlier plans were reviewed, together with the report of the royal commission, and a new design was adopted, using the same span distances and support locations. Two years were spent removing the massive twisted steel and debris from the 1907 collapse. New piers were built on the bedrock, and the new bridge was designed, manufactured, and erected by the St. Lawrence Bridge Company.

Unfortunately, a second tragedy occurred in 1916, during the final stages of construction, killing another 13 workmen. The new erection plan was to build both ends of the bridge part-way out from the shore, then to float the assembled centre span into the middle of the river and raise it into position. This plan worked smoothly in its initial stages but as the centre span was being lifted, it slid off its four corner supports into the river, carrying 13 men to their deaths and injuring 14 others. The cause of this failure was traced to material flaws. The centre span was rebuilt and successfully raised into place the following year. The Prince of Wales formally declared the bridge open in a ceremony on August 22, 1919.

Lessons learned   The 1907 Québec Bridge disaster made the public realize that only competent and ethical people should be permitted to practise engineering. In the decade after the opening of the Québec Bridge, the first provincial laws to license professional engineers were passed. A group of dedicated engineers instituted the Ritual of the Calling of an Engineer to encourage the ethical practice of engineering. Engineers who accept this oath are permitted to wear the iron ring discussed in Section 3.7. The iron rings used in the early ceremonies were allegedly made from steel scrap salvaged from the 1907 Québec Bridge collapse. Canadian professional geoscientists, following the engineering tradition, have established the Ritual of the Earth Science Ring. Some engineers in other countries, such as the United States, have copied the Canadian ritual, and hold a similar iron ring ceremony.

The scope and responsibilities of the engineering profession have expanded over the past century. Many current branches of engineering were unimaginable in 1907, and in most parts of Canada, engineering and geoscience are licensed as branches of the same profession. Provincial and territorial laws now permit only licensed professional engineers or professional geoscientists to make technical decisions "wherein the safeguarding of life, health, property or the public welfare is concerned."

2.2.2 The laws regulating engineering

To practise engineering in Canada, you must obtain a licence. For example, Figure 2.2 shows an engineering licence issued in Ontario, but licensing laws (or acts) exist in all provinces and territories. Each licensing law establishes an Association (an Ordre in Québec) to regulate the engineering profession and usually geoscience as well, as listed in Table 2.1. In this book, we use the term *Association* (with a capital letter, to distinguish its meaning from that of a generic association) to refer to any of these licensing bodies, including the Ordre.

Table 2.1 The provincial and territorial Associations that regulate the engineering profession in Canada, with their acronyms and web sites.

Association of Professional Engineers, Geologists and Geophysicists of Alberta	(APEGGA)	www.apegga.com
Association of Professional Engineers and Geoscientists of British Columbia	(APEGBC)	www.apeg.bc.ca
Association of Professional Engineers and Geoscientists of the Province of Manitoba	(APEGM)	www.apegm.mb.ca
Association of Professional Engineers and Geoscientists of New Brunswick	(APEGNB)	www.apegnb.com
Professional Engineers and Geoscientists of Newfoundland and Labrador	(PEG-NL)	www.pegnl.ca
Association of Professional Engineers, Geologists and Geophysicists of the Northwest Territories (representing NWT and Nunavut)	(NAPEGG)	www.napegg.nt.ca
Association of Professional Engineers of Nova Scotia	(APENS)	www.apens.ns.ca
Professional Engineers Ontario	(PEO)	www.peo.on.ca
Association of Professional Engineers of Prince Edward Island	(APEPEI)	www.apepei.com
Ordre des ingénieurs du Québec	(OIQ)	www.oiq.qc.ca
Association of Professional Engineers and Geoscientists of Saskatchewan	(APEGS)	www.apegs.sk.ca
Association of Professional Engineers of Yukon	(APEY)	www.apey.yk.ca

Figure 2.2   The engineering profession in Canada is regulated by provincial and territorial law. A typical engineering licence is shown above. (Courtesy of Professional Engineers Ontario)

The purpose of laws for licensing engineers is to protect the public, and the goal of the Associations established by the laws is to safeguard life, health, property and the public welfare, as discussed at length by Andrews in [3]. As mentioned above, most provinces regulate engineering and geoscience in the same act, but three provinces (Ontario, Québec, and Nova Scotia) regulate geoscientists separately from engineers, and Prince Edward Island and Yukon regulate engineers, but have not yet passed laws to regulate geoscientists.

Under the authority of the licensing acts, the Associations are empowered to monitor the standards of professional practice and to discipline practitioners where necessary. Each Association has independently developed detailed regulations, by-laws, and a code of ethics. In 1936, the Associations created an "umbrella" body called the Canadian Council of Professional Engineers (CCPE), which changed its name in 2007 to Engineers Canada. The role of Engineers Canada is to coordinate the engineering profession on a national scale by promoting consistency in licensing and regulation.

Consistent laws help engineers to move across Canada and to practise in different provinces and territories. Engineers Canada develops detailed policies, guidelines, and position statements at the national level, and the Associations are encouraged to review and adopt the documents, as appropriate. As a result, while each province and territory regulates the engineering profession independently, the laws across Canada are basically similar. The discussions of engineering law in this book therefore apply generally to all of Canada, although many examples are drawn from Ontario, which licenses almost half of Canada's engineers. All of the engineering and geoscience licensing laws are available on the Internet. They can usually be found on the Association or Engineers Canada web sites [8, 9].

The legislative act is the basic document that regulates the profession. In addition, regulations, by-laws, and codes of ethics have the force of law, since they are created under the authority of the act. For clarity, the difference between these documents is described in the following paragraphs.

Regulations   Although *regulations* require government approval, the Association usually prepares the regulations (not the legislature), under the authority of the act. Regulations provide more specific rules, details, or interpretations of clauses of the act. For example, the act typically states that academic and experience requirements for licensing will be set, but the regulations state these requirements in detail.

By-laws   *By-laws* are rules for running the Association itself. They concern the meetings of the council, financial statements, committees, and other internal matters. In some provinces, the by-laws are part of the regulations.

Code of ethics   A *code of ethics* is a set of rules of personal conduct to guide individual engineers. In some provinces, the code of ethics may be part of the regulations; in other provinces, it is a by-law. The code of ethics is discussed in more detail in Chapter 3.

In each province and territory, the engineering profession is self-regulating; that is, the engineers themselves regulate their profession by electing the majority of members to the Association council, which typically contains members appointed by the lieutenant-governor-in-council (the provincial government). Engineers also confirm, by ballot, the by-laws established by the council. For self-regulation to work effectively, engineers must be willing to participate in Association activities and to serve in the elected positions at the various council levels.

2.2.3   The legal definition of engineering

Engineers Canada has proposed a simple, national definition of the practice of professional engineering as follows. The "practice of professional engineering" means

> ... any act of planning, designing, composing, evaluating, advising, reporting, directing or supervising, or managing any of the foregoing, that requires

the application of engineering principles and that concerns the safeguarding of life, health, property, economic interests, the public welfare or the environment. [10]

This is an extremely broad definition. The term *engineering principles* is generally interpreted to mean those subjects defined in a university-level engineering curriculum: mathematics, basic science, engineering science, and complementary studies.

The purpose of this national definition is to promote uniformity of engineering law throughout Canada, thus assisting engineers to practise engineering in different provinces and territories. However, while every province has adopted a legal definition of engineering, only a few are similar to the Engineers Canada definition, and no two definitions are identical. You will find the legal definition for your province or territory by searching the provincial act on your Association's web site, listed above in Table 2.1. Andrews summarizes all of the provincial and territorial definitions in reference [3].

The provincial legislative acts define engineering practice and place a responsibility on the engineer to ensure that life, health, property, and the public welfare are protected. However, some engineering graduates accept temporary or permanent jobs that do not carry these responsibilities. Must these graduates be licensed as professional engineers? The following advice may be useful.

The key point is that almost every engineering job involves decisions that affect life, health, property, and the public welfare, including the environment, even if only indirectly or remotely. If the terms of a job truly do not meet the above definition, then licensing is not required, of course, but you should ask whether the job is really engineering. Moreover, in certain grey areas of activity where licences are rare, it is far better to have a licence that is not needed, than to break the law by practising illegally.

The lack of a licence may hinder your promotion, or limit future career advances. It is very awkward to apply for a job for which you are not legally qualified. Therefore, even graduates who work outside the mainstream of engineering may profit from being licensed, since the licence opens up many career opportunities.

Finally, your licence illustrates your respect for yourself and your profession, since unlicensed graduates may not legally use the title Professional Engineer or its abbreviation, P.Eng. Many engineers who leave the profession and are no longer responsible for engineering decisions still maintain their licences, as a matter of honour and pride in their training, experience, and accomplishments.

2.3   Admission to the engineering profession

The legislative acts in the provinces and territories define the admission requirements for the profession. The requirements are similar, but not identical, across the country. To qualify for a licence, an applicant typically must

- be a citizen of Canada, or have permanent resident status;
- have reached the age of majority (typically 18 years);
- satisfy the academic requirements;

- pass the professional practice examination;
- satisfy experience requirements;
- be of good character, as confirmed by referees.

The admission process is illustrated in Figure 2.3 for students in accredited university engineering programs. Regulations for admission change from time to time. In the past decade, the experience requirements were raised in most provinces from two years to four, with criteria added to define the quality of the experience. At the same time, an internship process was instituted to help prospective engineers through the process. Consult the information provided by your provincial Association for advice about preparing your personal information, academic qualifications, and engineering experience.

Most Associations have a student membership designed to promote communication between engineering students and their profession. A student membership may be very useful in the early stages of your professional career. Your Association is easily contacted through the Internet. The Engineers Canada web site (www.engineerscanada.ca) links to all Associations.

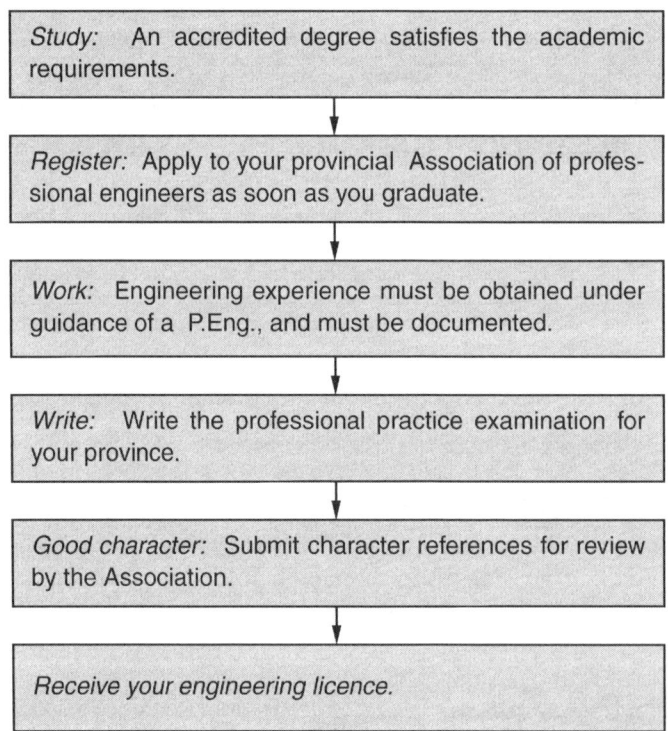

Study: An accredited degree satisfies the academic requirements.

Register: Apply to your provincial Association of professional engineers as soon as you graduate.

Work: Engineering experience must be obtained under guidance of a P.Eng., and must be documented.

Write: Write the professional practice examination for your province.

Good character: Submit character references for review by the Association.

Receive your engineering licence.

Figure 2.3   The steps to take to become an engineer through an accredited program.

2.3.1   Academic requirements

The provincial or territorial licensing Association evaluates each applicant's academic record and has final authority on all matters related to academic requirements. The qualifications of each applicant are evaluated according to the requirements given above.

Accredited degree programs
The Associations rely on the Canadian Engineering Accreditation Board (CEAB) for advice on Canadian engineering programs. The CEAB, a standing committee of Engineers Canada since 1965, evaluates engineering degree programs in Canada at the request of the universities. It also evaluates the accreditation process in countries where similar systems exist, and as a result of mutual recognition agreements, graduates of accredited programs in several countries are given reciprocal recognition of their academic degrees. The CEAB grants *accreditation* to Canadian programs but uses the term *substantial equivalency* for programs outside of the country. A mutual recognition agreement has been in effect with the Accreditation Board for Engineering and Technology in the United States since 1980 and, since 1989, an agreement sometimes called the Washington Accord has listed accreditation in the following countries as substantially equivalent: Canada, Australia, Ireland, New Zealand, the United Kingdom, and the United States. Accreditation agencies in other countries have since joined the Washington Accord and are listed in the annual CEAB Accreditation Criteria and Procedures report. The 2007 document [11] lists agencies in South Africa, Hong Kong, Japan, Singapore, Korea, and Taiwan (Chinese Taipei).

Engineers Canada concluded an agreement in 1999 to recommend that programs in France listed by the Commission des titres d'ingénieur be treated as substantially equivalent to Canadian programs. This agreement has been ratified by most Canadian licensing Associations.

The CEAB mandate was extended in 1997 to include the evaluation of programs outside of Canada. A few programs in Austria, Costa Rica, and Russia are listed in reference [11] as substantially equivalent to accredited Canadian programs.

Other engineering degrees
Applicants for an engineering licence without a CEAB-accredited engineering degree or one covered by international agreement must submit the details of their degree programs so they can be evaluated individually by the licensing Association. In recent years the PEO, for example, has received more applications and granted more licences to this category of applicants than to those with degrees from accredited programs. Technical examinations may be required in order to confirm that the degree meets minimum requirements.

Applicants without equivalent degrees
Applicants without a CEAB-accredited engineering degree or equivalent may be admitted to the engineering profession after writing examinations. The Canadian Engineering Qualifications Board (CEQB), another standing committee of Engineers Canada, advises licensing bodies on examinations, maintains a list of foreign engineering educational institutions, and promotes uniformity of treatment of applicants across the country.

The examination program depends on the applicant's previous education, but completing it is not an easy task, for several reasons:

- There are approximately 16 three-hour university-level admission examinations for each branch of engineering. An applicant may be exempt from some subjects, depending on his or her academic record or experience. Applicants who require more than nine examinations are not normally admitted to the examination program.

- The examination system is not an educational system. Persons applying for the examinations must study on their own and present themselves when they are prepared to write and pass the examinations.

- The licensing Associations do not offer classes, laboratories, or correspondence courses to prepare for the examinations, and there are no tutorial services to review examination results.

2.3.2 Experience requirements

To obtain a licence, you must have four years of experience, except in Québec, where three years are required. According to the national guidelines developed by the CEQB [12], the experience must be at the appropriate level, following the conferring of an engineering degree or the completion of equivalent engineering education. An applicant who has not satisfied the experience requirements is typically registered, temporarily, as an engineering internship trainee (EIT), member in training (MIT), or similar designation, depending on the province or territory. Most Associations have mentorship programs to help applicants to obtain advice from an engineer, who might also serve as one of the referees needed to vouch for character. Internship and mentorship programs are especially useful for graduates in work environments where there are few practising engineers.

General requirements Work experience acceptable for admission must possess the following characteristics, according to the Engineers Canada national guidelines for admission:

- The experience must be in areas generally satisfying the definition of engineering.

- The experience is normally in the same area as the applicant's engineering degree.

- The experience must be up to date.

- The application of theory must be included.

- There should be broad exposure to areas of management and the interaction of engineering with society.

- The experience must demonstrate progressively higher responsibility.

- Normally, experience is obtained under the guidance of a professional engineer.

In addition to the above criteria, the factors given below are typically considered for admission; however, provincial requirements may differ slightly, and criteria may change.

Canadian experience At least 12 months of the experience mentioned above must be obtained in a Canadian jurisdiction under the supervision of a person legally authorized to practise professional engineering.

Equivalent experience Appropriate credit will be given for equivalent experience, following the conferring of an engineering degree, for any of the following:

- experience obtained outside Canada;

- a post-graduate degree or combination of degrees;

- teaching at or above third-year university engineering level or the graduating year in a college;
- military experience.

Pre-graduation experience   Well-documented work-term or internship experience, up to a limit of one year, is usually also accepted as satisfying a portion of the experience requirements. The experience must be obtained after the midpoint of the undergraduate degree.

Documentation   Check the current requirements for the proper method to record the details of your experience. The following data is generally required:

- dates of employment, by month and year;
- names and locations of employers;
- detailed descriptions of the technical responsibilities and the services or products involved;
- dates of periods of absence from employment.

Experience in a different field   If your experience is not in the field of engineering for which your degree was awarded, the Association will evaluate the appropriateness of the experience.

2.3.3   The professional practice examination

Some provincial Associations create their own professional practice examination; others use the Alberta (APEGGA) exam, which is nationally available. The examination tests knowledge of ethics, professional practice, law, and liability. Every applicant for a licence must successfully complete the examination. Exemption is permitted only for applicants who are applying to be reinstated after an absence or for applicants with at least five years of membership in another Canadian Association.

Time limits   You must submit a licence application before you will be permitted to write the professional practice examination. In some provinces, you may not write the examination until one year after graduation from university, and you must pass the exam not later than two years following the date of submitting the application for membership (or the date of successful completion of all other examination requirements, whichever is later). The best advice is to obtain some experience after graduation, write the examination, and then complete the experience requirements for your application.

Exam schedules   The examinations are generally administered several times each year. The provincial Association should be contacted for precise locations and dates. These dates and other registration information are published on the Internet. The provincial Associations and web sites are listed above in Table 2.1.

2.3.4 Offering engineering services to the public

All provincial and territorial Associations impose licensing requirements on partnerships or corporations that offer engineering services to the general public. The purpose is to identify the licensed engineers in the corporation who are assuming responsibility for the corporation's engineering work. A corporate engineering licence is typically called a *permit to practise* or a *certificate of authorization*, depending on the province or territory.

Some Associations require individual engineers offering services to the public to satisfy additional licensing criteria, such as additional experience and professional liability insurance (also called *errors and omissions insurance*). Some acts exempt individuals (sole proprietorships) from the insurance requirements, providing that clients are informed, in writing, that engineering services are not covered by liability insurance. Typically, the client must acknowledge and accept this condition.

These additional licensing requirements are deemed necessary to achieve the Association's main responsibility, which is protecting the public. However, each province and territory interprets this responsibility differently, and no two acts are identical on all issues. Andrews [3] summarizes these additional licensing requirements. If you are planning to offer engineering services to the public, you must be licensed as an engineer, and you must contact your Association to determine what additional licensing, experience, or insurance requirements apply in your province or territory.

2.4 The purpose of provincial Associations

In concluding this chapter, the authors would like to stress that the primary purpose of licensing is to protect the public, not to help engineers. However, a structure that includes acts, Associations, regulations, by-laws, and codes of ethics makes engineering a very well-organized profession. In fact, public confidence in the engineering profession is very high.

Although some Associations, in addition to their licensing function, also provide fringe benefits—such as group insurance, retirement savings plans, employment advice, chapter-level technical meetings, and social events for members—this is not their primary purpose. Occasionally, people who see only these superficial aspects may believe that the licensing bodies are advocacy or service groups for engineers. They are not; they were set up to protect the public by regulating the engineering profession. The provision of member services is quite minor and does not conflict with regulating the profession.

In most self-regulated professions, there is a two-body organizational structure in which one organization regulates the members of the profession, and an independent organization works on behalf of its members by setting professional fees and organizing pension plans and other activities. The legal profession is a good example: in each province, the Law Society regulates members, and the Bar Association works on their behalf. Similarly, in medicine, the College of Physicians and Surgeons regulates medical doctors, and the provincial medical association works on behalf of its members. A similar structure has been proposed in engineering, but at present exists only in Ontario, where regulation is separated from service: the Association (PEO) regulates engineers, and the Ontario Society of Professional Engineers (OSPE) undertakes a role of advocacy, member services, and opportunities for professional development.

For the most current information on recent developments, academic and experience requirements, and admission procedures, consult the web site of your provincial Association, listed in Table 2.1. Additional information can be found through the Engineers Canada web sites [8–12].

2.5 Further study

1. Choose the best answer for each of the following questions.

(a) How does your university engineering program differ from most arts and science programs?

    i. The engineering program has more scheduled hours.

    ii. The engineering program is accredited by CEAB.

    iii. The engineering program satisfies the technical requirements leading to a professional designation (P.Eng.).

    iv. All of the above are correct.

(b) A provincial Association will require an applicant for an engineering licence who is a graduate of an accredited Canadian engineering program to write

    i. only the professional practice examination.

    ii. four confirmatory exams.

    iii. four technical exams from the applicant's branch of engineering.

    iv. all technical exams for the applicant's branch of engineering.

(c) The primary purpose of the provincial and territorial Associations that license engineers is

    i. to benefit engineers in general.

    ii. to exclude technologists and foreign engineers.

    iii. to protect the public.

    iv. to protect the government.

(d) How many years of relevant engineering experience are required by most provinces (except Québec) to qualify for a licence to practise engineering?

    i. one year   ii. two years   iii. three years   iv. four years

(e) The engineering profession is "self-regulating," as are all of the major professions, such as law and medicine. What does "self-regulating" imply?

    i. The laws and regulations are left entirely to the engineers themselves.

    ii. The laws are set by the government, but the engineers elect a council and monitor the profession and administer the laws.

    iii. The government establishes a licensing body and appoints engineers to oversee the profession.

iv. The "learned" engineering societies assess credentials of applicants and award "chartered" status to engineers when they reach a suitable level in the society.

(f) Geoscientists influence public health, safety, and welfare, particularly in oil, gas, and mineral exploration, environmental studies, mining activities, and construction of major engineering works such as dams and bridges. How are geoscientists regulated in most provinces and territories in Canada?

i. Geoscientists are regulated in the same acts as engineers.

ii. Geoscientists are regulated as are engineers, by separate (but equivalent) acts.

iii. Geoscientists are not regulated.

iv. Geoscientists are regulated by "learned" geoscience societies, such as the Canadian Geotechnical Society (CGS).

(g) The collapse of the Québec Bridge in 1907, killing 75 workmen, was a critical event that stimulated the drafting of laws to regulate the engineering profession. The main cause of the Québec Bridge collapse was traced to

i. the steamer Glenmont, which passed beneath the bridge shortly before the collapse.

ii. an overloaded train passing over the bridge.

iii. failure to deliver the telegraph message ordering workmen to leave the bridge.

iv. negligence of the design engineers to calculate the dead-load properly.

(h) The professional practice examination, which all applicants must normally write before obtaining a licence, contains questions on

i. ethics, professional practice, law, and liability.

ii. ethics, accounting, law, and liability.

iii. engineering and technology, ethics, and accounting.

iv. the environment, ethics, and accounting.

(i) In order to obtain an engineering licence, an applicant must have suitable experience. At least how much experience must have been obtained in Canada or in a Canadian engineering environment?

i. six months   ii. one year   iii. two years   iv. four years

(j) The main purpose of Engineers Canada is

i. to represent Canadian engineers in labour negotiations.

ii. to regulate the engineering profession at the provincial level.

iii. to coordinate the engineering profession on a national scale.

iv. to coordinate foreign engineering relief aid.

2. Inspect the web site of your provincial or territorial Association, in order to answer the following.

(a) Is there a student membership category? If so, how do you become a student member? What are the benefits of student membership?

(b) What title does your Association assign to engineering graduates in training for engineering practice?

(c) Does the Association require practising engineers to show evidence of continuing competence in order to maintain their licensed status?

3.   Is there a provincial engineering advocacy association in your province? If so, do the following.

(a) Determine and list activities that could, in principle, overlap with the activities of the regulatory Association. Do their terms of reference specify their relationship with the regulatory Association?

(b) Determine the conditions of membership.

(c) Is there a student membership category?

4.   Do any provinces or territories impose licensing requirements additional to those listed in Section 2.3? If so, what are the requirements, and where are they in force?

2.6   References

[1] H. MacKenzie, "Opening address for the debate on the *Professional Engineers Act 1968–69*, Bill 48," in *Legislature of Ontario Debates, 28th Session, 1969*, vol. 5, pp. 4791–4800, Toronto: Queen's Printer for Ontario, 1969.

[2] J. B. Carruthers, 1978. Personal communication to G. C. Andrews.

[3] G. C. Andrews, *Canadian Professional Engineering & Geoscience: Practice & Ethics*. Toronto: Nelson-Thomson, third ed., 2008.

[4] H. Petroski, *Engineers of Dreams: Great Bridge Builders and the Spanning of America*. New York: Vintage Books, 1995.

[5] H. Holgate, *Royal Commission: Quebec Bridge Inquiry*. Ottawa: Government of Canada, 1908. 7-8 Edward VII, Sessional Paper No. 154.

[6] G. H. Duggan, ed., *The Québec Bridge*. Montréal: Canadian Society of Civil Engineers, 1918. Bound monograph prepared originally as an illustrated lecture.

[7] Canada Department of Railways and Canals, *The Québec Bridge over the St. Lawrence River near the City of Québec*. Ottawa: Governor-General in Council, 1919. report of the Government Board of Engineers, Department of Railways and Canals Canada.

[8] Canadian Council of Professional Engineers, *P.Eng., The Licence to Engineer*. Ottawa: Canadian Council of Professional Engineers, 2002. <http://www.peng.ca> (March 9, 2008).

[9] Canadian Council of Professional Engineers, *The Four Steps to Becoming a P.Eng.* Ottawa: Canadian Council of Professional Engineers, 2002. <http://www.peng.ca/english/students/four.html> (March 9, 2008).

[10] Canadian Engineering Qualifications Board, *Guideline on the Definition of the Practice of Professional Engineering*. Ottawa: Canadian Council of Professional Engineers, 2002. <http://www.engineerscanada.ca/e/files/guideline_definition_with.pdf> (March 9, 2008).

[11] Canadian Engineering Accreditation Board, *Accreditation Criteria and Procedures*. Ottawa: Engineers Canada, 2007. <http://www.engineerscanada.ca/e/files/report_ceab.pdf> (March 9, 2008). Pre-publication data at <http://www.engineerscanada.ca/e/acc_programs_2.cfm> (September 10, 2008).

[12] Canadian Engineering Qualifications Board, *Guideline on Admission to the Practice of Engineering in Canada*. Ottawa: Canadian Council of Professional Engineers, 2001. <http://www.engineerscanada.ca/e/files/guideline_admission_with.pdf> (March 9, 2008).

Chapter 3

Professional Engineering Ethics

Students entering engineering programs have high ideals, and aspire to be highly skilled, accurate, dependable, ethical professionals. However, every profession attracts a few unscrupulous individuals who want to profit from unprofessional behaviour. Unprofessional behaviour harms the public, reduces public confidence in the profession, and may also end a promising career. Reports of malpractice in other professions, such as medicine, law, and accounting, appear fairly frequently in the news media. Fortunately, the engineering profession has fewer complaints of this type, but they do occur.

For example, a contractor may offer a bribe to the engineer supervising a construction project, if the engineer will allow the contractor to use cheap, substandard materials or methods. Offering a bribe and accepting a bribe are both illegal, but in addition to criminal charges, a professional engineer who accepts a bribe is also subject to disciplinary action by the licensing Association. Upon conviction, the discipline may include a reprimand, a fine, or even the loss of a licence to practise engineering. Ethical behaviour is therefore extremely important in professional activities.

In order to guide professional engineers, each provincial licensing Association publishes a code of ethics. These codes vary slightly from province to province, but they express the same ethical principles. This chapter discusses these principles, as well as

- legal significance of the code of ethics;
- applications of ethics in the workplace;
- definitions of professional misconduct;
- disciplinary powers of the Associations;
- ethical use of engineering computer programs and the engineer's seal; and
- significance of the iron ring, worn proudly by Canadian engineers.

3.1 Introduction to professional ethics

As explained in the previous two chapters, each province and territory of Canada has passed a law (act) to regulate the engineering profession. These acts establish engineering as a profession, and they require professional engineers to follow a code of ethics, which requires high standards of professional conduct. The code of ethics for your province or territory is found on your Association's web site. The web sites are listed in Table 2.1 of Chapter 2.

The main purpose of the code of ethics is to protect the public. The code of ethics is not a voluntary guide. To ensure that engineers are aware of the code of ethics, applicants for engineering licences are required to write a professional practice examination, in which the code of ethics must be applied to hypothetical cases in engineering practice. Examples of typical exam questions are given at the end of this chapter. Moreover, to ensure that engineers comply with the code of ethics, each act specifies disciplinary actions to be imposed on engineers who disregard the code.

Codes of ethics impose several duties on the practising engineer, including duties to society in general, to employers, to clients, to colleagues, to the engineering profession, and to oneself. The wording of these codes is typically based on common sense and natural justice. Therefore, it is not necessary for you to memorize the code; most engineers find that they follow it intuitively and never need to worry about charges of professional misconduct. However, a special case may arise if an engineer's employer acts unethically.

For example, consider the case where an engineer learns that the employer is illegally polluting waterways by discharging toxic waste, in violation of Canada's Environmental Protection Act. The engineer might be faced with agreeing to an unethical action (pollution) or confronting the employer and possibly losing his or her job. It should be emphasized that such serious situations are rare, but do occur. The Association is usually available to provide advice and to mediate. If the employer is an engineer, or a corporation licensed to practise engineering by the professional Association, then the code of ethics applies equally to the employer. In this case, toxic pollution is not only unethical but it is illegal, and if the employer will not stop it, both the code of ethics and the Environmental Protection Act require the pollution to be reported to Environment Canada.

An employee engineer should never have to choose between unethical behaviour and a disruption of employment. Such problems arise very infrequently, and the first step is to try to solve the problem by discussing it with the employer. If that attempt fails and internal solutions are clearly impossible, then a confidential contact should be made with the Association.

3.1.1 Codes of ethics: General principles

Codes of ethics [1] usually include statements of general principles, followed by instructions for specific conduct that emphasize the duties of the engineer to society, to employers, to clients, to colleagues, to subordinates, to the profession, and to himself or herself. Although the various codes express these duties differently, their intent and the results are very similar. The following paragraphs summarize what the codes of ethics have in common:

Duty to society A professional engineer or geoscientist must consider his or her duty to the public—or to society in general—as the most important duty. In other words, professionals have a duty to protect the safety, health, and welfare of society whenever society is affected by their work. In return, the professions receive the privilege of self-regulation. That is, the government delegates its authority to the Associations, which define standards of admission, discipline licensed members, and regulate the profession. This arrangement benefits society, because the Associations ensure that professionals are competent, reliable, up to date, and ethical.

Duty to employers
A professional engineer or geoscientist must act fairly and loyally to the employer, and must keep the employer's business confidential. Furthermore, a professional is obliged to disclose any conflict of interest.

Duty to clients
A professional engineer or geoscientist in private practice has the same obligations to clients as an employee has to the employer.

Duty to colleagues
A professional engineer or geoscientist must act with courtesy and goodwill toward colleagues. This golden rule is supported by all major ethical theories. Professionals should not permit personal conflicts to interfere with professional relationships. Most codes of ethics state specifically that fellow professionals must be informed whenever their work is reviewed.

Duty to employees and subordinates
A professional engineer or geoscientist must recognize the rights of others, especially if they are employees or subordinates.

Duty to the profession
A professional engineer or geoscientist must maintain the dignity and prestige of the profession, and must avoid scandalous, dishonourable, or disgraceful conduct.

Duty to oneself
Finally, a professional engineer or geoscientist must ensure that the duties to others are balanced by the individual's own rights. A professional person must insist on adequate payment, a satisfactory work environment, and the rights awarded to everyone through the Canadian *Charter of Rights and Freedoms*. The professional also has a duty to strive for excellence and to maintain competence in the rapidly changing technical world.

The provincial and territorial codes of ethics [1] contain the seven general duties described above, expressed with minor wording variations. Some codes of ethics impose additional duties. As previously mentioned, you can read the code of ethics for your Association on the Association's web site listed in Table 2.1 of Chapter 2.

3.2   Ethics in the workplace

Everyone wants to work in a fair, creative, productive, and professional environment, and the applicable code of ethics can help to establish such an environment. Every Association code of ethics states that professionals must show "courtesy and goodwill toward colleagues" or emphasizes the need for fairness, integrity, cooperation, and professional courtesy. These characteristics are essential in any professional organization, and one purpose of the code is to create a productive and professional workplace.

Although engineering students are not legally bound by the provincial code of ethics, it is a general rule in universities that any behaviour that interferes with the academic activity of others may be classified as an academic offence, which may cause the university's discipline code to be invoked. Academic offences include plagiarism, cheating, threatening behaviour, or other behaviour that has no place in the workplace or in a school for professionals. A professional atmosphere is equally important in the engineering student's workplace—the academic environment of classrooms, study rooms, libraries, residences, and the job sites where students work between academic terms. The following examples illustrate the importance of ethical conduct in the student workplace.

Professional behaviour

Engineering students are expected to act professionally toward employers, colleagues, and subordinates, even though they are not legally governed by a code of ethics. In particular, students on work-terms or internships have an obligation to act professionally on job sites and to keep employers fully informed. The required behaviour includes simple matters such as courtesy, punctuality, and appropriate clothing. For example, hard hats and steel-toed shoes are essential on a construction site but may be inappropriate in a design office.

As a second example, consider how unprofessional behaviour can undermine the student job-placement process. University placement staff put a great deal of effort into attracting employers to campus. Employers, in turn, invest time and money interviewing students for job openings. If a student accepts a job offer from one employer, whether verbally or in writing, then fails to honour it when a better offer arrives from a second employer, the student would clearly be behaving unprofessionally. If the commitment cannot be honoured because of circumstances beyond the student's control, then there is an obligation to inform the first employer immediately through the placement department and try to minimize the damage. In most cases, an alternative can be negotiated, but only if the student acts promptly and ethically.

Teamwork

As stated earlier, engineering has become a team activity because of the increasing complexity of projects. Engineering students must work effectively with engineers and other professional staff—and the most important goal is the successful completion of the project.

Engineering teams usually include persons with widely different abilities, interests, and education who cooperate by contributing their particular expertise to advance the project. The best engineering team will use each person according to his or her strengths and willingness to contribute to the whole effort. Professionals should not be judged by extraneous factors, such as race, religion, or sex. In addition to being unprofessional and contrary to codes of ethics, such discrimination is contrary to provincial and federal laws on human rights. Within the engineering profession, we must aim for a higher standard of conduct than the minimum set by law. This is the purpose of the code of ethics and is what we would expect of a rational profession like engineering.

Ethical problems

An engineer or an engineering undergraduate may occasionally be faced with an ethical dilemma. For example, suppose that you have a colleague or co-worker who has developed a serious personal problem such as drinking, drugs, or mental or emotional imbalance. You are faced with an ethical choice: do you help to conceal the problem and let the individual's health deteriorate, or do you try to get medical or other professional intervention, knowing that this may affect your friend's employment or end your friendship? Both choices are unpalatable.

The solution to an ethical dilemma depends on an evaluation of all factors of the case, using the code of ethics as a guide. You might find the design procedure in Chapter 15 of this book to be useful, since it is a problem-solving procedure. As the codes of ethics state, an engineer has a duty both to colleagues and to the employer, and this is the root of the dilemma. In view of these conflicting duties, it is usually impossible to give a simple resolution of such a dilemma. The solution requires gathering all of the pertinent information, evaluating the alternative courses of action, and selecting the one that achieves, or comes closest to achieving, the desired goal. When both alternatives are unpleasant, as in the above example, the least undesirable alternative must be chosen.

Resolving disputes In creative activities such as engineering, differences of opinion are common; in fact, diversity may help to achieve a team's goals. However, disputes must not be allowed to undermine the engineering team, and the best way to settle a dispute is through courteous, direct communication, with the expectation that a full review of the dispute will lead to its solution. In fact, this is the meaning of "good faith" in item 7(i) of the code of ethics. This straightforward technique will usually yield a good solution that will be accepted by everyone affected by it.

If you are involved in a dispute, you should never file a formal complaint until all possibilities for personal, informal resolution are exhausted. Sometimes, but very rarely, external agencies or authorities must be contacted to solve ethical problems in the workplace. This action, sometimes called "whistle-blowing," should be the absolute last resort. The professional Association is available for advice in these matters.

3.2.1   Whistle-blowing

Occasionally, but rarely in a professional career, it may be necessary to "blow the whistle" on unethical activity. Most engineers will likely never need to be concerned about the process. However, if an engineer observes unsafe, unethical, or illegal practices, the code of ethics requires that action be taken. The action always begins by assessing the degree of danger or illegality involved, and then following the chain of management within the organization to remedy the problem. Only after all internal routes have been unsuccessful should an engineer blow the whistle by going outside of the organization.

For example, consider the case mentioned earlier in this chapter, where an engineer discovers that the employer is illegally polluting waterways by discharging toxic waste in violation of Canada's Environmental Protection Act. The following steps might be useful to resolve such a case:

Get the facts and identify the urgency If the engineer has tested the effluent, and has documented proof of the toxicity, then the degree of danger is fairly high. Such waste disposal is not only unethical; it is illegal, so immediate action is necessary. It is also essential to identify the person in charge of the waste disposal, because that person will be most effective in resolving the problem.

Consider the solution The engineer must also decide on the simplest remedial action, in advance of any confrontation. That is, the engineer must not merely complain about a problem; the engineer must be prepared to offer a clear, simple, and preferably cost-effective solution to the problem.

Speak to the key person Usually, an informal, personal conversation with the person involved, describing the problem and proposing a solution, yields the best results. In many cases, the person responsible may not be aware of the problem, and will be eager to solve it.

Going higher   If the key person is not receptive to the engineer's presentation of the problem, and has no valid reason for refusing to remedy it, then the engineer must go over the head of this person and farther up the chain of management. This is not whistle-blowing, because the action is internal, and is the expected remedial action in a properly run organization. The engineer's goal is not to embarrass anyone, but to reduce the organization's liability for an unethical or illegal activity.

Blowing the whistle   When all internal avenues have been closed, and no action is forthcoming, then it may be necessary to go outside the organization. The licensing Association may be able to mediate in most cases, but illegal activities must eventually be reported to the regulatory bodies (such as Environment Canada), if the public health or welfare is involved. In fact, failure to take action may be considered to be professional misconduct or even complicity in the violation, as described in Chapter 15 of reference [1].

Although whistle-blowing is quite rare, several Associations have defined reporting procedures in publications available on their web sites. For example, PEO publishes a public information guide on the engineer's duty to report [2].

3.3   Professional misconduct and discipline

The duty of the professional Associations is to protect the public welfare and to act on complaints from the public. When complaints are made, the Associations investigate, try to mediate, and where necessary, take legal action. The action varies, depending whether the complaint concerns a licensed or unlicensed person or corporation.

Enforcement of the act   The purpose of the provincial or territorial licensing act is to protect the public, so only educated, experienced, competent professionals are allowed to practise. People or corporations that practise engineering or geoscience without a licence are breaking the law (the act). Associations are responsible for enforcing the act by prosecuting offenders in court. Each act typically states that it is an offence for an unlicensed person to

- practise professional engineering or professional geoscience, or
- use the title Professional Engineer, Professional Geoscientist, or the like, or
- use a term or title to give the belief that the person is licensed, or
- use a seal that leads to the belief that the person is licensed.

Most of the complaints are easily resolved, because unlicensed practitioners are frequently unaware that they are contravening the act, and, when informed, they promptly stop the offending behaviour, as described in Chapter 4 of reference [1].

Professional discipline   A second way to protect the public is to discipline professionals. The Association is granted a wide range of authority to discipline any licensed member who is shown to be negligent, incompetent, or guilty of professional misconduct. This authority is clearly defined in the provincial engineering act, and typically includes the power to fine the engineer, revoke or suspend the engineering licence, and, if appropriate, to monitor, inspect, or restrict the work done by the engineer. Results of discipline hearings may be published at the discretion of the discipline committee. The provincial and territorial acts are very similar, although not identical. They typically specify six causes for disciplinary action:

- professional misconduct, also called unprofessional conduct,
- incompetence,
- negligence,
- breach of the code of ethics,
- physical or mental incapacity, and
- conviction of a serious offence.

Although the Association staff receive and administer the complaints, discipline decisions are made by a discipline committee, composed mainly of professional engineers.

3.4   Common professional complaints

Conflict of interest   A common complaint to the provincial and territorial Associations concerns "conflicts of interest," which are mentioned prominently in every code of ethics. A conflict of interest occurs whenever a professional has a personal preference or financial interest that interferes with a duty to the employer, to the client, or to society. For example, an engineer who receives payments or benefits from two sources for the same service usually has a conflict of interest, because the engineer's duty to the client or employer may be compromised. As another example, if an engineer specifies that a certain product must be used on a project because the supplier paid the engineer a secret commission, then the engineer has a conflict of interest. In fact, accepting a secret commission is not only unethical but illegal.

A professional person must try to avoid conflicts of interest. In some cases, a conflict of interest may be unavoidable or may be very minor. When it cannot be avoided, then the conflict must be disclosed to the people involved, to ensure that everyone is aware of it and to ensure that the conflict does not affect professional decisions. An unavoidable conflict of interest is often acceptable if it has been fully disclosed.

Breach of standards   Another common complaint to the provincial and territorial Associations concerns the clauses in the code of ethics requiring the engineer to be competent. In design work, this requires the engineer to follow commonly accepted design codes and standards. Such standards are easy to obtain, thanks to the Internet (see Section 19.5.1). One of the first steps in the design process (see Chapter 15) is gathering information. Therefore, at the start of every design project, the design engineer should routinely search the Internet for appropriate design standards and safety regulations. A simple search for standards might pay immense rewards by contributing to a safe design and helping to avoid potential injury or financial loss.

3.4.1	**Case studies in ethics**

In each of the following situations, identify a course of action that is consistent with the code of ethics and avoids any appearance of professional misconduct. Try to prepare a response before referring to the authors' suggested answers, which follow the cases themselves.

Case study 3.1: Peak flow dilemma

You work for a consulting engineering company that specializes in the design of sewage treatment plants, and you have been assigned to study the sewage treatment plant in a small town. The plant was constructed about 20 years ago and generally operates very well. However, at least two or three times per year, during severe rainstorms, the plant is overloaded, and operators are forced to discharge untreated sewage into the river that flows by the town. You have measured the peak sewage flows and calculated that the plant capacity would need to be increased by about 40 % to cope with these occasional peak loads. However, during your work, you discovered drawings showing that several buried storm drains are improperly connected to the sewage system, and this extra flow explains why the overloads occur at the sewage plant only during rainstorms: half of the water treated during storms is rainwater. Obviously, if the storm drains were rerouted, the sewage plant would have adequate capacity.

You tell your boss about your discovery. Your boss explains that if you recommend that the town expand its sewage plant, then your employer would bid on the lucrative contract for this work and would likely get the contract. However, if you recommend that the town upgrade and reroute its storm-drain system, the contract cost will be much lower and would likely be won by a competing company that specializes in storm-drain work. Your boss suggests that you conceal the information about the storm drains and recommend that the town increase the sewage plant capacity. Either way, the problem will be solved, so your company might as well share in the profit. What should you recommend in your report?

Case study 3.2: Light bulb commission

You are the chief design engineer in an electrical design company. You have been awarded a contract to manufacture specialized lighting equipment for a nearby major airport extension. You contact several light bulb manufacturers for specifications and prices on the bulbs needed for the runway lighting. Some, but not all, of the information is available over the Internet, but one company representative visits you personally, provides all of the information that you need, and offers you a "confidential" cash commission for every hundred bulbs that you order. You review the sales literature and realize that the company manufactures the best bulbs available for brightness, shock resistance, and longevity, although its prices are a little higher than competitors' prices. You would probably select these bulbs anyway, so should you accept the commission from the sales representative?

Case study 3.3: Expert witness

You are in private practice as a consulting engineer in manufacturing processes. You are contacted by company A to give advice on a computer-controlled milling machine in its machine shop that does not cut metal within the guaranteed tolerances and is constantly malfunctioning. The manager at company A explains that the machine was purchased about a year earlier from the manufacturer, company B, but the machine has never operated properly. Company B is nearby, and its technicians have made many service visits, with no useful results. When you see the machine, you remember that you were involved in the design of the control software, several years earlier, at an early stage of development of the milling machine. You provided advice to company B, and the development was carried out by its engineers, but you know the machine fairly well. However, the engineers from company A tell you that they want to hire you to analyze the machine and to testify as an expert witness when they sue company B. What ethical issues arise here, and what should you do?

Case study 3.4: Surveyor's assistant

Your employer has assigned you the task of hiring an assistant to help the survey crew. The assistant carries equipment and clears sight-lines for the surveyors in rough bush country. You place an advertisement on the company web site. The best-qualified person who answers the advertisement coincidentally happens to be your cousin, whom you met a few times at family gatherings. You do not know the cousin very well, but your last names are the same, so other employees may know that you are related and may think you are favouring your relatives. Should you hire your cousin?

Case study 3.5: Whistle-blowing dilemma

Assume that you are a recently licensed professional engineer, working for a consulting engineering company that is supervising the construction of a major building. You assist your boss in monitoring the delivery of materials (sand, gravel, concrete, and steel) and components (doors, windows, roofing, etc.) to the job site, which is almost the size of a city block. On behalf of the client, you routinely count the material and components delivered, and ensure that they are installed according to the plans. Occasionally, you notice small discrepancies. The invoices do not agree with the materials delivered, and you report the shortages to your boss, who listens, but tells you to ignore each report because, "In a project this large, some shrinkage occurs." However, one day you notice a truck leaving the site with a few doors and windows that should have been unloaded. When you stop the truck, the driver tells you: "This part of the load is going to your boss's new cottage." You refuse to let the truck leave the site, but your boss intervenes and overrules you, on the basis that the components are the wrong size and the paperwork to return them to the manufacturer is in order. You suspect that your boss might be stealing from the job site by redirecting materials for personal use. What should you do?

3.4.2   Suggested solutions to case studies

Ethical courses of action, for each of the cases above, are suggested below.

Suggested solution 3.1: Peak flow dilemma

Under the code of ethics, you have an obligation to put the public interest ahead of narrow personal gain. Moreover, you have an obligation to complete your task honestly, competently, and professionally. To conceal the real problem with the sewage system would be dishonest and unprofessional. If you follow the boss's dishonest suggestion and the true facts concerning the storm drains should later become known, you could be charged with concealing a conflict of interest, unprofessional conduct, or both; and if found guilty, you could be subject to a fine, reprimand, or even loss of your professional engineering licence. Therefore, without hesitation, you would recommend the reconnection of the storm drains as the best choice in this case.

Suggested solution 3.2: Light bulb commission

Accepting a secret commission is both unprofessional and illegal. The code of ethics requires the engineer to act as a faithful agent of the employer. In fact, Ontario specifically defines secret compensation from more than one party, for the same service, to be professional misconduct. You therefore have a conflict of interest; to avoid any appearance of professional misconduct, this conflict must be disclosed to your employer. When the employer knows all the facts, a fair decision can be made. It is possible, but unlikely, that your employer would allow you to accept the commission. Most likely, the employer would insist on purchasing bulbs from a company with more honest salespeople, or, if the bulbs are purchased from the representative anyway, the employer would expect a price reduction equal to the commission.

Suggested solution 3.3: Expert witness

Although your prior knowledge of the defective machine may seem to be an asset, your previous work for company B may conflict with your work for company A as an expert witness. When you worked for company B, you undoubtedly gained knowledge about the company's technical methods, processes, and business affairs, which might become part of the lawsuit planned by company A. Under the code of ethics, you have a continuing obligation to your former client to keep such matters confidential. Moreover, you do not know whether your earlier advice may be part of the problem with the machine. You therefore have a conflict of interest on two levels: your possible personal responsibility for the design problem, and your continuing obligations to company B. An expert witness is expected to provide an impartial assessment of the facts of the case, regardless of which side pays the expert. You are not likely to be impartial when you are investigating your own work, and your previous work for company B creates obligations and gives an appearance of possible bias. Therefore, the offer of employment as an expert witness for company A should be declined.

Suggested solution 3.4: Surveyor's assistant

You have advertised properly and your cousin has received no favouritism in the hiring process, so you have acted as a "faithful agent" for your employer, as required by the code of ethics. However, because the best candidate is your cousin, other employees may perceive that some favouritism was given. In cases such as this, which occur fairly often, the best procedure is to make a list of the applicants, and write a brief summary explaining what is needed for the job, the reasons why you are recommending the person selected, and disclose that this person is your cousin. You would then pass this information and all of the applications on file to your boss for a final decision. If you have made the right choice, your boss will confirm it; if you have unwittingly been biased in favour of your cousin, then your boss will correct the error.

Suggested solution 3.5: Whistle-blowing dilemma

In this case, whistle-blowing would be unjustified with such flimsy evidence, and might be a career-limiting trap. You have only a verbal allegation, which might have been a joke by the truck driver, and it has been contradicted by your boss. Stealing materials from a job site is a serious offence, particularly for a professional, and action should never be taken on the basis of mere suspicions. You brought the discrepancies to your boss's attention, your boss has taken an apparently valid action, and you should not proceed unless you have more information. You would, of course, be more diligent in watching for evidence that corroborates or disproves the allegation. For example, if you checked the paperwork for the order, and found that the components were, indeed, returned to the manufacturer, then you would likely be glad that you had not rushed to judgment. However, if you learned that the components are missing, that your boss is indeed building a cottage, and that materials were taken from the workplace at the cottage site, then you would have to act. Otherwise, you might be considered to be an accomplice in the theft. Assuming that you obtain clear evidence of theft, even by misdirection, you would follow the process suggested earlier in this chapter, and suggest a simple remedial action, perhaps returning the materials or reimbursing the client. Then you would meet with your boss and lay out your evidence. If your boss should refuse to cooperate, you would go to your boss's boss, or to the client, who would likely turn the matter over to police. Obviously, you would be a key witness, so you must have solid evidence. It is extremely unlikely that involving the news media would benefit the course of justice.

3.5   The professional use of computer programs

Computer software is essential to all aspects of engineering design, testing, and manufacturing. Several ethical concerns are created by the use of software, and some of these cross the boundary of ethical responsibility to become legal issues. Concerns to be discussed briefly include: liability for computer errors, software piracy, plagiarism from the Internet, and protection against computer viruses.

Liability for computer errors The engineer is responsible for decisions resulting from his or her use of computer software. In future, this responsibility may be shared, as software development becomes the responsibility of the software engineering profession, but at present, the engineer, not the software developer, is legally responsible for engineering decisions.

Significant engineering programs are never provably defect-free. However, if an engineering failure results from faulty software, legal liability cannot be transferred to the software developer, any more than it can be transferred to an instrument manufacturer because of faulty readings from a voltmeter. To put it even more clearly: *the engineer cannot blame the software* if the engineer makes decisions based on incorrect or misunderstood software output.

In the event of a disaster, the software developer may be liable for the cost of the software, but the engineer will likely be legally liable for the cost of the disaster. Therefore, when software is used to make engineering decisions, the engineer must:

- be competent in the technical area in which the software is being applied,
- know the type of assistance provided by the software,
- know the theory and assumptions upon which it was prepared,

- know the range and limits of its validity,
- test the software to ensure that it is accurate.

The extent and type of tests will vary, depending on the type of software, but the tests must be thorough. Usually, the engineer must run test examples through the software and compare the output with independent calculations. In some safety-critical applications, computations are performed by two independently written programs and accepted only if the results agree. However, the engineer must never use software blindly without independent validation.

Software piracy

Piracy is the unauthorized copying or use of computer software and is, quite simply, theft. Software piracy occurs more frequently than other types of theft because software can be copied so easily and because the risks of piracy are not very well known.

Software is protected by copyright and trademark laws, as discussed in more detail in Chapter 17. Piracy is an infringement of the developer's rights and is clearly unethical. Engineers who commit software piracy are subject to civil litigation and, in some cases, criminal charges, as well as professional discipline if the conduct of the engineer is deemed to be professional misconduct.

In addition, it is short-sighted to use unauthorized software in engineering, because an engineer who uses such software runs a much greater risk of legal or disciplinary action if a project runs into problems. In any legal proceeding, the use of unauthorized software would be convincing evidence of professional misconduct.

Organizations such as the Canadian Alliance Against Software Theft (CAAST) have been formed to combat software piracy. CAAST is an alliance of several software publishers, created with the goal of detecting and prosecuting cases of software copyright infringement in Canada. CAAST acts very aggressively through the Internet to reduce software piracy by educating the public, detecting infringers, and enforcing the copyright laws.

The ethical use of software stimulates software development and increases creativity, productivity, and job opportunities. Don't get involved in software piracy!

Plagiarism from the Internet

Plagiarism is another ethical problem that has been aggravated by the software explosion. Plagiarism is defined as taking intellectual property, such as words, drawings, photos, artwork, or other creative material that was written or created by others, and passing it off as your own. Plagiarism has become more common in recent years, mainly because of the convenience of cutting and pasting information from the Internet. Plagiarism is always unethical and is specifically contrary to several sections in engineering codes of ethics.

Plagiarism could lead to disciplinary action if the actions of the engineer are deemed to be professional misconduct.

Engineering students should be aware that plagiarism can result in severe academic penalties and can delay or abort a promising engineering career. All universities expect students to know what plagiarism means and to avoid it.

When you submit a document with your name on it, such as an engineering report, you take responsibility for everything in the document except for material that is specifically identified as coming from other sources. If material created by others is not clearly and completely cited, then you are liable to be charged with plagiarism.

The format for citing material created by others is explained in Chapter 8 of this book.

Computer viruses   Like other professionals, engineers depend on their computer systems to run effectively and efficiently. The proliferation of computer viruses that damage or overload computer systems is an ever-growing threat. Virus vandalism can cause expensive and dangerous failures. Individuals who participate in creating or disseminating computer viruses are subject to criminal prosecution, and every engineer has a duty to expose such conduct.

3.6   Proper use of the engineer's seal

The provincial law (act) provides for each professional engineer to have a seal, such as that shown in Figure 3.1, denoting that he or she is licensed. All final drawings, specifications, plans, reports, and other documents involving the practice of professional engineering should bear the signature and seal of the professional engineer who prepared and approved them. This is particularly important for services provided to the general public. The seal has legal significance, since it implies that the documents have been competently prepared. In addition, the seal signifies that a licensed professional engineer has approved the documents for use in construction or manufacturing. The seal should not, therefore, be used casually or indiscriminately. In particular, preliminary documents should *not* be sealed. They should be marked "preliminary" or "not for construction."

The seal signifies that the documents have been prepared or approved by the person who sealed them, implying an intimate knowledge and control over the documents or the project to which the documents relate. An engineer who knowingly signs or seals documents that have not been prepared by the engineer or under his or her direct supervision may be guilty of professional misconduct. The engineer may also be liable for fraud or negligence if misrepresentation results in damages.

Engineers are sometimes asked to "check" documents, then to sign and seal them. The extent of work needed to check a document properly is not clearly defined, but such a request is usually not ethical. The engineer who prepared the documents or supervised their preparation should seal them. If an engineer did not prepare them, then perhaps the preparer should have been under the supervision of an engineer. The PEO *Gazette*, for example, reports many disciplinary cases involving engineers who improperly checked and sealed documents that later proved to have serious flaws.

Figure 3.1   A typical professional engineer's seal. On an actual seal, the name of the licensed engineer appears on the crossbar. (Courtesy of Professional Engineers Ontario)

3.7 The iron ring

The engineering codes of ethics enjoin each engineer to act in an honest, conscientious manner. However, there is a much older voluntary commitment, written by author and Nobel Prize winner Rudyard Kipling and first used in 1925, called the Obligation of the Engineer, which is the focus of a ceremony known as the Ritual of the Calling of an Engineer. Those who have participated in the ceremony and made such a commitment can usually be identified by the wearing of an iron ring on the small finger of the working hand.

The ceremony is conducted by the Corporation of the Seven Wardens, which does not seek publicity although it is not a secret society. The corporation is totally independent of the provincial licensing Associations and the universities. Its members are volunteers.

The wardens usually conduct the ceremony in the late winter or early spring of each year, and invite students who are eligible for graduation to participate. The wardens allow attendance by previously ringed engineering graduates, who are often parents, grandparents, or others who wish to present a ring to a graduating student. The ceremony is not open to the public although, at a few institutions, students are allowed to invite a small number of family members.

Following the long-standing success of the uniquely Canadian ritual, ceremonies with similar intent but different obligation and ring have been introduced in the United States, beginning in 1970.

The engineering iron ring does not signify that a degree has been awarded or that the wearer is a Professional Engineer, and it carries no direct financial benefit. However, it is a reminder that the wearer has participated in the ceremony and has voluntarily agreed to abide by the obligation. The obligation is brief, but it is a solemn commitment, in the presence of practising engineers and other graduating students, to maintain high standards of performance and ethics.

3.8 Further study

The questions following the Quick Quiz illustrate ethical dilemmas that are found in engineering practice. Questions 5, 6, and 7 have been adapted from Ontario professional practice examinations. These questions should be answered by citing and explaining the code of ethics sections and the professional misconduct sections that apply.

1. Choose the best answer for each of the following questions.

(a) The principal purpose of a code of ethics is to

    i. raise public esteem of the profession.

    ii. control admission to the profession.

    iii. protect the public by enforcing high standards of professional conduct.

    iv. protect the public by discouraging unlicensed practitioners.

(b) Almost all provincial codes of ethics state that the engineer must consider the duty to public welfare as paramount. What does *paramount* mean in this context?

    i. The public safety must be placed ahead of personal gain.

    ii. The engineer must pay all appropriate income, business, and sales taxes.

    iii. Public projects (for provincial or federal governments) must be safer than private projects.

    iv. Public projects must be scheduled for completion before private projects.

(c) Although provincial codes of ethics state that public welfare is paramount, the codes also impose a duty to keep the employer's business affairs confidential. Therefore, a dilemma may occur. For example, consider the case where an employer has an unsafe work site, or plans to use substandard materials in a bridge support structure, thereby creating a public-safety hazard. If the engineer cannot convince the employer to follow safe construction practices, then the engineer is encouraged by the code of ethics to

    i. follow the employer's instructions and keep the information confidential.

    ii. report the situation to the news media.

    iii. try to resolve the problem with the employer, and ask the Association to mediate.

    iv. report the situation immediately to the police.

(d) The term *conflict of interest* means (when used in a code of ethics) that the engineer

    i. is working with a person whose personality conflicts with the engineer's personality.

    ii. has a personal or financial interest that conflicts with the employer's interests.

    iii. has committed an unprofessional act.

    iv. has lost interest in a project.

(e) An engineer is assigned, by an employer, to purchase materials for a construction project and to supervise the construction. In collaboration with a partner, the engineer creates a delivery company that purchases the materials, delivers them to the construction site, and charges the costs to the engineer's employer, with an added profit for the engineer and the partner. The engineer's employer is not aware that the engineer is the half-owner of the delivery company. Has the engineer infringed the code of ethics? If so, what should be done?

    i. No infringement has occurred; the engineer is performing the required duties.

    ii. No infringement has occurred if the delivery company is in the partner's name.

    iii. A conflict of interest has occurred, but no action need be taken if the delivery company is in the partner's name.

    iv. A conflict of interest has occurred; the engineer must disclose this conflict to the employer and ask whether the employer will allow the arrangement to continue.

(f) Which of the following is *not* grounds for charging an engineer with professional misconduct?

    i. gross negligence

    ii. incompetence

    iii. conviction of a traffic offence

    iv. conviction of a serious criminal offence

(g) Which of the following is grounds for charging an engineer with professional misconduct?

    i. failing to correct or report a dangerous work site under the engineer's control

    ii. sealing a drawing prepared by a technologist under the engineer's direct supervision

    iii. accepting a secret commission from a supplier for buying the supplier's products

    iv. sealing a drawing by an unlicensed person not under the engineer's direct supervision

    v. making public statements that are not based on firm knowledge and conviction

(h) Which of the following is grounds for charging an engineer with professional misconduct?

    i. An engineer signs and seals a report as a favour to a friend, even though the report was not prepared or checked by the engineer.

    ii. An engineer undertakes work outside the engineer's area of competence.

    iii. An engineer fails to inform an employer or client of a conflict of interest.

    iv. An engineer fails to correct or to report a situation that may endanger the public.

    v. All of the above.

(i) An engineer designs a machine using software purchased from a commercial software company. The machine later fails, and it is discovered that the engineer relied on the output from the software, which was grossly in error. What is the likely outcome, concerning the liability for the cost of replacing the failed machine?

    i. The software developer is fully liable; the software was faulty, and this is the root cause.

    ii. The liability would be shared equally between the engineer and the software developer.

    iii. The liability would be shared, but the software developer would likely have to pay more.

    iv. The engineer is fully liable; the engineer cannot blame the software developer, who is liable only to reimburse the cost of the software.

(j) An engineer designs a circuit-board layout for a client, using pirated electrical circuit design software, obtained from the Internet. The circuit-board design fails, because spacing is inadequate for installation of the chips. The client has spent much money in purchasing unusable materials and in preparing prototype circuit boards, which must now be scrapped. Obviously, the design software was grossly inadequate for the task. What is the likely outcome?

i. No one is to blame; the pirated software was faulty, and this is the root cause.

ii. The engineer is fully liable for the client's direct loss and the costs of any delay.

iii. The engineer is fully liable for the client's direct loss plus the costs of any delay, and may also be reported to the provincial Association for disciplinary action based on negligence, incompetence, or unprofessional conduct.

2.   John Jones is a professional engineer who works in the engineering department for a medium-sized Canadian city. He has been assigned to monitor and approve, on behalf of the city, each stage of the construction of a new sewage treatment plant, since he was involved in preparing the specifications for the plant. The contract for construction has been awarded, after a competitive bidding process, to the ACME Construction Company. About 10 days before construction is to begin, he finds a gift-wrapped case of rye whiskey on his doorstep, of approximate value $600. The card attached to the box says, "Looking forward to a good professional relationship," and is signed by the president of ACME Construction. Is it ethical for Jones to accept this gift? If not, what action should he take?

3.   Alice Smith is a professional engineer with several years of experience. Ima Turkey, who is a graduate of a school of engineering but has never registered as a professional engineer, approaches her. Turkey offers Smith $1000 if she will put her seal and signature on documents that Turkey has prepared to make them acceptable to the city official who issues building permits. Is it ethical for Smith to do this?

4.   René Brown is a professional engineer who has recently been appointed president of a medium-sized dredging company. Executives of three competing dredging companies approach him. He is asked to cooperate in competitive bidding on dredging contracts advertised by the federal government. If he submits high bids on the next three contracts, then the other companies will submit high bids on the fourth contract and he will be assured of getting it. This proposal sounds good to Brown, since he will be able to plan more effectively, if he is assured of receiving the fourth contract. Is it ethical for Brown to agree to this suggestion? If not, what action should be taken? If Brown agrees to this suggestion, does he run any greater risk than the other executives, assuming that only Brown is a professional engineer?

5.   As chief engineer of the XYZ Company, you interviewed and subsequently hired Mr. A for an engineering position on your staff. During the interview, Mr. A spoke of his engineering experience in Québec, where he worked previously, and stated that he was "a member of the Order of Engineers of Québec." You assumed that he was a licensed professional engineer in your province, also. You had business cards printed for his use describing Mr. A as a professional engineer, and he accepted and used these cards without comment. Some months later you received a call from a client, complaining that

Mr. A was calling himself a professional engineer when, in fact, he was not licensed to practise in your province. Upon investigation, you found this to be true. You fired Mr. A immediately. Was your action ethical in this matter? Was Mr. A's action ethical or legal? Refer to the code of ethics, the act, or both in your answer.

6. You are a professional engineer employed by a consulting engineering firm. Your immediate superior, who is also a professional engineer, is the project manager and prepares the invoices for work done on your projects. You accidentally find an invoice for recent work that you performed for a client. You are surprised to see that your time has been reported incorrectly. You decide to check further into this by reviewing your time sheets. The time sheets show that time charged to other work has been deliberately transferred to a different job. You try to raise the subject with your boss but are rebuffed. You are quite sure something is wrong, but are not sure where to turn. You examine the code of ethics for direction. What articles are relevant to this situation? What action must you ethically take?

7. You are an engineer with XYZ Consulting Engineers. You have become aware that your firm subcontracts nearly all of the work associated with the setup, printing, and publishing of reports, including artwork and editing. Your wife has some training along this line and has some free time. You decide to form a company to enter this line of business together with your neighbour and his wife. Your wife will be the president, using her maiden name, and you and your neighbours will be directors.

Since you see opportunities for subcontract work from your company, you believe that there must be similar opportunities from other consulting firms. You are aware of the existing competition and their rates charged for services and see this as a nice little sideline business. Can you ethically do this? If so, what steps must you take?

3.9 References

[1] G. C. Andrews, *Canadian Professional Engineering & Geoscience: Practice & Ethics.* Toronto: Nelson-Thomson, third ed., 2008. Sections 3.1.1 and 3.1.2 have been adapted, with minor changes, from this book, with permission of the publisher.

[2] Professional Engineers Ontario, *A Professional Engineer's Duty to Report: Responsible Disclosure of Conditions Affecting Public Safety.* Toronto: Professional Engineers Ontario, 2001. <http://www.peo.on.ca/complaints/duty_to_report.pdf> (March 9, 2008).

Chapter  Engineering Societies

Most undergraduates learn about engineering societies through design contests held on campus. For example, the Society of Automotive Engineers (SAE) organizes an annual Formula SAE race (as shown in Figure 4.1) for students who design, build, and drive the vehicles. However, the main role of engineering societies is to develop and distribute technical and professional information, including technical standards and research results. This key role is vitally important to practising professional engineers. This chapter describes:

- the purpose and history of engineering societies,

- differences between engineering societies and licensing Associations,

- the importance of engineering societies,

- some criteria for selecting and joining engineering societies.

Figure 4.1 A Formula SAE race, in which engineering students design, build, and race cars under specified rules (Photo courtesy of Formula SAE team, University of Waterloo).

4.1   The purpose of engineering societies

The general term *engineering societies* includes *technical societies, engineering institutes,* or even *learned societies.* Engineering societies are voluntary organizations, but they provide technical information that is essential to most practising engineers. This information is often too specialized to be provided by the provincial licensing Associations.

To avoid confusion, it should be emphasized that engineering societies are significantly different from the provincial professional engineering licensing Associations discussed in Chapters 2 and 3. The provincial Associations are created by law and have the legal authority to license engineers and to regulate engineering practice. Membership in (or registration by) the provincial Association is not optional, because a licence is required to practise engineering in Canada. Conversely, membership in an engineering society is voluntary.

Engineering societies often advocate on behalf of their members, and this is another distinctive difference from the provincial licensing Associations. The primary goal of licensing Associations is to protect the public, so they cannot advocate for engineers—it could be a conflict of interest. The roles of engineering societies and licensing Associations are therefore different and complementary.

The role of engineering societies in Canada

Engineering societies usually focus on an engineering discipline, such as computer, electrical, or mechanical engineering; on an industry or specialization, such as the nuclear, mining, or manufacturing industries; or on a specific membership, such as engineering students, engineers in management, or consulting engineers. Some well-known engineering societies are listed in Section 4.5.

The purpose of engineering societies may vary, depending on the society, but the main goals usually involve activities of vital interest to their members, such as:

- publishing technical information,
- developing technical codes and standards,
- encouraging engineering research,
- organizing engineering meetings and conferences,
- organizing engineering design competitions for students,
- organizing short courses on specialized topics,
- advocating on behalf of their members, when appropriate.

These activities have an immense impact on the engineering profession. They stimulate the creation of knowledge and innovative new products. Codes and standards guide engineering design to assure safety and product quality. Meetings, conferences, and short courses are extremely useful for the rapid dissemination of new ideas.

In addition, all engineering societies advocate on behalf of their members, and this is the most distinctive difference from the provincial professional engineering Associations. The primary mandate of the provincial Associations is to protect the public, so the Associations would have a conflict of interest if they took a leading role in advocating for engineers.

The role of engineering societies in other countries

The role of engineering societies varies among countries. The licensing laws in other countries are less comprehensive than in Canada, so the engineering societies play a role in regulating the engineering profession. Discipline-based engineering societies assess education and training programs and evaluate the qualifications of individuals who apply to be registered or "chartered."

In the United Kingdom, for example, registration as a chartered engineer (C.Eng.) is a desirable qualification, but is not generally required by law to practise engineering. However, British regulations have started to specify the C.Eng. for some safety-related work.

4.2 The history of engineering societies

The Industrial Revolution stimulated the need to disseminate technical information. Britain led the way in establishing engineering societies, with the Institute of Civil Engineers in 1818, followed by the Institution of Mechanical Engineers in 1848. In the decade following 1848, many additional societies were established [1].

In the United States, the first engineering society was the American Society of Civil Engineers, founded in 1852. Several others were established later, such as the American Society of Mechanical Engineers in 1880 and the American Society of Heating and Ventilating Engineers in 1894.

The American Institute of Electrical Engineers was established in 1884 as a national society, primarily for electric power engineers, but later electronics and communications engineers were included. In 1912, the Institute of Radio Engineers was formed as an international body for both professionals and non-professionals in the new field of radio communications. These two organizations merged in 1963 to create the Institute of Electrical and Electronics Engineers (IEEE), which is now the world's largest engineering society. Within the IEEE are about 40 specialty societies, with names ranging from aerospace electronic systems to vehicular technology, and with members in all parts of the world. The IEEE publishes over 130 specialized periodicals and magazines, and arranges or co-sponsors 450 technical conferences annually.

History of engineering societies in Canada

Several engineering societies were formed in Canada shortly after Confederation, between 1867 and 1900. Examples are the Canadian Institute of Surveying in 1882 and the Engineering Institute of Canada in 1887, originally the Canadian Society of Civil Engineers. The Canadian Institute of Mining and Metallurgy was formed in 1898.

The formation of student societies began in 1885 with the Engineering Society of the University of Toronto; surprisingly, the "Society was, indeed, a 'learned society,' and published and disseminated technical information [...] in addition to looking after the University undergraduates in engineering" [1].

In keeping with trends toward greater specialization, several Canadian engineering societies have recently been reorganized and new societies created. The Engineering Institute of Canada (EIC), which had served all disciplines for many years, recognized that it could not serve the many diverse specialties of engineering within a single organization; the EIC is now a federation of member societies, as listed in Section 4.5.

4.3	The importance of engineering societies

When you leave university and begin to practise engineering, your technical ability will be enriched by participation in (one or more) engineering societies, and your competence will be maintained. Moreover, maintaining your competence—through conferences, courses, workshops and similar professional activities—is not voluntary. Every provincial and territorial law (act) defining professional engineering (and some codes of ethics) requires the practitioner to maintain professional competence, and this requirement is also stipulated in the Engineers Canada guideline on continued competency assurance [2].

The provincial Associations have begun programs to monitor the continuing professional development (CPD) of their members, starting with the Alberta Association (APEGGA) with a mandatory program in 1998 and the British Columbia Association (APEGBC) with a voluntary program in 2001. As of 2007, all provincial and territorial Associations have adopted the Engineers Canada guideline, or are in various stages of adoption. Some programs are voluntary and some are mandatory. Fortunately, whatever CPD requirements exist in your jurisdiction, engineering societies can help you to maintain your professional competence.

Engineering societies hold a vast storehouse of useful knowledge that they readily offer to members. Societies also stimulate and encourage research into new areas, and provide a wide range of conferences, publications, courses, and workshops to disseminate the information to the engineers who need it and can apply it.

4.4   The relationship of the engineer to laws and organizations

The relationships between the practising engineer and the laws, agencies, and organizations that regulate and assist engineers are shown graphically in Figure 4.2. In each province, the government has passed a law to regulate the engineering profession, and under the law, an Association has been created. The council, staff, and committees of each Association enforce the relevant act and admit, regulate, and discipline members of the profession.

As described in Chapter 2, the provincial Associations are linked by Engineers Canada, which assists the Associations in establishing national policies, and represents the Associations in negotiating international issues. Engineers Canada has two key standing committees (as shown in Figure 4.2), which perform important tasks for the engineering profession. The Canadian Engineering Accreditation Board (CEAB) evaluates the quality of engineering degree programs, and the Canadian Engineering Qualifications Board (CEQB) advises the Associations on standards for admission to the profession, standards of practice, and ethical conduct.

The CEAB is particularly relevant to engineering students because it visits established engineering degree programs to verify that academic standards are met. New undergraduate engineering programs are evaluated by CEAB in the final year of the first graduating class. Although accreditation is rarely denied entirely, some programs are accredited for less than the full term of six years. This limited accreditation usually stimulates rapid remedial action by the university. The list of CEAB-accredited programs is published annually, but does not distinguish between limited and full-term accreditation.

Although the provincial Associations regulate the engineering profession, they cannot provide the lifelong learning that all engineers require to keep their knowledge up to date. The simplest way for you to keep up with the rapid changes in your profession is to join and participate in the activities of an engineering society specializing in your discipline. Most practising engineers are members of both a Canadian engineering society and an international engineering society.

Several Canadian societies exist to advocate for engineers in a more personal way. Although the members of these societies are licensed engineers or engineering students, the societies are not, strictly speaking, "learned" societies. These advocacy societies focus on improving the welfare of the individual, rather than the advancement of a specific discipline.

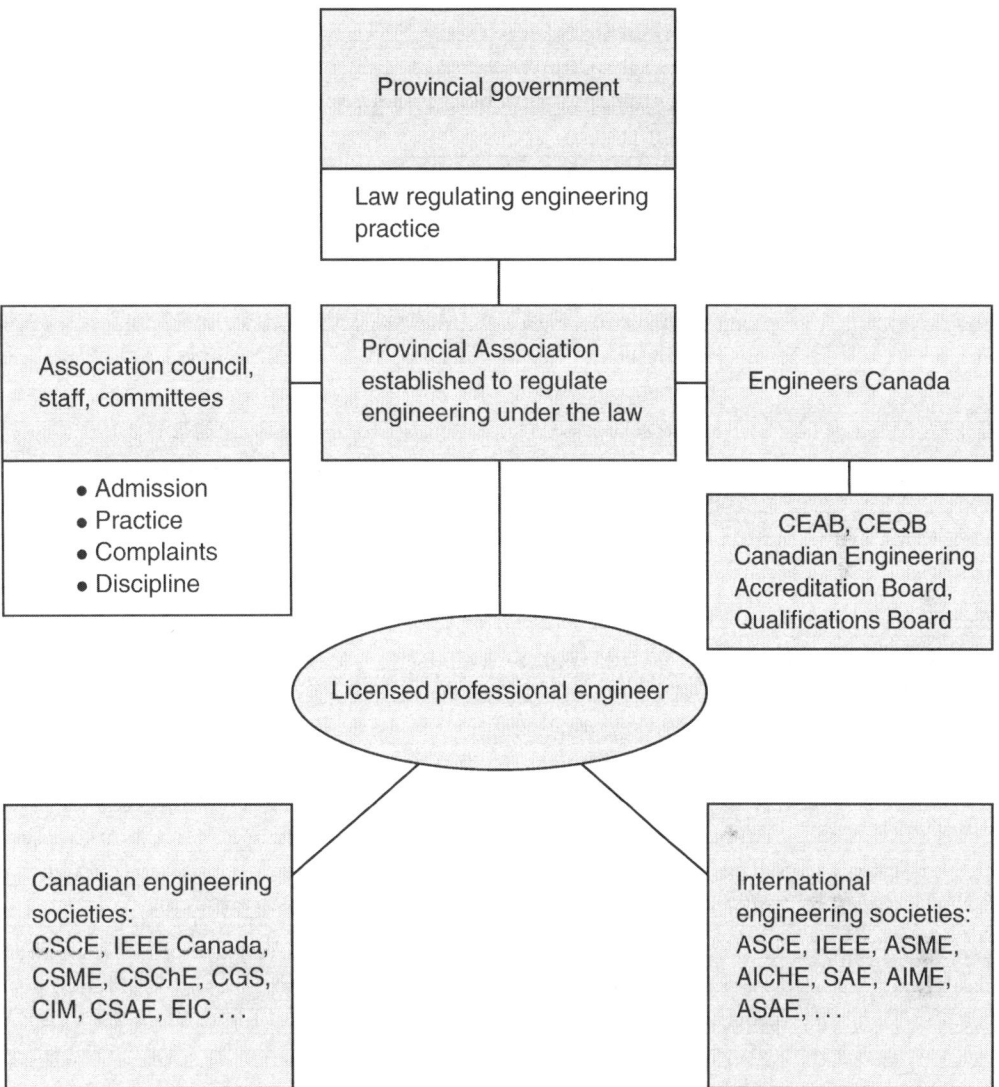

Figure 4.2   The relationship of the licensed engineer to the provincial engineering Association, Engineers Canada, and Canadian and international engineering societies.

For example, the Canadian Federation of Engineering Students (CFES) is an advocacy group that works on behalf of engineering students, mainly by improving communication between students at different universities, and acting as a liaison with Engineers Canada and with the National Council of Deans of Engineering and Applied Science. The CFES is the main organizer of the Canadian Engineering Competition, a prestigious annual design competition that attracts the top engineering students from across Canada.

Similarly, the Canadian Society of Professional Engineers (CSPE) is an umbrella body for provincial societies that promote the welfare and working conditions of individual licensed engineers. The first provincial society formed to advocate for individual engineers is the Ontario Society of Professional Engineers (OSPE), which describes itself as "a member-interest society for Ontario's professional engineers." Self-employed consulting engineers have had provincial advocacy groups for many decades. A federal "umbrella" group, the Association of Consulting Engineers of Canada (ACEC), coordinates the activities of these provincial consulting advocacy groups.

In the United States, the National Society of Professional Engineers (NSPE) advocates for the welfare of individual U.S. engineers, and its web site (www.nspe.org) is of interest to Canadian engineers. Self-employed U.S. consulting engineers also have advocacy groups, coordinated by the American Council of Engineering Companies (ACEC).

Finally, honorary engineering societies exist, such as the Canadian Academy of Engineering, which was established in 1987. The academy exists to promote engineering, to recognize important service, and to speak as an independent voice for engineering. It cannot be joined; the fellows of the academy elect new members, from all disciplines, based on their record of distinguished service and contribution to society, to Canada, and to the engineering profession. The academy is distinguished and exclusive; membership is limited to a maximum of 250 fellows.

4.5 Choosing your engineering society

As a professional engineer in a rapidly changing world, you have an obligation to maintain your competence. Engineering societies are one of the best sources of up-to-date technical information, and the societies are as important today as they were during the Industrial Revolution. Each professional engineer should be a member of at least one society. For tax purposes, engineering society dues are deductible from personal income, for practising engineers, under Canadian income tax laws.

Your choice of society depends mainly on your engineering discipline. Most major societies have student chapters, and you should, if possible, participate in their activities to learn about the society. However, if student chapters do not exist in your area, then you can learn about engineering societies easily through the Internet or by speaking with a practising engineer.

Internet addresses are listed on page 61 for some of the larger societies, and a search of the Internet for "engineering societies" will yield thousands of organizations, but not all are equally well established and trustworthy. A good guide to societies is the *International Directory of Engineering Societies and Related Organizations* [3], published by the American Association of Engineering Societies (AAES). This directory should be in your university library. The directory lists the purpose, membership, address, dues, and other information about each society.

4.5.1 Canadian engineering societies

Engineering Institute of Canada (EIC) As mentioned previously, the long-established EIC is now a federation of engineering member societies that cooperate to advance their common interests, such as maintaining engineering competence through continuing education courses, recognizing engineers by special awards, preserving Canadian engineering history, and creating opportunities for Canadian engineers to participate in disaster relief activities.

Table 4.1 lists the web site addresses of the EIC, its main constituent societies, and several other well-known Canadian engineering societies. The addresses were current at the time of publication of this book. In addition to the engineering societies listed in Table 4.1, several Canadian engineering societies have been established for advocacy, honorary, or charitable purposes. A sampling of these societies is given below, with their web addresses:

Advocacy societies
- Canadian Federation of Engineering Students (CFES), www.cfes.ca
- Canadian Society of Professional Engineers (CSPE), www.cspe.ca
- Ontario Society of Professional Engineers (OSPE), www.ospe.on.ca
- Canadian Military Engineers Association (CMEA), www.cmea-agmc.ca

Consulting engineering advocacy organizations
Every Canadian province and territory except Prince Edward Island and Nunavut has a voluntary advocacy organization devoted to consulting engineers. The names, locations, and web sites of these consulting advocacy organizations are available from the web site of their umbrella group: the Association of Consulting Engineers of Canada (ACEC), www.acec.ca.

Honorary society
- The Canadian Academy of Engineering, www.acad-eng-gen.ca

Charitable societies
- Engineers Without Borders (EWB), www.ewb-isf.org
- Registered Engineers for Disaster Relief (RedR), www.redr.ca

Table 4.1   Major Canadian engineering societies

Society		Web address
EIC	Engineering Institute of Canada	www.eic-ici.ca
CGS	Canadian Geotechnical Society	www.cgs.ca
CSCE	Canadian Society for Civil Engineering	www.csce.ca
CSME	Canadian Society for Mechanical Engineering	www.csme-scgm.ca
CSChE	Canadian Society for Chemical Engineering	www.chemeng.ca
CSEM	Canadian Society for Engineering Management	www.csem-scgi.ca
IEEE	Institute of Electrical and Electronics Engineers, Canada	www.ieee.ca
CNS	Canadian Nuclear Society	www.cns-snc.ca
CIM	Canadian Institute of Mining, Metallurgy and Petroleum	www.cim.org
CSBE	Canadian Society for Bioengineering	www.bioeng.ca
CMBES	Canadian Medical and Biological Engineering Society	www.cmbes.ca
CIC	Chemical Institute of Canada	www.cheminst.ca
CSSE	Canadian Society of Safety Engineering	www.csse.org
MTS	Canadian Maritime Section of the Marine Technology Society	www.mtsociety.org
CDA	Canadian Dam Association	www.cda.ca

Memorial Foundation The Canadian Engineering Memorial Foundation (CEMF) (www.cemf.ca), was created with the help of Engineers Canada, following the tragic murder of 14 young women in an engineering class at the École Polytechnique, Montréal, in 1990. CEMF is funded entirely by donations, and it awards scholarships to outstanding female engineering students.

4.5.2 American and international engineering societies

In 1904, five U.S. societies cooperated to create the United Engineering Trustees, Inc. The five founding societies have undergone changes over the years, and in 1998 the United Engineering Trustees, Inc., was reorganized as the United Engineering Foundation, Inc. More recently, some of the duties of the foundation were undertaken by the American Association of Engineering Societies, which claims a membership of over 25 organizations, most of them devoted to general disciplines.

U.S. advocacy society The principal American advocacy organization is the National Society of Professional Engineers (NSPE), www.nspe.org.

International societies Thousands of societies exist to develop and disseminate information about specific interests. Major American and international societies are listed in Table 4.2. Most of the large engineering societies include subgroups devoted to specialized topics. In fact, the IEEE, the world's largest engineering society, includes approximately 40 special-interest groups. The more-established organizations publish periodicals or other material that is available to members or is in your university library. However, the most convenient way to find most societies is by a simple web search. A search for the name of almost any engineering device, theory, or topic will often reveal a society devoted to the subject.

Table 4.2 Major American and international engineering societies

Society		Web address
ASCE	American Society of Civil Engineers	www.asce.org
AIMBE	American Institute for Medical and Biological Engineering	www.aimbe.org
AIME	American Institute of Mining, Metallurgical and Petroleum Engineers	www.aimeny.org
ASME	American Society of Mechanical Engineers	www.asme.org
IEEE	Institute of Electrical and Electronics Engineers	www.ieee.org
AIChE	American Institute of Chemical Engineers	www.aiche.org
ASAE	American Society of Agricultural Engineers	www.asae.org
ASEE	American Society for Engineering Education	www.asee.org
AIPG	American Institute of Professional Geologists	www.aipg.org
ANS	American Nuclear Society	www.ans.org
AGMA	American Gear Manufacturers Association	www.agma.org
EAMT	European Association for Machine Translation	www.eamt.org
SAE	Society of Automotive Engineers	www.sae.org

A word of caution is in order: Not all of the information obtained from the Internet is true. In particular, numerical data obtained from the Internet is not usually reliable enough to be the basis for engineering decisions. Only publications from reputable sources, such as the engineering societies listed in this chapter, may be depended upon to be accurate, because their publications must stand the test of peer review, debate, and validation. The established engineering societies, therefore, provide the extra measure of reliability needed for engineering decisions. Many American engineering societies also publish codes of ethics, although formal enforcement of these codes is extremely rare.

4.6   Further study

1.   Choose the best answer for each of the following questions.

(a)   In Canada, engineering societies were first formed

    i.   by French army engineers in Québec, prior to 1700.

    ii.   shortly after Confederation, between 1867 and 1900.

    iii.   during the Canadian centennial celebration, in 1967.

    iv.   during the centennial of engineering in Canada, in 1987.

(b)   Engineering societies are important, because they

    i.   develop and publish technical codes and standards.

    ii.   encourage engineering research and the exchange of results at conferences.

    iii.   organize engineering design competitions for students.

    iv.   assist engineers in remaining competent through specialized short courses.

    v.   perform all of the above functions.

(c)   The Canadian Engineering Accreditation Board (CEAB) is part of Engineers Canada, and is

    i.   the committee that evaluates and accredits university engineering programs.

    ii.   the national honorary engineering society, with limited membership.

    iii.   a national engineering advocacy society.

    iv.   the committee that advises on qualifications for admission and standards of practice.

(d)   Engineering societies are distinctly different from the provincial and territorial engineering licensing bodies (the Associations and the Ordre in Québec), because the societies

    i.   are created by acts of the provincial or territorial legislatures.

    ii.   exist primarily to assist engineers by developing and disseminating engineering information.

    iii.   exist primarily to regulate the engineering profession as required by law.

    iv.   require compulsory membership in order to practise engineering.

(e) The provincial and territorial engineering licensing bodies (or Associations) have agreed to act jointly on national interests through an "umbrella" body, called

    i. the World Federation of Engineering Organisations (WFEO).

    ii. the American Association of Engineering Societies (AAES).

    iii. the Canadian Society for Professional Engineers (CSPE).

    iv. Engineers Canada.

(f) The main goal of most engineering societies is to

    i. license professional engineers.

    ii. assist practising engineers.

    iii. publish books.

    iv. advertise the engineering profession.

(g) The Canadian Academy of Engineering is

    i. a national university for engineering.

    ii. a national honorary engineering society with limited membership.

    iii. a national engineering advocacy society.

    iv. a national engineering fraternity, open to all professional engineers.

(h) The main purpose of the provincial Associations (including the Ordre in Québec) is to

    i. encourage engineers to become entrepreneurs.

    ii. regulate engineering practice and protect the public.

    iii. publish engineering standards and codes of practice.

    iv. advocate on behalf of engineers.

(i) The Engineering Institute of Canada (EIC) has a long history, but has changed roles in the past decade or so. The role of EIC is now

    i. to license engineers who satisfy the education and experience requirements.

    ii. as a national honorary society, with limited membership.

    iii. to monitor and accredit university engineering programs.

    iv. as a federation of Canadian engineering societies, with a mission to assist continuing professional development.

(j) Information is increasingly obtained through the Internet. The established engineering societies play a leading role in this revolutionary access to information, because they

    i. have Internet addresses and well-developed web sites.

    ii. encourage research in applying newly developed techniques to engineering problems.

    iii. provide engineering information that can be trusted.

    iv. perform all of the above functions.

2.   Consulting engineers have an advocacy group, the Association of Consulting Engineers of Canada. What is the role of ACEC, and how does it differ from CSPE and the Ontario Society of Professional Engineers (OSPE)?

3.   Find at least five student design competitions currently underway or held in the last five years, such as the Canadian Engineering Competition sponsored by CFES or the Formula SAE race sponsored by SAE.

4.   Determine the purpose (mission), activities, criteria for admission, and financial support of the Canadian Academy of Engineering in more detail than described in this chapter.

5.   Find and list the requirements for continuing professional development (CPD) for professional engineers licensed in your provincial or territorial Association. Is the CPD requirement voluntary or compulsory? How is the CPD activity reported, and what amounts of CPD are recommended or required?

6.   A long-standing debate considers whether Canadian engineering societies are needed or whether Canadian engineers should simply join foreign-based societies that are already in existence and have a longer history, more publications, and a larger membership. Write a brief summary listing the pros and cons of these two alternatives. Does Canada need distinct engineering societies? Are engineering societies truly apolitical, or do national interests influence the research published in their journals and periodicals? Are dues paid to international engineering societies ever used to lobby governments in favour of a specific nation's interests? Are there uniquely Canadian conditions that would justify uniquely Canadian societies? What is your conclusion or recommendation?

7.   Find and list the requirements, if any, of your provincial Association for engineers to demonstrate continuing professional development (CPD) after receiving a licence. Is the CPD requirement voluntary or compulsory? If no CPD program is in place, what does your act or code of ethics state about the need for continuing competence or ensuring your competence to practice?

8.   Find your nearest advocacy association devoted to consulting engineers. What is its stated purpose?

9.   Visit the web site (www.cemf.ca) of the Canadian Engineering Memorial Foundation (CEMF) and read the qualifications and eligibility requirements for its scholarship program. Determine if you or members of your class are eligible to apply for a CEMF scholarship.

4.7	References

[1] L. C. Sentance, "History and development of technical and professional societies," *Engineering Digest*, vol. 18, no. 7, pp. 73–74, 1972.

[2] Canadian Engineering Qualifications Board, *Guideline on the Definition of the Practice of Professional Engineering.* Ottawa: Canadian Council of Professional Engineers, 2002. <http://www.engineerscanada.ca/e/files/guideline_definition_with.pdf> (March 9, 2008).

[3] American Association of Engineering Societies, *International Directory of Engineering Societies and Related Organizations.* Washington, D.C.: American Association of Engineering Societies, sixteenth ed., 1999.

Chapter 5

Advice on Studying, Exams, and Learning

Studying to be an engineer is not a spectator sport—success requires participation. Even if you were at the top of your class in high school, you are in a new league now, with classmates who know how to set priorities, schedule their time, take notes, and organize their studies. This chapter will help you to excel academically, but for best results, you must "read it before you need it." That is, you should read this chapter early in the academic year and apply these skills regularly. This chapter tells you:

- how much time you should expect to spend studying,

- how to organize yourself to study effectively,

- how to prepare for examinations and write them,

- what to do if things should go wrong in your studies.

If you follow this advice (and a first hint is shown in Figure 5.1), you will almost certainly succeed and still have the free time to enjoy university life. However, this chapter discusses only the main strategies for university success—for more help, your university probably provides study skills workshops. To find them, start with an Internet search of your university web site, using key words such as *study skills*, *note-taking*, *time management*, *test* or *exam* or *learning strategies*, *writing techniques* or *strategies*, or *study groups*. The university groups, publications, and workshops in references [1–9] are typical. If these aids do not help, speak to someone immediately: start by speaking to your professors, but senior students are also a good source of advice, and every university has counselling staff who are hired to provide personal help and confidential guidance. Finally, many books on study skills are available in your university library, and may be found by a simple search of the library catalogue, using the key words suggested above. A few examples of such books are references [10–12].

Figure 5.1 Resist the urge to sit in the rear rows of large university lecture halls. Sit at the front and participate.

5.1   The good and bad news about university studies

The good news is that you have survived a very competitive engineering admission process, so you almost certainly have the academic ability to graduate. The bad news is that your high-school study skills probably will not be good enough for university. Professors report that in some programs, the average grade obtained in the first year by a typical engineering student is significantly below that student's high-school average. A partial explanation for this phenomenon is that university programs typically admit only the top fraction of high-school graduates, so almost everyone in your university class was a top student in high school. If you want to cope with the higher university standards, you must study effectively.

Moreover, engineering courses are difficult. This should not be surprising; courses are difficult in all professional schools, including law, medicine, optometry, accounting, and others. Professional schools cannot risk losing their accreditation because of low standards. However, there is no required quota of student failures, as some students may believe, and the greatest problem for students is not high standards; it is coping with newfound freedoms.

5.2 How much study time is required?

An informal survey of university students showed that, in addition to lectures and laboratories, students typically spent 28 to 32 hours per week completing assignments and studying. That is five hours per day for a six-day week! You might need a little more or a little less study time, depending on your capabilities. In any case, organize your timetable so that, after time is allotted for lectures, clubs, sports, and entertainment, about 30 hours per week are free for assigned work and studying.

At a university, free time tends to get lost! Universities generally treat students as responsible adults. Some students may be tempted simply to relax and to neglect their studies, with tragic consequences. There are no watchful parents or high-school teachers to worry about, and university professors do not check attendance or remind students to submit assignments. However, students who lose sight of academic priorities get a rude awakening when they learn that universities promote students only on the basis of demonstrated performance, not on future potential. Don't be tempted to skip classes, ignore assignments, or procrastinate.

5.3 Managing your time

Every university student has heavy time commitments, many of which are beyond the student's control. You must attend lectures, laboratory sessions, tutorials, and meetings. Professors set deadlines for assignments and projects. However, you also need a personal life, to meet people, develop friendships, and enjoy university. Problems arise when these activities conflict or deadlines are missed. You need to manage your time!

You likely developed a technique for time management in high school. However, if not, the basic ideas of time management are very simple. Time management requires a little basic planning and discipline—but not much. Planning is simply a four-step process, from general plans to specific action, as follows:

- Term calendar: First, you need a general plan of the term. This is simply a calendar showing the dates of key events during your academic term.

- Weekly timetable: Second, you need a weekly timetable. Your classes are usually scheduled weekly, so this is a key document. Most importantly, it can be used to calculate your free time.

- "To do" list: Third, you need a master "to do" list of tasks to be done, and their deadlines, so that you can set priorities for your work.

- Synchronizing: Finally, you must synchronize these documents, usually every few days, to avoid conflicts, to set priorities, and to allocate your free time effectively.

The four steps are explained in detail below, but you may adapt these recommendations to suit your personal preferences.

Step 1:
Prepare a term
calendar

A calendar is a very useful map of the months ahead. Get a calendar that covers your academic term, and insert all key academic events, such as project due dates, field trips, mid-term tests, and final exams. Include major social events, also. If you know that you will be attending a special concert or watching the Grey Cup game or the hockey playoffs, put the event on the calendar and work around it. Keep your calendar up to date!

Every office-software program has a calendar template, and the university bookstore sells calendars of all sizes, from pocket size to huge white-board calendars that you bolt to the wall. (But don't confuse size with ease of use.) Wall calendars give you the big picture; however, they do not show routine activities, so that takes us to the next step: your weekly timetable.

Step 2:
Prepare a weekly
timetable

A weekly timetable shows your routine activities (like classes and labs), and you may already have this—universities usually give students a printed class schedule. Check that your lecture times and locations are correct. Your timetable is usually set for the term, but if your classes change drastically, revise your timetable. For example, when your final exams start, prepare a new weekly timetable (or exam timetable, as explained later in this chapter).

Write all your scheduled weekly courses and labs on your timetable, and block out time for normal mealtimes, regular sports or physical exercise, sleeping, social events, and spiritual attendance. The timetable should still show about 30 hours per week of "free" time. "Free" time means unscheduled time, free of lectures, labs, meals, meetings and social commitments. Most of it will likely be evenings or weekends. This is your study time. If you do not have 30 hours per week of "free" time (or fairly close to it), including evenings or weekends, then you should re-examine your timetable critically, and see if all the activities are essential.

Post a copy of your final timetable on your bulletin board with your wall calendar, and keep another copy with your notebooks, for use in class. To achieve your goals, you must apply your free time to the academic studies with top priority. However, to set priorities, you need a "task list" or "to do" list.

Step 3:
Keep a "to do" list

Make a task list or "to do" list, and review it daily. Carry a pad or notebook, and when your professors assign a task, add it to the list, along with the due date. If possible, make a rough estimate of the time needed for each new assignment. Add personal tasks to the list, also. When tasks are completed, cross them off the list, or start a new list.

Step 4:
Synchronize, set priorities, and schedule your "free" time

Every few days, you must "synchronize" your calendar, timetable, and "to do" list. The key purpose of synchronizing is to avoid conflicts between your calendar and your timetable, to set priorities for the tasks on your "to do" list, and to decide how to use your unscheduled "free" time.

Examine each task on your "to do" list, and decide its priority. The priority depends on urgency (or due date), time available, and work required. Number your tasks in order of priority. At this point, you can see what needs to be done first (on your "to do" list), and you can see when you can do it (from your calendar and timetable), so insert the tasks in the unscheduled "free" time on your weekly timetable. Your time is now properly managed, so get working on your tasks!

Some final advice

Time management always involves the four steps described above, but many students simplify the paperwork by buying planners. For example, many bookstores sell personal planning books, usually in leather ring binders. These planners contain all three documents in one binder (calendar, timetable sheets, and "to do" list).

Most planners also include an appointment book, and many students use the appointment book in place of (or in addition to) the calendar. This condenses the time management into a smaller format, which you can carry with you. It is usually convenient to use both an appointment book and a calendar, although it is a minor duplication. Conversely, if you condense everything (calendar, timetable sheets and "to do" list) into a single binder, it is very portable, but don't lose it!

Some students simplify the paperwork differently, by photocopying the timetable for each week of the academic term. This permits the calendar and the "to do" list to be synchronized, right on a copy of the timetable, each week. This method is simple, and works effectively. Many other modifications are possible. Use what works for you!

Other students go the electronic route, and use a Blackberry or a personal digital assistant (PDA) that fits in a pocket. However, the authors of this text recommend that students start with paper, to keep time management as simple and as cheap as possible. When you master the effective use of your time, your method will adapt easily to a computer or PDA. The four basic steps in effective time management are the same, regardless of the electronic complexity.

5.4 Preparing for the start of lectures

A few simple tasks can get your term off to a good start.

Prepare a timetable

As mentioned previously, schedule your lectures, clubs, sports, and entertainment, but leave at least 30 hours per week free for reading and studying.

Buy the specified course textbooks

Many university bookstores encourage textbook orders by Internet or telephone, so that bookstore line-ups can be avoided. Obtain the texts as soon as possible; sometimes texts sell out and must be restocked. Used books might be available, but make sure that you obtain the correct edition.

Skim through the texts before lectures start

As soon as you get your texts, skim through them. If you read even a little bit about a subject before a lecture, you will be astounded how simple and logical it becomes. If you cannot do this for every subject, choose only the most difficult courses. You will be able to ask good questions, clear up doubts, and avoid panic at the end of the term.

Obtain a detailed outline for each course

Each of your professors will post a course outline on the Internet or distribute a detailed course outline in the first lecture. If no outline is provided, speak to your professor, because you need to know what to expect. In particular, the professor must define the course content, the marking scheme, and the main assignments, early in the term.

5.5   Developing a note-taking strategy

Note-taking during lectures is a personal matter and most students have their own preferences. Before the term starts, you should decide what your note-taking strategy will be for each course. The authors of this text recommend a moderate note-taking strategy that is suitable for most courses, provided that a textbook has been specified and the professor is following it. Your class notes should include headings, derivations, and quotations that can be reviewed later and compared with the textbook. For this strategy, the general rules are:

- Make notes for every lecture.
- Include the course number and date on each page of notes.
- Make your notes brief but complete, with appropriate headings, textbook page references, the topics discussed, derivations that are difficult, and points that are controversial.

Regardless of your note-taking strategy, organize your notes in a file or binder, and review them, along with the textbook or other course aids, to learn the concepts. If you have questions, raise them at the next lecture or tutorial session. Reviewing your lecture notes for a few minutes before the next lecture will greatly improve your understanding of the material. Lecture notes are also an excellent resource for review prior to examinations.

5.6   A checklist of good study skills

The checklist below lists 10 basic points; they are common sense but are worth repeating. The checklist is arranged in order of importance, with the most important rules first. The first six points are absolutely essential—if you are not following all of them, you will likely have academic problems.

Attend the lectures

1. Attend lectures regularly, and pay attention. This is extremely important; a few skipped lectures may seem innocuous, but in a fast-moving course, they can begin a vicious cycle of losing interest and skipping more lectures, which ends when the entire class is dropped.

Submit all assignments

2. Even if assignments are not counted as part of the term grade, the final exam will undoubtedly include topics from the assignments. Work on assignments in tutorial sessions where you have the help of a teaching assistant.

Find a place to study

3. If your residence is not suitable for studying, use the study tables in the university library or unused classrooms, or even the local coffee shop.

Don't take part-time employment

4. Outside work is a common cause of poor performance. You may have had a part-time job in high school, but time is much more limited in university. Speak to a university student-loan counsellor if you have financial problems.

Don't procrastinate

5. Major assignments can turn into disasters if left to the last minute. Even an hour spent early in the project, organizing and scheduling how and when you will complete the assignment, will pay off dramatically as the deadline gets closer.

Work weekends

6. Don't schedule every weekend for trips, parties, or travel home. Some weekend work is essential, whether it is studying, catching up, reviewing, writing reports, or preparing for examinations.

Don't stay up all night

7. Late nights, whether for work or play, disrupt your life and can lead to serious health problems. To study effectively, you need a minimum of seven to eight hours of sleep each night, or at least 50 hours per week.

Concentrate on your work

8. When you study, do it! Work will be done more quickly if you concentrate on it. Don't study in bed or try to do two things at once, such as watching TV with your books open; you're just wasting time.

Read ahead

9. A brief preview of your texts every week gives you an overview of each subject and greatly aids the learning process. If you cannot do this for every course, just preview your most difficult courses. If a recommended course text is not very readable, ask your professor, librarian, or bookstore to suggest an alternative text, or seek information on the Internet.

Reward yourself

10. When you are studying, take a five- to ten-minute break every hour. Stretch, walk, exercise, or reward yourself with coffee or a snack. If you have a deadline for a major project, promise to reward yourself with an entertaining night off when the deadline is met.

5.7 Collaborating on assignments

Most assigned problems are intended to apply the principles introduced in lectures, and you should submit them regularly, on time. However, avoid the extremes of spending too little or too much time on them. An engineer must strive for an optimal return for the time invested. Occasionally, enthusiastic students spend an excessive amount of time on interesting projects. Remember the "law of diminishing returns": as the time spent on a project increases, the incremental benefit decreases.

Many students collaborate when working on assignments, and this joint work is usually beneficial. However, there is a fine line between collaboration and copying. When you submit an assignment with your name on it, it is assumed that you have prepared all of the material in the assignment, except where you indicate that material has been taken from other sources and the sources are cited. To put this point in clearer terms: exchanging ideas by talking to one another is collaboration; exchanging written materials is copying, and exchanging computer files is serious copying. Although copying is easy to do, professors and teaching assistants watch closely for it.

Another word for copying is *plagiarism,* which is defined as taking any intellectual property, such as words, drawings, photos, or artwork, that was written or created by others and presenting it as your own. If you submit an assignment with your name on the front, and include material inside that has been taken from others but not cited, then you have committed plagiarism, and you may be subjected to severe disciplinary action, including dismissal from the university. Plagiarism is discussed at several places in this book; in particular, the proper method of using material provided by others and citing the source is described in Chapter 8.

5.8   Preparing for examinations

Few people like exams, but they are essential in the university. Exams were originally devised, centuries ago, to prevent favouritism, and they are used today for the same purpose. Exams ensure that students are promoted on knowledge and ability, and not on apple-polishing, bribery, or luck. There are no limits on the number of students who can pass. Exams, therefore, are only an impartial metre-stick, applied to see that everyone measures up. Try to view them in a positive way, as an achievable challenge.

Examinations are also a learning experience; in fact, the effort put into summarizing and organizing the course material in preparation for an exam is usually very efficient learning.

As soon as the exam schedule is known, usually a few weeks before the end of the term, you should begin your exam routine, as described in the following paragraphs.

Make a timetable
- Prepare a personal exam schedule, showing your scheduled exam dates, plus any remaining lectures, tutorials, interviews, or important social events. Identify the uncommitted days remaining until exams begin. This is the time that you can control.

Review each subject
- Include a block of time in your exam schedule to review each subject before the exams begin. If your assignments interfere with your review schedule, talk to your professors; maybe an assignment can be delayed or shortened.

Organize your notes
- At this point you will appreciate having a course outline and dated notes for each lecture. Old exams are also very useful references; obtain them and list the topics covered by the old exams.

Review systematically
- Ideally, you should review each course in three stages: first, review the outline and purpose of the course; second, reacquaint yourself with the main topics; third, review derivations, assignments, and problems from previous exams. Prepare brief review notes as you go, containing definitions, summaries, and lists of equations. These notes will be very valuable for a final review the day before the exam.

Chemical stimulants

- Avoid the use of mood-altering drugs of any kind. Even coffee, in excess, can cause trouble. Collaborating with fellow students is useful, but avoid "all-nighters."

Final review

- Save the afternoon before the exam for a final review. You should have started studying early enough that you can spend these last few hours rereading your notes, trying old exam problems, and reviewing key points. Then relax, and make sure you get a good sleep.

5.9 Writing examinations

Even if you are well prepared, it is normal to feel slightly tense before an exam. Don't let it bother you; everyone else feels the same, even if you can't see it. These suggestions may help you.

Take a walk

- Take a brisk walk before the exam to reduce anxiety and clear your mind.

Arrive early

- Arrive a little early; check your pens and pencils, and visit the toilet.

Read the questions carefully

- *Read the exam paper!* Many students give excellent answers to questions that were not asked.

Easiest question

- Always solve the easiest question first.

Defer the tough questions

- If you are faced with a really tough question, read it thoroughly and go on to the next question. Your mind will work on it subconsciously, and you may have the answer when you come back.

Write clearly

- Write clearly and solve problems in a logical order. This shows a methodical approach to problem-solving, and almost always will get you a higher grade.

Describe how to solve the problem

- If time is running out, describe how you would solve the problem if you had more time. A blank page gets a zero, but a description of how to proceed might get partial marks. Remember that the exam is a communication between you and your professor, so you may include any comments, references, or explanations that you would make orally.

5.10 When things go wrong

Everyone has bad luck occasionally: sickness, a car accident, family problems, and legal, social, or other problems. Every university has a procedure for helping students with serious personal problems at exam time. However, the student must take the initiative! If you have a serious problem that clearly interferes with your ability to write an examination, then tell someone! The appropriate person may be your medical doctor, counsellor, professor, department chair, or the exam-room supervisor (proctor). However, the earlier you speak out, the easier it is to remedy the problem. Don't wait until you receive your examination results to say that you were seriously ill, for example.

After the examination, if you feel that there is an error in your grade, it can be reviewed. This causes some inconvenience and should not be done casually. A formal letter to the university registrar explaining your reasoning (and there must be a reason) will set the process in motion. Refer to your university calendar for more information about the examination review process.

Finally, if you find that the subjects you are studying are not really the topics that you expected when you enrolled in engineering, you may have enrolled in the wrong program. Your professors and faculty counsellor may be able to help you define your career objectives more clearly. This book, particularly Chapters 1 to 4, is intended to help you clarify your career choice. However, if you still have doubts, get more advice and guidance. Every university has a counselling service available to assist students, such as shown in references [1–9]. If you are not sure how to contact your university counselling service, try searching your university web site.

| 5.11 | **Your professional career and lifelong learning** |

There are three major learning tasks in the lifetime of every engineer. The first is to succeed as a student, the second is to obtain and profit from practical experience, and the third is to maintain continuing professional competence in the fast-changing world of engineering.

Take charge of your timetable

This chapter explains how to succeed at university. Since the university demands most of your time, you must identify the free time that is left over, and use it effectively. If you apply the advice in this chapter, you will succeed as a student and still have time to enjoy the friendship, diversity, and good times of university life.

Document your work experience

Remember to document your work experience, particularly if you are in a cooperative (work–study) program or an internship program. Don't delay, or the details may slip from your mind. Your student work experience is valuable, because work-terms and internships link theory to practice. At the end of each project, or at the end of each work-term, prepare a summary of your activities, so that it can be inserted into your résumé.

Updating your résumé is a minor task, but is very useful—it gives you a sense of pride and achievement, and an up-to-date résumé will be essential for interviews when you apply for your next job.

Documenting your work experience may shorten the time it takes to get your engineering licence. Most provincial Associations require four years of documented experience to obtain an engineering licence. However, they will typically accept up to one year of experience obtained after the midpoint of a bachelor's degree program if the work is well documented, so keep your résumé up to date.

Prepare for lifelong learning

University studies are a springboard to a professional career that involves life-long learning. As explained in Chapter 2, every provincial and territorial professional engineering act (or code of ethics) contains a clause that requires the practitioner to maintain professional competence. Every provincial and territorial licensing Association has adopted either a mandatory or voluntary continuing professional development (CPD) program. In fact, several licensing Associations monitor and audit the continuing professional development of professional engineers and geoscientists.

In summary, remember that your university engineering program is not just a set of related courses—it is an entry into an important and dynamic profession that looks to the future and builds on new developments. Your task is to prepare as an undergraduate, profit from your work experience, and maintain your competence as new knowledge becomes available.

5.12 Further study

1. Choose the best answer for each of the following questions.

(a) Engineering attracts some of the very best students. The university grades achieved by most first-year engineering students are typically

    i. much higher than their high-school grades.

    ii. about the same as their high-school grades.

    iii. much lower than their high-school grades.

(b) The informal survey of engineering students reported in this chapter revealed that engineering students typically spend how much time per week completing assignments and studying (outside of lectures and labs)?

    i. 12 to 24 hours per week

    ii. 28 to 32 hours per week

    iii. 32 to 40 hours per week

    iv. 40 to 64 hours per week

(c) The main objective in managing your time is to

    i. identify your free time and then schedule this time to achieve your goals.

    ii. prepare a term calendar.

    iii. prepare a weekly timetable.

    iv. maintain a "to do" list.

(d) In this chapter, the authors provide a strategy for writing examinations. Which of the following is *not* recommended?

    i. Take a brisk walk before the exam.

    ii. Read the exam paper.

    iii. Always solve the hardest question first.

    iv. If you are faced with a tough question, go to the next question and come back later.

(e) Studying for an examination by reviewing old exams is

    i. unfair and not recommended.

    ii. fair but not recommended.

    iii. an obvious step in a well-organized study plan.

(f) Skipping lectures

    i. is normally not permitted in universities.

    ii. has been found to be occasionally useful, since it increases student interest in the course.

    iii. can begin a vicious cycle of losing interest and skipping more lectures.

    iv. is acceptable, as long as you have paid your tuition fees.

(g) Collaboration with other students on course assignments is

    i. strictly forbidden.

    ii. allowed but not recommended.

    iii. encouraged, provided that the collaborator is identified on the assignment.

    iv. encouraged, providing that only ideas are exchanged (no written matter).

(h) If you have a serious personal problem that affects your academic performance, you should

    i. wait until examinations are over to speak to your professors.

    ii. keep the situation a secret, since problems may affect your assigned grades.

    iii. contact a counsellor or (in an exam) speak to the exam proctor.

    iv. tell a relative, as soon as possible, to put the situation on record.

(i) Staying up all night is

    i. occasionally necessary before an examination and is recommended.

    ii. not recommended at any time and may lead to health problems.

    iii. recommended for studying but not for partying or examinations.

    iv. recommended for partying or studying but not for examinations.

(j) Engineering, like almost every profession, involves lifelong learning. Which of the following will *not* help you to maintain and demonstrate your continuing competence?

    i. documenting your experience at the end of each project (or work-term)

    ii. joining the student branch of your engineering society (if one exists)

    iii. becoming a student member of your provincial Association (if permitted in your province)

    iv. selling your engineering textbooks to pay tuition fees

2.   Using your computer, prepare a calendar for the end of your current academic term, including an end-of-term exam schedule, as suggested in Section 5.3. Insert dates of known university or class events. Compare and exchange templates with other class members to ensure that all essential events are included. When you receive your final exam schedule, insert the exam dates as suggested in Section 5.3, and schedule your exam review.

3. Search your university web site using the key word *plagiarism.* Search also using the key words *academic offence.* What is your university's policy on plagiarism? Does your university classify plagiarism as an academic offence?

4. Search your university web site for documents including the phrase *study skills,* as suggested on page 67. Compare the study skills on the web site with the suggestions in this chapter. Do the suggestions that you found on the Internet substantially agree or disagree with those in this chapter?

5.13 References

[1] UBC Student Services, *New to UBC, Welcome.* Vancouver, BC: University of British Columbia, 2007. <http://www.students.ubc.ca/newtoubc/index.cfm> (March 9, 2008).

[2] University Student Services, *Academic Support Centre.* Edmonton, AB: University of Alberta, 2007. <http://www.uofaweb.ualberta.ca/academicsupport> (March 9, 2008).

[3] University of Manitoba, "Learning assistance centre," 2005. <http://umanitoba.ca/student/u1/lac> (March 9, 2008).

[4] Counselling Services, *Study Skills Package.* Waterloo, ON: University of Waterloo, 2005. <http://www.adm.uwaterloo.ca/infocs/study/index.html> (March 9, 2008).

[5] University of Toronto, "Counselling and learning skills service," 2007. <http://www.calss.utoronto.ca> (March 9, 2008).

[6] First Year Office, *Reading and Study Skills Workshops.* Montréal, PQ: McGill University, 2007. <http://www.mcgill.ca/firstyear/workshops> (March 9, 2008).

[7] Concordia Counselling and Development, *Groups & Workshops—Learning and Study Skills Strategies.* Montréal, PQ: Concordia University, 2007. <http://cdev.concordia.ca/workshops/learning.html> (March 9, 2008).

[8] Student Services, *Study Skills Counselling.* Saint John, NB: University of New Brunswick at Saint John, 2007. <http://www.unbsj.ca/studentservices/study> (March 9, 2008).

[9] Student Affairs and Services Counselling Centre, *Study Skill Links.* St. John's, NL: Memorial University of Newfoundland, 2006. <http://www.mun.ca/counselling/academic/studyskill.php> (March 9, 2008).

[10] D. H. O'Day, *How to Succeed at University.* Toronto: Canadian Scholars' Press, 1990.

[11] M. N. Browne and S. M. Keeley, *Striving for Excellence in College: Tips for Active Learning.* Upper Saddle River, NJ: Prentice Hall, 2001.

[12] P. Schiavone, *How to Study Mathematics.* Scarborough, ON: Prentice Hall Canada, 1998.

Part

II

Engineering Communications

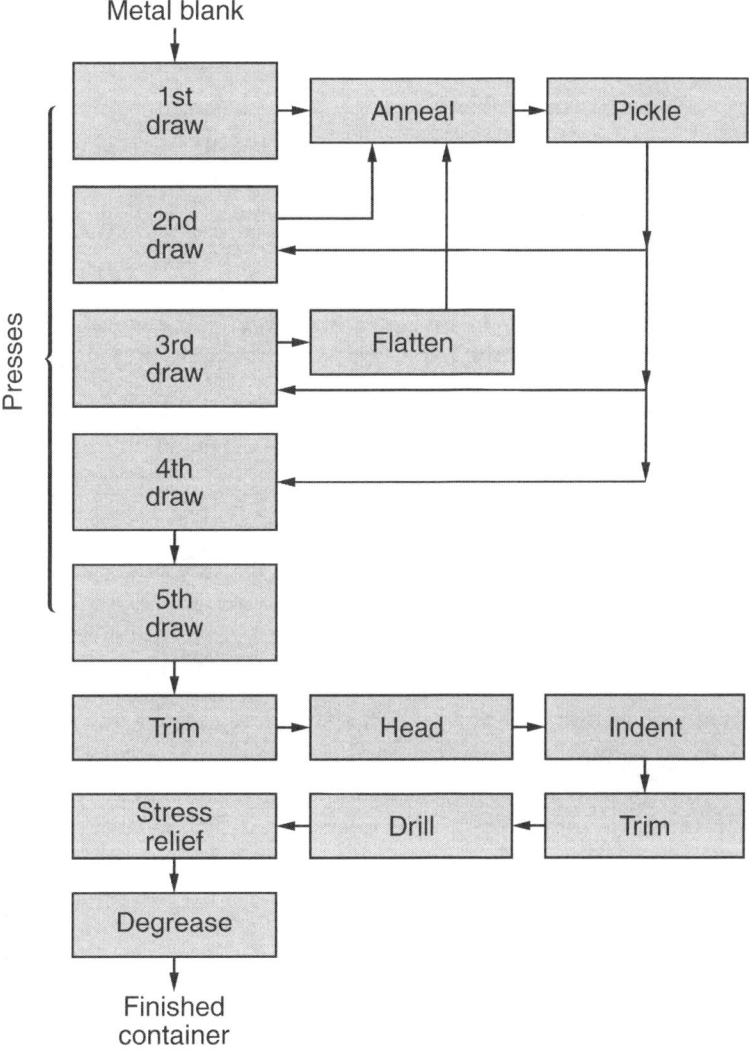

Figure II.1 Flowcharts are used in all engineering disciplines to communicate the essentials of a sequence of steps or events. The example above shows the steps in the manufacture of seamless metal containers by a process called deep-draw forming.

Engineers are creative people! However, creative ideas must be conveyed clearly to others. You will need effective communication skills in order to inform and persuade others, and to record the progress and results of your work. In addition to normal business documents, engineers must produce clear and correct technical reports, drawings, diagrams, graphs, letters, memoranda, and electronic communications.

Engineers must also interact with the business world, and the ability to write and speak effectively in business transactions is important. On rare occasions, you may have to defend your actions and conclusions to business associates or even under the scrutiny of a lawyer in court. Most engineers seek to avoid legal confrontations, but some act regularly as expert witnesses.

Effective communication skills, which are essential for successful engineering practice, should be learned early and refined throughout your career. Part II covers the basics of effective written and oral communication, in the following four chapters:

Chapter 6   **Technical documents:** What types of documents do engineers write? This chapter describes documents that are typically found in an engineering project or are common to engineering and business, and explains briefly how to prepare them. An introduction to information gathering and hints for effective oral presentations are also given.

Chapter 7   **Technical writing basics:** Are you familiar with the basic rules of English grammar? This chapter gives you hints on writing style and punctuation and discusses some errors to avoid. The chapter concludes with hints for the use of the Greek alphabet in technical writing and a grammar self-test.

Chapter 8   **Formal technical reports:** This chapter concentrates on the technical report, which is one of the most common engineering documents, and one of the most important. Technical reports may discuss a wide range of topics and are normally organized according to the basic rules explained in this chapter. Knowing how to begin is also important in report writing, and this chapter explains six simple steps in writing an engineering report. The chapter concludes with a report checklist to help you review your report before submitting it.

Chapter 9   **Report graphics:** The term *graphics* includes diagrams, charts, graphs, sketches, artwork, and even engineering calculations. Graphics often convey the central message in engineering documents, where you frequently must describe trends, patterns, geometric concepts, or complex shapes. This chapter describes basic principles of creating good graphics and the rules for including them in documents such as technical reports.

Chapter

Technical Documents

Technical writing is distinguished by an emphasis on objectivity, clarity, and accuracy. Graphics are included as necessary to explain and clarify. The information must be presented logically and objectively, and all conclusions must be supported by the document contents or by reference to other documents. Considerable skill and creativity may be required to present complex material clearly and concisely, but artistic licence is not allowed when presenting conclusions.

Technical writing is an art and requires talent and intelligence, but it is also a craft, the tools (see Figure 6.1) and strategies of which can be learned. In this chapter, you will learn about the kinds of documents you may have to write as an engineer and acceptable formats for them. The chapter also contains an introduction to research methods and to technical presentations and associated written or visual aids.

There are many reference books containing advice for writing technical documents; a selection is given by references [1–9] at the end of this chapter.

A technical document, whether produced at work or for academic credit, is intellectual property and subject to copyright, as discussed in Chapter 17.

Figure 6.1 The computer has done to the typewriter what the automobile did to the horse and buggy. In the recent past, an engineer would instruct a typist equipped with a modern version of the antique shown, but now engineers are expected to participate in computerized publishing. Higher-quality documents result, but expectations of quality are correspondingly higher.

6.1 Types of technical documents

The range of technical documents produced by engineers is as large as the set of organizations and situations in which engineers find themselves. Nevertheless, the most common document types are easy to list. Most technical documents are variations of the documents described here.

Documents are generally produced with the aid of computers and word-processing or graphics software. Many documents are meant to be read on a screen or in print format, but the overriding requirement for clarity is the same in both cases. The principles involved in the creation of the two formats are nearly the same, except that very long or complicated documents are more easily comprehended when they have been printed on paper.

Organizations may have their own requirements for the style and format of technical documents. Follow the general guidelines in the following sections, unless your organization has specific rules.

6.1.1   Letters

Letters are used to communicate with people outside the company or organization of the sender. The distinguishing characteristic of a technical letter is not found in its format, but in the requirements for clarity and objectivity.

A typical letter, shown in Figure 6.2, has eight basic parts:

1. company letterhead including the sender's return address,

2. date that the letter was signed,

3. inside address of the recipient,

4. subject line,

5. salutation,

6. body, organized as required by the content, but beginning with an introduction that presents the purpose of the letter and desired outcome if any, and ending with a closing statement of follow-up actions desired or to be taken,

7. complimentary closing,

8. signature of sender with printed name and title.

When attachments are included with the letter, a note is made at the bottom of the page, as shown in the example.

Date formats   Format the date as shown in the figure, unless you are required to do otherwise. Never use the format 09/12/2009: this is ambiguous and typically means 12 September 2009 in the United States, but 9 December 2009 in Europe.

6.1.2   Memos

The word *memo* is a short form of *memorandum* and literally means "something to be remembered." However, normally the word simply means a written communication between people within a company or organization. Memos are used for all informal and formal written communication, except that letters may be preferred for very formal internal communication such as the establishment of a contractual obligation.

As shown in Figure 6.3, a memo contains the following components:

1. heading: the title "Memorandum" or "Memo,"

2. "To" line: name or names of the recipient or recipients,

3. "From" line: name of the sender,

4. optional "Copies" line: persons to whom copies of the memo are sent for information; sometimes the archaic "cc," meaning "carbon copy," is substituted for "Copies,"

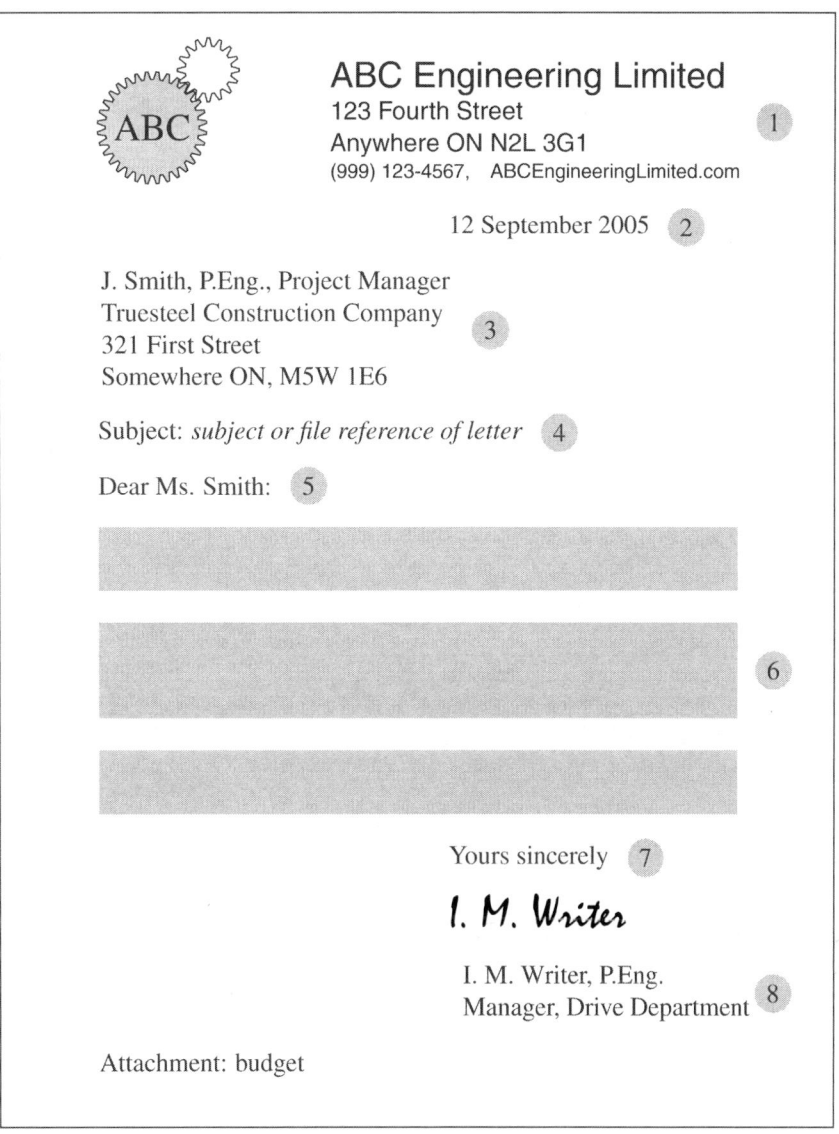

ABC Engineering Limited
123 Fourth Street
Anywhere ON N2L 3G1   ①
(999) 123-4567,   ABCEngineeringLimited.com

12 September 2005   ②

J. Smith, P.Eng., Project Manager
Truesteel Construction Company   ③
321 First Street
Somewhere ON, M5W 1E6

Subject: *subject or file reference of letter*   ④

Dear Ms. Smith: ⑤

⑥

Yours sincerely   ⑦

I. M. Writer

I. M. Writer, P.Eng.
Manager, Drive Department   ⑧

Attachment: budget

Figure 6.2   The format of a typical business letter. The numbered circles denote standard letter components. Reference to an attachment is included.

5. "Date" line,

6. "Subject" line: subject of the memo,

7. body,

8. signature, which may be optional, depending on company policy.

Unlike letters, informal memos typically need not be signed. The format shown is common, but any neat format that contains all the necessary information is usually acceptable. Most word processors provide letter and memo templates.

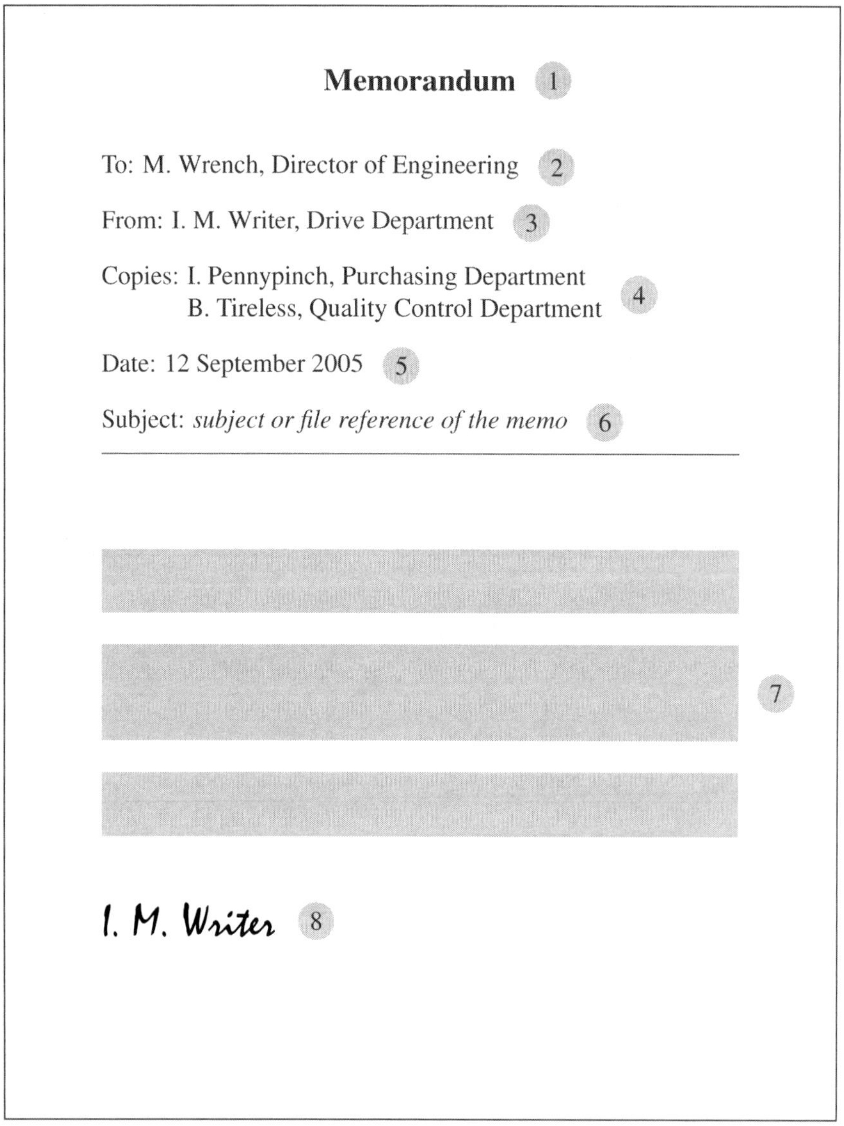

Memorandum ①

To: M. Wrench, Director of Engineering ②

From: I. M. Writer, Drive Department ③

Copies: I. Pennypinch, Purchasing Department
        B. Tireless, Quality Control Department ④

Date: 12 September 2005 ⑤

Subject: *subject or file reference of the memo* ⑥

⑦

I. M. Writer ⑧

Figure 6.3 A typical memo. The "Copies" item is included when necessary. The signature may be optional, depending on company policy.

6.1.3   Email and electronic messages

Electronic mail (email) sent and received by computers or portable electronic devices is pervasive. An email message is like a memo that has not yet been printed. Thus, email can be informal or formal, although documents that must be signed will have to await the widespread adoption of encrypted digital signatures. The great convenience of electronic messaging raises several issues, which are discussed in the following paragraphs.

Privacy Many companies reserve the right to inspect any message sent using their facilities, and some strictly prohibit the use of their facilities for personal purposes. On some systems, the number and size of messages sent and received by an individual are available to any other user of the mail system. Never send an email that you would not want your boss or colleagues to know about or read.

Security   Business email may be sent through the Internet, but many large companies maintain their own internal networks for security reasons, and other companies make arrangements for encrypting messages sent through public channels. Never send an unencrypted email containing material that would cause difficulty if disclosed.

Disagreement   Anyone who has witnessed the animosity that is common in Internet discussion groups will be aware that debate quickly tends to become hardened or acrimonious from lack of eye contact, body language, and voice tone, which are important parts of face-to-face communication. If you need to resolve a difference of opinion with someone, speak to him or her personally or use the telephone, in preference to debating by email. The exchange of memos can also be potentially acrimonious, but email can be sent much more quickly and can lead to interpersonal difficulties that should not exist in a professional environment. Therefore, if you must debate with someone using email, pay special attention to being polite, wait a day if possible before returning messages, and carefully reread your replies before sending them.

Text messaging   Like email, electronic text messaging is widely used, and messages sent through company facilities may be recorded like any other business document or communication. The ability to send quick messages can be a great convenience but can also be cause for sober reflection about remarks that were sent too hastily. Never send a message that you would not want your fellow employees or other colleagues to see.

6.1.4   Internet postings

The Internet provides a means by which anyone can post opinion, commentary, fantasy, practical information, or indeed, anything that can be written. To say that something has been "posted" is an apt description, since placing a document on the Internet is analogous to posting a piece of paper on a bulletin board. The posting may be available worldwide to anyone or to a more restricted audience, depending on reading habits or restrictions placed on membership of reader groups.

It is essentially impossible to remove or suppress a document that has been posted on a web page or in a news group. You should assume that anything you post may be read by a future prospective employer. Some companies seek out employees by looking for postings by people who have used and understood their products. Others search for postings of job applicants in order to find information about them prior to an interview. It goes without saying that if you have been unhappy with a company or individual, then venting your frustration on the Internet could lead to great trouble, since the laws of libel apply as for any other publication. Restricting the posting to a social networking forum may provide no effective protection. Post on the Internet only material that you would be willing to print and pin to the wall of your work space.

6.1.5   Specification documents

In order to obtain an engineering product or service, a client or purchaser usually writes a specification document containing the criteria that must be satisfied by the product or service. The specification document is used by the engineers who design, build, or otherwise provide the product. Thus a specification contains the criteria that the desired product or service must satisfy, and may form part of a contract between the client and provider. The client may be an external supplier or another department within the same organization.

A specification document is basically a list of criteria or tests that determine the characteristics required of a desired product, component, process, or system. Thus, for example, an airline company might produce a set of specifications for an airplane that it wishes to purchase, listing the required payload, range, speed, operating environment, internal fixtures, delivery schedule, payment constraints, and other factors. Manufacturers would then bid on the proposed contract, adapting their existing airplane types to meet the desired specifications, and would likely have to go through a similar process of writing specifications to obtain the engines, electronics, landing gear, hydraulic systems, and the many other components required. The process is repeated at increasing levels of detail. Thus, specification documents may be used from the most general level to describe complete airplanes, down to the detailed level to describe the size and strength of individual bolts, for example.

A specification document generally contains the following.

1. Introduction and scope: the general purpose of the product or service required, with an overview of the range of application.

2. List of requirements: each requirement that must be satisfied is listed, possibly with the procedures to be used to test whether the requirement is satisfied.

6.1.6   Bids and proposals

Bids and proposals are offers from engineers to provide services. A *bid* is an offer to provide specified services; a *proposal* typically suggests a means of meeting a need that has not been specified precisely and offers to provide the required service.

Thus, a potential client may send a "request for quotation" to eligible bidders, including specifications for the job to be done. The response to the client is a bid document, which states how the engineering work will comply with the specifications and any exceptions that are proposed, together with a price at which the company is willing to do the work. Other information may be required by the client or offered by the engineer, such as the qualifications of the individuals who will perform the work. Government agencies tend to have strict rules for the format and timing of bid documents.

When required engineering services are not precisely defined, and there are several possible solutions, then a "request for proposal" is issued. The engineer is then free to outline a solution and to offer to provide it. Since a proposal is in part intended to convince a prospective client of the competence of the engineering provider, the proposal document is produced using a permanent binding, in a formal but attractive style, and is accompanied by a letter of transmittal as discussed for engineering reports in Chapter 8.

Bid and proposal documents, when accepted, may form part of a legal contract that binds the client and engineering group together for the project.

6.1.7	Reports

Reports may be produced in many forms. Usually the production of the document has been commissioned by a superior or client, either to answer specific questions, to investigate a situation in the interest of the client, or simply to provide a record of an event or situation. Therefore, reports may be technical or non-technical, formal or informal, depending on the circumstances.

Formal technical reports may be the official record of a company's conclusions or actions, and may have to withstand the scrutiny of a legal proceeding, for example. Chapter 8 discusses formal technical reports typically required within a company, by a client contracting for engineering services, or from a student returning from an industrial internship or work-term.

In the following, the logical structure of reports is described, together with their physical structure. This section then discusses experimental and laboratory reports, evaluation reports, and progress reports.

Logically, there are three main components in a report.

Introduction
1. The introduction gives the purpose and background of the work presented in the report. The questions being investigated are posed, together with background information required by the intended readership.

Detailed content
2. The main body contains the investigations, results, or analysis satisfying the purpose of the report, in as much detail as the intended readers are expected to require. The material is presented in a clear, logical sequence, using sections and subsections as necessary.

Conclusions
3. The conclusions section gives the answers to the questions posed in the introduction. In a long report, these answers may have appeared in the detailed content and are summarized in this section; in a shorter report, the conclusions supported by the analysis are stated. It should be possible to pair each conclusion with a question posed in the introduction.

The physical components of many reports are similar to those of a book.

Front matter
1. The front matter contains components such as the title page and table of contents, which introduce and index the document. In the past, the front matter could be composed only after the body of the document was finished and the page numbers of the body had been determined. Therefore, the front matter had to be given a separate set of page numbers, which were printed in lowercase roman numerals. This numbering tradition persists, even with the use of computers for report production.

Body
2. The body material is divided into numbered sections, beginning with the introduction section and typically ending with the conclusions section, with other sections and

subsections as necessary to present the material clearly and logically to the intended audience.

Back matter   3. At the back of the report there may be material that supplements the contents of the body but is not essential for reading the report. The back matter may also contain material that is too voluminous to put in the body or that would otherwise break up the flow of the body. Typically the back matter contains a list of references and possibly appendices. Appendices are structured like report sections and are normally labelled with the letters A, B, ... , rather than numbered.

Report formats vary according to their formality, purpose, and content. However, specific formatting rules may be imposed by the client or by the company producing the report. It is always wise to check for special requirements, and when several documents of the same type are to be produced, to develop a document template in the correct format, thus ensuring uniformity.

The format of a formal engineering report is discussed further in Chapter 8, but some other common report types are briefly described below.

Experimental and laboratory test reports   All engineering students will have to write laboratory reports during their degree programs. The purpose of the laboratory is typically to illustrate theoretical concepts studied in class or to teach a test procedure, and the report is written according to the principles of the "scientific method," which requires the objective presentation of the work and the conclusions that follow from it. A list of typical headings is shown in Table 6.1. The main part of the report consists of the numbered sections, beginning with the introduction section and ending with the conclusions section.

In business and industry, the purpose of the technical report is rarely to verify newly learned theory; the report is usually intended to determine

- the feasibility of a new process, product, or device to be used, for example, in the design of a new product,

- the results of a standardized test of performance or safety, or

- exploratory data, such as soil tests for building foundation design.

A test report generally must contain all of the details of the apparatus, procedure, and test conditions. However, many reports are the result of routinely conducted standard tests, such as the verification of the properties of concrete or tests of the hardness of a metal alloy. In such cases, the report may consist only of an introduction, reference to the industry standards defining the tests, a statement of any variances from or exceptions to normal procedure, the measured data with its analysis, and resulting conclusions.

Evaluations   Product or process evaluations may be required for product improvements, to investigate unexpected performance, or to judge the qualities of an alternative or competing product. In all cases, the introduction must carefully state the criteria by which the evaluation is being made, and the conclusions must compare performance to the stated criteria.

Feasibility reports   Feasibility reports are similar to evaluations but discuss projected or hypothetical products or processes, rather than those that are available. Prototypes of proposed products

Table 6.1   Components of a typical laboratory report, with numbered sections in the body.

Title page	Department name, course name and number, professor name, student name and number or other required identification, report title, date.
Summary	A summary in less than one page of the report contents and conclusions.
Contents	A list of the report section and subsection titles and numbers, with page numbers.
List of figures	The figure numbers and captions, with page numbers.
List of tables	The table numbers and captions, with page numbers.
1. Introduction	The purpose and circumstances under which the experiment was conducted and the report written. The questions to be answered in the report must be posed.
2. Theory	The theoretical concepts required to understand the experiment and the hypotheses that are to be tested.
3. Equipment	A description of the equipment required, given in sufficient detail to duplicate the setup in equivalent circumstances.
4. Procedure	The methods used to obtain the observations and data recorded in the report.
5. Results	The data recorded with qualitative observations as necessary.
6. Analysis	Computations with graphs or other information required to compare the data with the hypotheses tested.
7. Conclusions	The answers to the questions posed in the introduction.
References	Author, title, and publication data in a standard format for the documents cited in the report.
Appendices	Large tables, diagrams, computer programs, or other material that is not meant to be read as part of the body of the report.

are sometimes the subject of feasibility reports, in order to predict the performance of the final product.

Evaluations and feasibility reports have a common structure: the performance of the whole is determined by examining its parts, comparing their performance with respect to stated criteria, and reaching global conclusions which are presented logically, for example, by a table such as described in Section 15.5.1.

Progress reports   During an engineering project, it may be necessary to send reports to company management or to clients, with the object of recording what parts of the work have been completed or are underway. These reports may also be written as an internal record of steps undertaken and accomplished. The primary question to be answered is whether the actual progress corresponds to the budgeted time, financial and physical resources, and human expertise. The content of these reports can form a partial draft of the final report.

6.2 Finding information

The content of communications such as letters, memos, and email is usually known in advance, but other documents, such as reports, often require research during the writing process. Technical projects typically require repeated searches for information throughout their progress toward a final report. You should cultivate basic research skills; this section lists some places where you can begin.

Suppose, for example, that you are a member of a team that has to design and test a simple prototype control system containing a pump, switch, and level sensor to maintain an approximately constant fluid level in a tank from which fluid is independently drawn. A similar situation might be the design and assembly of parts for an electronic circuit, or many others involving finding and assembling new components. During the project, you may have to verify the basic physics of the devices, find equations that model their behaviour, identify possible vendors of the parts required, and choose from among available devices. The decisions involved in such a design project are described in more detail in Chapter 15. Here, sources of relevant information will be discussed.

Libraries The primary source of information for engineering undergraduates is the university library. The catalogues of major libraries continent-wide are also accessible over the Internet, and material can be borrowed through interlibrary loan arrangements. University collections concentrate on academic material relevant to the teaching and research in the institution. Material of direct relevance to commerce is emphasized less, but there usually is a service that assists in the search for relevant material wherever it may be found. For the example project, expect to find textbooks or research papers related to the basic physics or chemistry required for device modelling. There may also be research papers describing similar projects.

The collections in public libraries tend to be less academically specialized than those of university libraries, but may contain industry-related periodicals not found in typical university libraries.

Many companies maintain libraries of technical material relevant to their practice or business, particularly of material relating to codes, standards, industry publications, and often government publications.

In searching library catalogues, you should understand the main classification systems: the Dewey decimal classification and the American Library of Congress classification. Table 6.2 shows the major subject categories of the latter and Table 6.3 the subclasses of the T (technology) class.

Manufacturers and vendors For parts and equipment specifications, manufacturer-supplied material must be consulted. The Internet is a good place to search, because vendors have an interest in providing ready access to their catalogues and tend to maintain up-to-date web pages. Finding vendors may require searching the web site of industry associations or of compendia specializing in particular product types. For components, consult the catalogues of wholesalers and component vendors. To find vendors in your area or preferred component sources, consult the stores or purchasing department of your organization, where current catalogues are often kept. In fact, your colleagues may be the best source of information relevant to your project.

Table 6.2   Major subject categories in the Library of Congress classification.

Code	Category	Code	Category
A	General works	**M**	Music
B	Philosophy and religion	**N**	Fine arts
C	History and auxiliary sciences	**P**	Language and literature
D	Universal history and topography	**Q**	Science
E/F	American history	**R**	Medicine
G	Geography, anthropology, folklore	**S**	Agriculture
H	Social sciences	**T**	Technology
J	Political science	**U**	Military service
K	Law	**V**	Naval services
L	Education	**Z**	Library science and bibliography

Standards   Many design projects require reference to technical standards. See Chapter 19 for a discussion of where they may be found.

Patents   For information about intellectual property, particularly to find the name of persons or companies active in development, consult the membership list of industry associations or the Canadian patents index at the Canadian Intellectual Property Office (CIPO), as discussed in Chapter 17.

Reference books   There are many encyclopedic engineering handbooks available. Each attempts to summarize the current state of the art in an engineering discipline or subdiscipline. Two examples that illustrate the range of possibilities are *The Electrical Engineering Handbook* [10], which covers a broad discipline and is produced by a commercial publisher, and the *SME Mining Engineering Handbook* [11] in a more specialized field, published by a technical society. Not only do such books provide a ready source of ideas, they often

Table 6.3   Technology subcategories in the Library of Congress classification.

Code	Category	Code	Category
TA	General engineering, including general civil engineering	**TK**	Electrical engineering, nuclear engineering
TC	Hydraulic engineering	**TL**	Motor vehicles, aeronautics, astronautics
TD	Sanitary and municipal engineering	**TN**	Mining engineering, mineral industries, metallurgy
TE	Highway engineering	**TP**	Chemical technology
TF	Railroad engineering	**TR**	Photography
TG	Bridge engineering	**TS**	Manufactures
TH	Building construction	**TT**	Handicrafts, arts and crafts
TJ	Mechanical engineering	**TX**	Home economics

include extensive lists of detailed references. To find relevant technical handbooks, consult the catalogue of a major library and the web sites of the major technical publishers, which often also publish university course textbooks.

The Internet
The growth of the Internet has allowed anybody with a computer and service provider to become a publisher. Vast quantities of information are available, published by persons and organizations with motives that are personal, altruistic, educational, commercial, or criminal, to mention only a few possibilities. Specialized discussion groups allow questions to be posed and answered. While this information and the search engines that simplify its sifting are great conveniences, there is often little guarantee of correctness. Always determine the identity and goals of the publisher before trusting documents obtained over the Internet. For technical material, archival documents published on paper by reliable commercial publishers, engineering societies, standards organizations, or governments is to be preferred to unverifiable electronic material.

Organize your search
Searching for information is like other kinds of research: the number of documents available may be enormous, and you must exercise discipline in searching and classifying it for your purposes. Don't read the full text of everything you encounter, but read the abstracts, make notes, and keep the electronic equivalent of a card file to index what you have found. You may have to refer to several documents in detail during the course of a project, and a preliminary classification of the material to narrow your search to the most relevant possibilities can be a time-saver.

6.3 Technical presentations

Engineers are called on often to speak about their work, and surveys show that the ability to speak well to a group is a major advantage for career advancement. A presentation may be informal, such as your response to a request to "say a few words about your work" to colleagues. It may also be very formal, such as a project proposal to senior engineers or an oral summary of a report to a client. This section concentrates on technical presentations that can be supplemented by visual aids including projected graphics.

The structure of a technical presentation is like a technical document, but the circumstances require special attention. You will have the advantage of dialogue with the audience and the disadvantage of limited audience time and attention.

The essential steps toward good presentations are the following:

Identify your audience
The information contained in a weekly presentation to your co-workers will differ greatly from material presented to a visitor or client. Ask yourself, "What questions would the audience wish to ask?" and concentrate on those questions.

Identify your message
Before preparing the details of your talk, complete the sentence "I want to tell my audience . . ." and concentrate on this sentence when defining the title of your talk, the introduction, and the conclusions.

Repeat three times
At the beginning, summarize what you are going to say. Then give the details, and finish your presentation with a summary of the main message.

Vary your presentation Television producers and good speakers know that audience attention will begin to wander after about 10 minutes unless it is attracted by a change of pace. Introducing or changing visual aids, pausing for questions, or changing subtopics will vary your presentation.

Present professionally Your appearance and comportment count for much. Dress neatly and appropriately, perhaps by dressing up a little. Be enthusiastic. Speak clearly and look directly at individuals in different parts of the audience.

Time yourself If you will be given a limited time to speak, practise giving your talk and carefully time yourself. Beginners often underestimate the time it will take them to present technical material.

Use visual aids Prepare visual aids as described below, but do not simply read them; rather, talk about them to your audience. Apply the formula "touch, turn, talk"—first point to the slide item of interest, turn and make eye contact with your audience, then talk about the item. Print your slides, several to a page, and make notes for yourself on each page to guide your presentation.

Prepare handouts A preprinted handout can reinforce your message in the memory of the audience. Distribute handouts at the beginning of the talk only if you refer to them and if they will not distract the audience; otherwise distribute them at the end.

Practise Like any skill, oral presentation can be improved by careful preparation and repeated practice. There are many books that give good advice [12, 13], but public speaking is like swimming: it cannot be learned completely from a book.

Don't present too much One of the most common and significant errors is to present too much material for the audience to grasp in the time allowed. For example, the NASA Columbia Accident Investigation Board reported that a presentation with visual aids was a contributing factor in the decision to allow the space shuttle to attempt to land. The presentation showed wing-damage findings in a confusing slide that was crammed with bullet points and symbols [14]. The Columbia disintegrated while descending through the atmosphere, killing all on board. The board stated, "It is easy to understand how a senior manager might read this [. . .] slide and not realize that it addresses a life-threatening situation."

6.3.1 Visual aids

A good set of slides (projected images) can emphasize and amplify your oral message and assist the audience to focus on the important points. However, a slide-based presentation may not be the best way of delivering your message. Slides generally communicate best when the room and audience are large and the essential information is visual but, in a small setting, they can distract your audience and divert attention from you. The most appropriate supplement for your talk may be at most one or two slides of graphical material. A poor set of slides distracts attention and emphasizes your disorganization.

All office-software packages include tools for easy preparation of visual aids to accompany oral presentations. Employ such an aid when appropriate, but do not let it oversimplify your message or clutter your presentation.

Introduce the talk and the slides   Begin your talk with a title slide that includes your name and the date. Place a short heading above or below each slide. Include an introductory slide that lists or maps out the components of the talk. If your presentation will exceed 10 minutes, summarize your main message near the beginning.

Check readability   Make sure that each slide can be read from the back of the room. Inevitably this means using large type and at most 100 characters per slide.

Simplify, but not too much   Avoid elaborately decorated templates and overuse of distracting transitions between slides. Include nothing that is not essential to your message, but do not oversimplify to the point of invalidity.

Check mathematics   The correct treatment of units (discussed in Chapter 10) and mathematics poses a special problem: presentation-graphics programs were not originally designed for the precise choice of fonts, symbols, spacing, and symbol placement that is required. Recent add-ons improve the situation, but make sure that all technical notation appears correctly.

Be careful of lists   Avoid overuse of the lists that presentation-graphics programs tend to induce. Not all technical talks have a list-like structure. A set of well-prepared graphics is often much clearer.

Create custom diagrams   Technical presentations are often centred on at most a few charts or diagrams. Re-design them to be viewed on a screen; in most cases, even good illustrations from printed material contain too much detail for a slide.

   Figure 6.4 contains a draft of a set of illustrations for an oral progress report. The illustrations have been developed approximately according to the following discussion.

Title slide   Identify the subject of the talk and the speaker. Unobtrusively including the company logo gives a subtle message. Slides should be formatted consistently.

Slide 2   Audiences need an overview of the presentation. The conclusion should not be left to the end if the talk will last a significant time.

Slide 3   An introduction giving the background of the talk can be in point form, but the points should only introduce what you say. Discuss the points; do not simply read them.

Slide 4   After introducing the topic, you can narrow the focus to the essentials of the body and the conclusions that will result.

Slide 5   Diagrams are better than lists for topics or questions that do not follow one another. Two or more options for consideration can be presented this way.

Slide 6   The audience needs the essential data that will allow them to make decisions. Allow comparisons to be made easily by showing the options in similar structures.

Slide 7   Clearly show the choices that you have made or that the audience will have to consider.

Slide 8   Your conclusions should follow from the body of your presentation or from cited information that you have summarized.

Slide 9   When you give recommendations, be as specific as you can about what action should be taken and by whom.

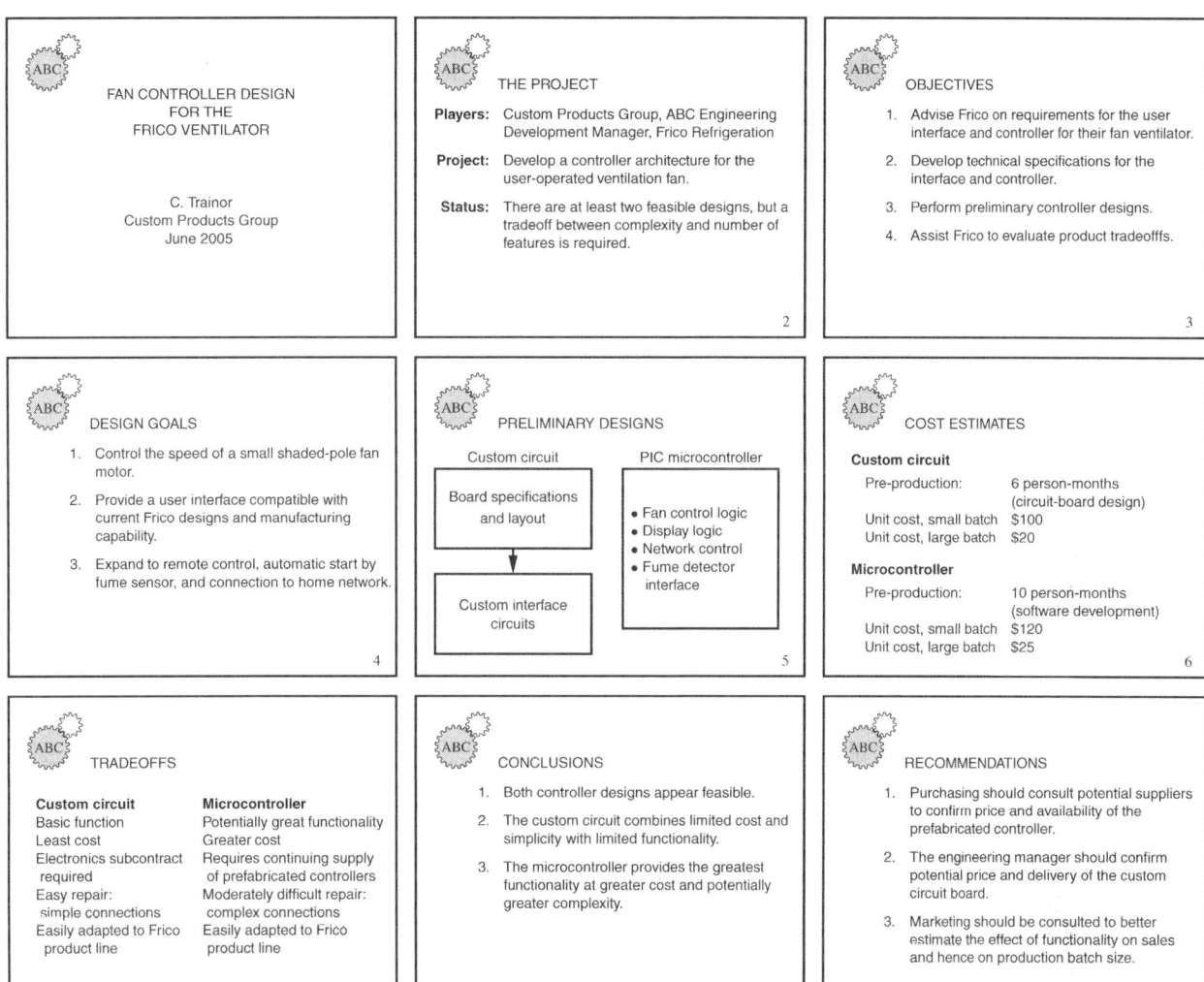

Figure 6.4   Draft illustrations for a talk on the status of a small design project are shown. The Custom Products Group of ABC Engineering has been asked to advise Frico Refrigeration on the design of a fan controller for a ventilator product. The project is progressing well, but a basic decision must be taken about the customer requirements: is it more important to have a sophisticated display and control features, or should simplicity and the resulting cost savings have priority?

6.4   Further study

1.   Choose the best answer for each of the following questions.

(a) The purpose of an experimental report in business or industry is typically

     i. to test feasibility, performance, or to obtain exploratory data.

    ii. to verify newly developed theory.

   iii. to be a vehicle for learning a test procedure.

(b) There is only one correct format for a technical letter.

 i. true ii. false

(c) Feasibility reports typically

 i. evaluate the projected performance of a product or process by analyzing its components.

 ii. analyze the performance of competing products.

 iii. test products to destruction to determine their feasibility for safety-critical applications.

(d) A specification document

 i. contains an introduction, detailed analysis, and conclusion.

 ii. does not form part of a contract.

 iii. analyzes the performance of the whole by analyzing its parts.

 iv. contains an introduction, together with a list of criteria that a product or service must satisfy.

(e) The main purpose of a progress report is usually

 i. to compare accomplished results to budgeted time, resources, and expertise.

 ii. to form a draft of the final project report.

(f) Check the incorrect statement or statements, if any.

 i. A technical memorandum may contain a horizontal line separating the header from the body.

 ii. Technical memoranda are always written on company letterhead.

 iii. A memorandum normally includes the date and "To," "From," and "Subject" lines.

 iv. A technical letter includes a date, a salutation, a complimentary closing, and a signature.

(g) Normally the difference between a bid and a proposal is that

 i. a proposal may suggest a means of satisfying a need as well as offering to perform the work at a given cost.

 ii. a bid suggests a means of satisfying an incompletely specified need as well as defining the cost.

 iii. a proposal is not concerned with cost but suggests a way of satisfying an imprecisely specified need.

(h) The main logical components of a report are

 i. sections containing an introduction, the detailed analysis, and conclusions.

 ii. the front matter, body, and back matter.

 iii. the title page, appendices, and references.

 iv. the introduction, summary, and conclusions sections.

(i) Normally a memorandum has at least the following components:

    i. heading, date, "Subject" line, signature, body

    ii. heading, date, "To" line, "From" line, "Subject" line, signature, body

    iii. heading, date, "To" line, "Subject" line, signature, body

    iv. heading, date, "To" line, "From" line, "Subject" line, body

(j) Email sent by students or company employees

    i. is normally encoded so that it cannot be read by others.

    ii. has copyright protection and should never be read by others without the permission of the author.

    iii. may be subject to inspection by others.

2.   Check your word processor to see if it has a standard format or template for writing technical letters and memoranda. How do the templates differ from the formats suggested in this chapter?

3.   Prepare word processor templates of (a) a laboratory report for use in your own classes and (b) a formal engineering report that you could use to report during an industrial internship or work-term.

4.   Evaluate Figure 6.4 critically to see if you can suggest improvements to the slides shown.

6.5   References

[1] D. Beer and D. McMurrey, *A Guide to Writing as an Engineer*. New York: John Wiley & Sons, second ed., 2005.

[2] M. Northey and J. Jewinski, *Making Sense, A Student's Guide to Research and Writing: Engineering and the Technical Sciences*. New York: Oxford, 2005.

[3] J. N. Borowick, *Technical Communication and its Applications*. Upper Saddle River, NJ: Prentice Hall, 2000.

[4] K. W. Houp, T. E. Pearsall, E. Tebeaux, S. Cody, A. Boyd, and F. Sarris, *Reporting Technical Information*. Scarborough, ON: Allyn and Bacon Canada, 1999.

[5] D. Jones, *The Technical Communicator's Handbook*. Toronto: Allyn and Bacon, 2000.

[6] W. S. Pfeiffer and J. Boogerd, *Pocket Guide to Technical Writing*. Toronto: Prentice Hall Pearson, second Canadian ed., 2006.

[7] D. G. Riordan and S. E. Pauley, *Technical Report Writing Today*. New York: Houghton Mifflin Company, 1999.

[8] J. S. VanAlstyne, *Professional and Technical Writing Strategies*. Upper Saddle River, NJ: Prentice Hall, 1999.

[9] K. R. Woolever, *Writing for the Technical Professions*. New York: Addison-Wesley Educational Publishers, Inc., 2002.

[10] R. C. Dorf, ed., *The Electrical Engineering Handbook*. Boca Raton, FL: CRC Press, second ed., 1997.

[11] A. B. Cummins, I. A. Given, and H. L. Hartman, eds., *SME Mining Engineering Handbook*. Littleton, CO: Society for Mining, Metallurgy and Exploration, 1992.

[12] R. E. Burnett and J. S. McKee, *Technical Communication*. Scarborough, ON: Thomson Nelson, 2003.

[13] P. Reimold and C. Reimold, *The Short Road to Great Presentations*. Hoboken, NJ: Wiley-Interscience, 2003.

[14] C. Thompson, "PowerPoint makes you dumb," *The New York Times, December 14,* 2003.

Chapter

7

Technical Writing Basics

When graduate engineers are surveyed about key skills for career success, communication skills are inevitably mentioned. You must be a competent writer. You must be able to use and learn from a dictionary, a thesaurus (Figure 7.1), and a technical writing reference kept near your desk, or their software equivalents. You must also be proficient with software tools for word processing, for creating diagrams, and for managing bibliographies. Like an athlete, you should exercise your skills regularly and continually attempt to improve them.

In this chapter, you will

- learn hints for improving your writing style,

- learn some of the most common writing errors and how to avoid them,

- review basic English punctuation,

- review parts of speech,

- learn hints for using Greek letters in technical writing.

The questions at the end of this chapter include a self-test for you to gauge your writing ability.

516 Intelligibility

N. *intelligibility*, knowability, cognizability; explicability, teachability, penetrability; apprehensibility, comprehensibility, adaptation to the understanding; readability, legibility, decipherability; clearness, clarity, coherence, limpidity, lucidity 567n. *perspicuity*; precision, unambiguity 473n. *certainty*; simplicity, straightforwardness, plain speaking, plain speech, downright utterance; plain words, plain English, mother tongue; simple eloquence, unadorned style 573n. *plainness*; easiness, paraphrase, simplification 701n. *facility*; amplification, popularization, haute vulgarisation 520n. *interpretation.*

Adj. *intelligible*, understandable, penetrable, realizable, comprehensible, apprehensible; coherent 502adj. *sane*; distinguishable, audible, recognizable, unmistakable; discoverable, cognizable, knowable

Figure 7.1 Part of an entry from a thesaurus [1]. An English dictionary, a technical dictionary, and a thesaurus, or their electronic equivalents, are essential tools for producing correct technical documents.

7.1   The importance of clarity

Engineers often work in teams, and effective teamwork depends on effective communication. Moreover, engineering work normally must be recorded for future reference. Decisions affecting worker or client health, property, or public welfare must be carefully recorded. In the event that your company has to demonstrate that it has followed standard professional practice, your documents might be scrutinized in a court of law. For these and many other reasons, you must write unambiguously and clearly.

The remainder of this chapter contains some points to keep in mind as you analyze the clarity and correctness of your writing. However, you must also know how to describe and interpret inexact quantitative data, using the methods of Part III, and to design graphs, drawings, and other artwork as described in Chapter 9, since a picture can explain complex relationships that are difficult to express verbally. The logic-circuit diagram in Figure 7.2, for example, shows at a glance how a particular combination of logic elements is constructed. Similarly, a surveyor's plan shows the location and shape of a piece of land accurately and concisely, but a lawyer's written description of the same piece of land might require several pages of legal jargon. Although a full discussion of engineering drawing is beyond the scope of this book, you should develop your ability to sketch and to make technical drawings using software and other tools.

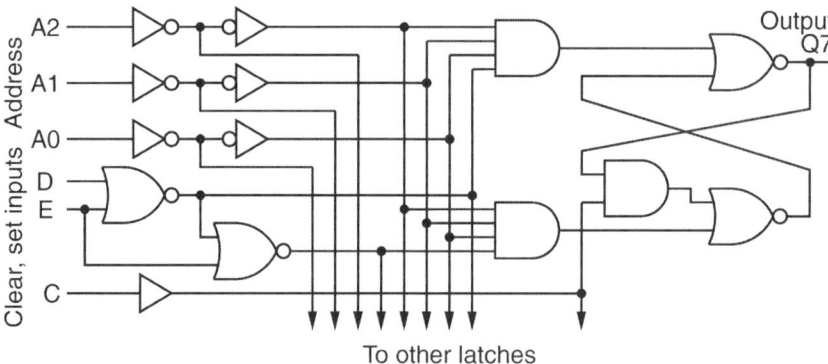

Figure 7.2   A diagram shows the connections of a binary logic circuit more clearly than could be done using words. This circuit is called a general-purpose latch. Inputs A0–A2 are address lines, and the inputs C, D, E are used for setting and clearing the output Q7.

7.2   Hints for improving your writing style

Practise your skills consciously whenever you have to write, consulting a dictionary, thesaurus, university-level grammar handbook, or electronic equivalents of these to hone your skills. Those who suggest that they have no need for aids or for improving their writing skills are like athletes who suggest that they have no need to train. Keep handy an English reference such as [2] or [3]. If your writing includes significant amounts of mathematical analysis or formulas, consult reference [4].

Clarity and brevity   In technical writing you will be mainly concerned with communicating information, so your style should be clear and concise. Say what you mean, as briefly and clearly as possible. However, while short sentences are concise, a report containing only short sentences is tedious to read; vary the sentence length to avoid monotony.

Correct terms   Define any specialized terms, either in the text or in a glossary, that your intended reader may not understand. Avoid slang, jargon, or compound words that do not appear in a dictionary. Consult a technical dictionary, as required, to verify your usage. General technical dictionaries such as [5] may suffice, but you may also have to consult a specialized dictionary, such as for construction [6] or electronics [7]. In some fields there are also on-line reference dictionaries that may be found using a web search.

Specific terms Use specific, concrete words rather than generalities. For example, in describing a manufacturing operation, the statement "A hole was made in the workpiece" may be true, but *made* is a weak, general word. It is more informative to say "A 12 mm hole was drilled through the workpiece." Similarly, "The machinery cannot be moved because the floor is bad" may be true, but be specific: "The wooden floor has several rotten planks and is unlikely to support the machinery adequately." Avoid generalities and overuse of weak verbs such as *to be, to have, to make,* and *to use.*

Spelling and punctuation Poor spelling and punctuation will destroy the reader's confidence in your work. If you are in doubt about the meaning or spelling of a word, consult a dictionary. If you have difficulty choosing precisely the right word, consult a thesaurus. Your computer may have a thesaurus or spell checker; however, these tools cannot identify correctly spelled words substituted for others, such as in the pairs

- *form* and *from,*
- *there* and *their,*
- *dough* and *doe,*
- *ferry* and *fairy,*

or the triples

- *your, you're,* and *yore,*
- *for, four,* and *fore.*

You must proofread what you have written. See Section 7.3 for a basic summary of punctuation.

Tense Inconsistency of tense and excessive use of the passive voice are common report-writing errors. The tense of a verb indicates the time of the action or state. For most practical purposes, there are six tenses, as illustrated in Table 7.1, although purists double this number to 12 by considering certain auxiliary forms [8]. Since most technical reports are written after measurements or other work is completed, the present or past tenses are most appropriate. Write in the future tense only when you are discussing an event that will occur in the future, such as in a section discussing suggestions for future work. Do not use the future tenses to refer to something appearing further along in the report.

Voice There are two voices: active and passive, illustrated in Table 7.1. The active voice is the more emphatic and interesting, as seen by comparing the following two examples:

Active: "The machine operator repaired the damaged workpiece."

Passive: "The damaged workpiece was repaired."

Table 7.1   Tenses for the verb *to see* in both active and passive voice. All tenses are technically correct, but the less desirable wording has been shaded.

Tense	Active voice	Passive voice
Present	I (We) see. You see. He (She, It) sees. They see.	I am (We are) seen. You are seen. He (She, It) is seen. They are seen.
Past	I (We) saw. You saw. He (She, It) saw. They saw.	I was (We were) seen. You were seen. He (She, It) was seen. They were seen.
Future	I (We) will see. You will see. He (She, It) will see. They will see.	I (We) will be seen. You will be seen. He (She, It) will be seen. They will be seen.
Present perfect	I (We) have seen. You have seen. He (She, It) has seen. They have seen.	I (We) will be seen. You have been seen. He (She, It) has been seen. They have been seen.
Past perfect	I (We, You) had seen. He (She, It, They) had seen.	I (We, You) had been seen. He (She, It, They) had been seen.
Future perfect	He (She, It, They) will have seen.	He (She, It, They) will have been seen.

Use the passive voice if the doer of the action is irrelevant to the discussion and the active voice to produce shorter, clearer sentences. For example, use the active voice

- to bring a main idea to the front of a sentence: "Figure 1 shows ... ,"
- to emphasize an identity: "Onitoba Manufacturing provided the specially designed clamps ..."

Person   The three choices of person, first (I, we), second (you), and third (he, she, it, they), are illustrated in Table 7.1.

Use the first person

- to take responsibility, for example, "We recommend ... ,"
- to place yourself in an active role, for example, "By the third interview, we observed that the attitude of staff members had changed."

The second person is often implicit in instructions; for example, the subject *you* is implied in, "Close the feed valve when the liquid level reaches the red line." The word *you* should mean the reader, not people in general.

Most technical writing should be objective and independent of the observer. Therefore, the third person is sometimes most appropriate, particularly in descriptive parts of formal technical reports. However, the third person leads to passive, complicated sentence structure; do not overuse it.

Questions of tense and person may be arbitrarily resolved if your employer (or university course) requires you to write in the third person; if so, then follow the rule but include plenty of active sentences. The authors discussed voice and person while writing this book. We decided that active sentences are more interesting, so you will find that we frequently use the first and second person in this edition.

7.3   Punctuation: A basic summary

Punctuation is a code to tell you where stops, pauses, emphasis, and changes in intonation occur. The 13 basic punctuation symbols are explained briefly below.

Period   The period (.) is a "full stop," which ends a sentence. It is also used after most abbreviations, for example, *Mr., Ms., Dr., Prof.*

Comma   The comma (,) indicates a slight pause to separate ideas, such as to separate two independent clauses joined by a conjunction; for example, "School was hard, but life was full." The comma is also used as follows:

- to give a pause after transitional words such as *however, nevertheless, moreover, therefore;*

- to separate a long introductory phrase from a principal clause; and

- to separate words, phrases, or clauses in a list of three or more like items.

Commas are often used in pairs, like parentheses, to enclose words, phrases, or clauses that add non-essential information. The meaning of the sentence must not change if the commas and enclosed material are deleted. As a general rule, a single comma should not separate the subject from the verb or the verb from the object.

Semicolon   The semicolon (;) indicates a longer pause than a comma but shorter than a period. It is used between two main clauses that express a continued thought but are not joined by a conjunction; for example, "Intellectuals discuss ideas; gossips discuss people." It is also used between two clauses that are joined by a conjunctive adverb; for example, "It rained; fortunately, the wedding was indoors." The semicolon is also used in place of a comma in a list if the phrases or clauses in the list contain commas.

Colon   The colon (:) usually introduces a list of like items, as a substitute for *namely,* or *for example.* It should not be used unnecessarily; for example, after a verb or after *of.* The colon is also used to introduce a formal statement, quotation, or question; for example, "The valve was marked: Do not open!"

Hyphen The hyphen (-) joins compound adjectives; for example, *natural-gas producer, two-month vacation.* A long word that does not fit on a line is hyphenated between syllables and completed on the next line. Hyphens are also used in some compound nouns, for example, *passer-by, dry-cleaning.* There are no clear rules about the construction of compound nouns; consult a reference when in doubt.

Dash The dash (—) shows a sudden interruption in thought. For example, "It stopped raining—I think!" The dash is common in dialogue but is rare in formal writing.

Question mark The question mark (?) is a full stop at the end of a direct question; for example, "May I help you?" However, an indirect question ends with a period; for example, "He asked if he could help you."

Exclamation point The exclamation point (!) is the full stop at the end of a sentence that expresses surprise or strong feeling; for example, "We won! I can't believe it!" The exclamation point is used rarely in formal writing.

Apostrophe The apostrophe (') has two uses. The first is to indicate possession. It is used at the end of a noun, followed by the letter *s*; for example, "He has writer's cramp." The apostrophe also replaces the deleted letters in abbreviations; for example, *doesn't; it's.* Note that *it's* always means *it is* and does not mean *belonging to it.*

Quotation marks Quotation marks (" ") enclose direct quotes and, when the style requires it, article titles in a list of references.

Brackets Brackets ([ ]) enclose material added editorially to clarify meaning; for example, "The boss [Mr. George] was described as being sick."

Parentheses Parentheses "( )" enclose ideas of secondary or alternate importance. An idea placed in brackets or parentheses must not be essential; the meaning of the sentence should not change if the parentheses and the enclosed material are deleted.

Ellipses Ellipsis points (. . .) are three spaced points used in place of omitted words.

7.4 The parts of speech: A basic summary

Proper sentence structure is the key to clear writing. Sentences are created from eight common parts of speech: nouns, pronouns, verbs, adjectives, adverbs, prepositions, conjunctions, and interjections. These terms signify not what a word is, but how it is used, since the same word may be used in several ways. To refresh your memory, the parts of speech are defined below:

Noun A noun names a thing or quality; for example, *a box, beauty.* A proper noun names a specific person or place and has an initial capital letter; for example, *Fred, Waterloo.*

Pronoun A pronoun takes the place of a noun; for example, *it, her, him, anyone, something.* Each pronoun refers to a noun, which is called the antecedent of the pronoun. The pronoun and its antecedent must agree, and the antecedent must be clear and unambiguous; for example, "The box broke; burn it," and "Fred called; please call him."

Verb	A verb is often called an action word, since it indicates action, although it can also indicate a state of being; for example, *run, jump, make, be, exist.*
Adjective	An adjective modifies nouns; for example, a *big* box; a *tall* building, a *small* raft.
Adverb	An adverb modifies verbs, adjectives, or other adverbs; for example, *rapidly, softly.*
Preposition	A preposition is a linking word, for example, *in, on, of, under, over, to, across.* A preposition is used to show the relationship of a noun or pronoun to some other word in the sentence, for example, *depth of water, voltage across the terminals, on a shelf.*
Conjunction	Conjunctions are also linking words, for example, *and, but,* and *or;* they join words, such as *bat and ball.* They may also join phrases, for example, *date of arrival and length of stay,* or clauses.
Interjection	Interjections express emotion, for example, *Oh!, Ouch!, Wow!,* and are rarely used in technical writing.

7.5   Avoid these writing errors

This section reviews some basic errors that you should avoid. Such a brief discussion cannot cover every topic; the handbooks in the references should be consulted for further explanations and additional topics.

Sentence fragments   The sentence is the basic building block of written communication and contains a complete thought. However, in the fury of composition, sometimes writers leave fragments in their work that have partial meaning but are not complete sentences. For example, "The new, advanced gear-box. Delivering over 200 HP!" In spoken dialogue, the speaker's tone of voice and emphasis help us to fill in any missing information. However, in a formal written report, sentences must be complete.

Dangling modifiers   The dangling modifier is a phrase, usually at the start of a sentence, that fails to identify what it modifies. A typical example of this common error is, "Although only 16 years old, the policeman treated him like an adult." This sentence implies that the policeman was only 16 years old, whereas the writer probably meant that the subject of the policeman's attention was 16 years old. The reader expects the phrase to modify the subject of the sentence, in this case, the policeman. Avoid this common error by asking yourself, whenever you write a modifying phase, "What does it modify?"

Comma splices and run-on sentences   Sometimes writers splice two principal clauses into a single sentence, using a comma. For example, "The voltage limit was exceeded, the capacitor failed." Correct this error by splitting the sentence or inserting a semicolon or conjunction. For example, "The voltage limit was exceeded. The capacitor failed," or "The voltage limit was exceeded; the capacitor failed," or "The voltage limit was exceeded, and the capacitor failed." A run-on sentence is an even more basic error; it is a comma splice in which the writer left out the comma.

Superfluous commas   Sometimes writers add superfluous (excess) commas. For example, "The data were read but, not organized in charts. Perhaps, the data were written down incorrectly."

Extra commas confuse the reader. Although every comma indicates a pause, not every pause requires a comma. For example, "The data were read but not organized in charts. Perhaps the data were written down incorrectly." It is worth spending the time to review the reasons for using the comma, defined in detail in Section 7.3.

Subject–verb disagreement The appearance of an incorrect verb may be a sign of an excessively complicated sentence, as well as a basic error of grammar. The most common error is the incorrect plural verb form illustrated by "The insertion of zener diodes stabilize the circuit."

Adverb and adjective confusion Adverbs and adjectives are both modifiers. Do not confuse them. Adverbs frequently end in *ly,* for example, *rapidly, softly,* and *quickly.* Adjectives modify only nouns, but adverbs may modify verbs, adjectives, or other adverbs. It is correct to say "a really big box" but not correct to say "a real big box," since *really* is an adverb, but *real* is an adjective and cannot modify the adjective *big.* If you are not sure whether a word is an adverb or an adjective, consult your dictionary.

7.6 The Greek alphabet in technical writing

Technical writing often requires the use of mathematical formulas that contain Greek letters. Use these letters sparingly, particularly if the report is intended for readers without mathematical expertise. In addition, do not use the Greek letters of identical appearance to English letters. Consult reference [4] for detailed advice about writing mathematics. The Greek alphabet is shown in Table 7.2.

Table 7.2 The Greek alphabet, showing English equivalents. The meaning attached to each symbol varies with the context.

Greek letter		Greek name	English equivalent	Greek letter		Greek name	English equivalent
A	α	alpha	a	N	ν	nu	n
B	β	beta	b	Ξ	ξ	xi	x
Γ	γ	gamma	g	O	o	omicron	ŏ
Δ	δ	delta	d	Π	π	pi	p
E	ϵ	epsilon	ĕ	P	ρ	rho	r
Z	ζ	zeta	z	Σ	σ	sigma	s
H	η	eta	ē	T	τ	tau	t
Θ	θ	theta	th	Υ	υ	upsilon	u
I	ι	iota	i	Φ	ϕ	phi	ph
K	κ	kappa	k	X	χ	chi	ch
Λ	λ	lambda	l	Ψ	ψ	psi	ps
M	μ	mu	m	Ω	ω	omega	ō

| 7.7 | **Further study** |

1. Choose the best answer for each of the following questions.

(a) Choose the best sentence.

    i. The engineer will keep a weekly log showing the work accomplished and the schedules met, and how the material deliveries have been coordinated.

    ii. The engineer will keep a weekly log showing the work accomplished, the schedules met, and the coordination of material deliveries.

    iii. The engineer will keep a weekly log showing the work accomplished, the schedules met, and how the material deliveries have been coordinated.

(b) Choose the correct sentence.

    i. The other engineer and I were called to help.

    ii. Me and the other engineer were called to help.

(c) Choose the best of the following sentences.

    i. Remember the words of Wilfrid Laurier: "The twentieth century belongs to Canada."

    ii. Remember: the words of Wilfrid Laurier "The twentieth century belongs to Canada."

    iii. Remember the words of Wilfrid Laurier "The twentieth century belongs to Canada."

    iv. Remember the words of Wilfrid Laurier, "The twentieth century belongs to Canada."

(d) Choose the correct sentence.

    i. Its assumed that the seatbelt was correctly manufactured, since its inspection certificate was signed.

    ii. It's assumed that the seatbelt was correctly manufactured, since it's inspection certificate was signed.

    iii. Its assumed that the seatbelt was correctly manufactured, since it's inspection certificate was signed.

    iv. It's assumed that the seatbelt was correctly manufactured, since its inspection certificate was signed.

(e) "The social life in residence has a major affect on the academic performance of students."

    i. The sentence is correct as written.

    ii. The word *students* should be capitalized.

    iii. This conclusion is obviously untrue.

    iv. This conclusion is obviously true.

    v. The word *affect* should be changed to *effect*.

(f) Choose the correct sentence.

    i. The car that won the race was the newest model.

    ii. The car which won the race was the newest model.

(g) Choose the best sentence.

    i. Although not yet 16, her co-workers treated her like an adult.

    ii. Although she was not yet 16, her co-workers treated her like an adult.

(h) Choose the best of the following sentences.

    i. The criteria for specifying this steel was it's tensile strength.

    ii. The criteria for specifying this steel was its tensile strength.

    iii. The criterion for specifying this steel was it's tensile strength.

    iv. The criterion for specifying this steel was its tensile strength.

(i) Choose the correct sentence.

    i. Engineering programs include mathematics, physics, and chemistry.

    ii. Engineering programs includes: Mathematics, Physics, and Chemistry.

    iii. Engineering programs include Mathematics, Physics, and Chemistry.

    iv. Engineering programs includes: mathematics, physics, and chemistry.

(j) Choose the correct sentence.

    i. The foundations specifications were three pages long.

    ii. The foundations specification was three pages long.

    iii. The foundations specification were three pages long.

    iv. The foundation specifications were three pages long.

2. The simple test below has 30 questions in total. Give yourself about 20 minutes to complete it. Then check the answers in the appendix.

Scoring: 27–30: Excellent   24–26: Good   20–23: Fair   < 20: Poor

A. Vocabulary In the following sentences, select the appropriate word:

(a) The kids are (all right, alright).

(b) In his fall, he narrowly (averted, escaped) death.

(c) The committee voted to (censure, censor) him for his poor behaviour.

(d) The engineering faculty is (composed, comprised) of six departments.

(e) The police siren emitted a (continuous, continual) scream.

(f) I am indifferent to her wishes; in fact, I (could, could not) care less.

(g) Her hint (inferred, implied) that she had a gift for him.

(h) We (should of, should have) removed the specimen before it broke.

(i) This machine is more reliable, since it has (fewer, less) parts to fail.

(j) The conference is (liable, likely) to be held at the university.

(k) He rewrote the subroutine (his self, himself).

(l) He asked me to (loan, lend) him my lawnmower.

(m) She bought a small (momento, momentum, memento) as a souvenir.

(n) It was so hot that I (couldn't hardly, could scarcely) breathe.

(o) Asbestos is (non-flammable, inflammable).

B. Grammar and usage   In each of the following sentences, identify whether the grammar and usage are correct or incorrect. If incorrect, write the corrected version.

(a) Americans speak different.

(b) She hit me hard on the arm.

(c) The technician took a real big wrench out of the toolbox.

(d) My sister, always bugging me.

(e) The four students were quarrelling between themselves.

(f) What about those Blue Jays!

(g) Driving down the road, the poles went past us quickly.

(h) Check you're spelling thoroughly.

(i) Whom do you recommend?

(j) Torquing is when you tighten a nut with a wrench.

C. Punctuation   In each of the following sentences, state whether the punctuation is correct or incorrect. If incorrect, write the corrected version.

(a) She said, "I am sure the fuse is broken; it happens frequently."

(b) He shouted, "The transmission lost it's oil!"

(c) They all drove over here in John's and Jean's cars.

(d) The crowd was mad, it strung him up.

(e) He bought ten dollars' worth of candy for the children's party.

3.   Correct or improve each of the following statements.

(a) Never use no double negatives.

(b) In his defence, he respectively told the magistrate that the auto was stationery at the time of the accident.

(c) A verb have to agree with its subject.

(d) Your represented by your writing.

(e) Clichés should be as scarce as hen's teeth.

(f) Proof-reed everything, especially watch you're spelling.

(g) Eschew obfuscation in whatever manifestation.

(h) Make sure colloquial language ain't used.

(i) When dangling, subjects are not properly modified.

(j) Throughly check you're spelling and punctiliousness.

4. The following sentences contain word usage or clichés that should be avoided. In each case, suggest a simpler or alternative way to state the idea.

(a) The machine is not functioning at this point in time.

(b) The project is inoperative for the time being; it's on the back burner.

(c) In the stock market, fear and greed dominate; this is the name of the game.

(d) You know, basically, I'll have the chocolate ice cream.

(e) The three alternatives caused him a dilemma. By and large, the bottom line is consumer driven.

5. Find a grammatical error or punctuation error in the body of this book, and write a letter of reprimand to the authors. Be polite but firm.

7.8 References

[1] R. A. Dutch, *Roget's Thesaurus.* London: Longmans, 1963.

[2] J. Buckley, *Checkmate: A Writing Reference for Canadians.* Scarborough, ON: Nelson Thomson Learning, 2003.

[3] K. R. Woolever, *Writing for the Technical Professions.* New York: Addison-Wesley Educational Publishers, Inc., 2002.

[4] N. J. Higham, *Handbook of Writing for the Mathematical Sciences.* Philadelphia: Society for Industrial and Applied Mathematics, second ed., 1998.

[5] R. Ernst, *Comprehensive Dictionary of Engineering and Technology: With Extensive Treatment of the Most Modern Techniques and Processes.* Cambridge: Cambridge University Press, 1985.

[6] L. F. Webster, *The Wiley Dictionary of Civil Engineering and Construction.* New York: John Wiley & Sons, 1997.

[7] IEEE, *IEEE Standard Dictionary of Electrical and Electronics Terms.* New York: Institute of Electrical and Electronics Engineers, 1977.

[8] D. Hacker, *A Canadian Writer's Reference.* Scarborough, ON: Nelson Thomson Learning, 2001.

Chapter

8

Formal Technical Reports

The formal technical report is the definitive presentation of the results of engineering work. The designation "formal" refers primarily to the logical structure and writing style, rather than to the purpose or application. A report might record the investigation of an accident, the analysis of a failure, or the feasibility of a design. Indeed, the subject matter of formal reports is as broad as all of engineering.

Above all other considerations, a formal report must be correct, clear, objective, and complete. At every stage, the skeptical reader should be able to ask, "Why should I trust this?" and find an explicit answer in the report. This is a new style of writing for anyone whose reading experience has been confined to textbooks and various forms of literature. The report should be carefully and artistically produced, with professional-quality layout and graphics (Figure 8.1), taking advantage of modern desktop publishing or other more elaborate facilities.

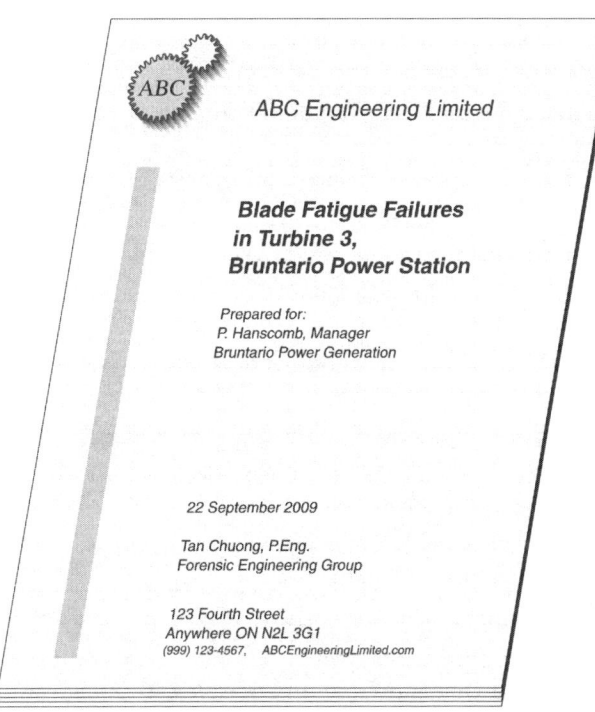

ABC Engineering Limited

Blade Fatigue Failures in Turbine 3, Bruntario Power Station

Prepared for:
P. Hanscomb, Manager
Bruntario Power Generation

22 September 2009

Tan Chuong, P.Eng.
Forensic Engineering Group

123 Fourth Street
Anywhere ON N2L 3G1
(999) 123-4567, ABCEngineeringLimited.com

Figure 8.1 A well-produced technical report makes a subtle statement about the quality of its content. For details on cover components, see page 113.

In this chapter, you will learn

- the logical purpose and structure of a report,
- the purpose and format of the physical components of a report,
- a strategy and checklist for writing reports.

The great majority of report documents can be produced by following the suggestions given in this chapter. However, nearly every definition and description is subject to variation or exception according to the report contents, the overriding need for clarity, and company policy. The purpose of the chapter is to present *one* description of a solidly defensible report, together with an acceptable format. For more comprehensive possibilities, consult books that have been written about technical communication, such as references [1–3]. For advice on writing material containing mathematics, see reference [4]. Consult your library for other references.

The clarity of reports is so important that standards have been developed for their organization. In addition to defining a standard format, reference [5] includes samples of report elements and a bibliography of style manuals, specialized dictionaries and handbooks, symbol standards, library reference aids, and graphic arts handbooks.

With your word processor you can and should create templates of the main kinds of reports that you may have to produce in the future.

8.1 Components of a formal report

The main logical components of essentially all reports are the introduction, detailed content, and conclusions as described in Section 6.1.7. The physical presentation of these and supplementary components can greatly affect the clarity and emphasis of the report. Readers will most easily understand material presented in a logical and familiar format. Therefore, the rules for the layout and physical design of reports are based both on logic and on tradition or familiarity. The main physical components are the front matter, report body, and back matter; Table 8.1 lists generic headings for technical reports. However, the section headings in the body should be changed to be more descriptive of the subject at hand.

Your company or university department may have specific rules about the format of formal reports, and the headings in Table 8.1 may have to be modified to conform. You may be given a report template that specifies formatting rules such as heading and text font, document margins, and other details.

In many ways, a report is similar to a book, although reports are not normally divided into parts and chapters as this book is, but into sections and subsections as are the book chapters. You will probably not copy the heading formats of this book, but we specifically chose its section and subsection numbering to serve as a model for a report. You can also refer to the front matter of this book as a model, although a report typically does not include a preface.

8.1.1 The front matter

The number of pages and the details of the front matter can be known only when the rest of the report is complete. Traditionally, to avoid complete renumbering of the document pages, front-matter pages were numbered independently and distinctly, using lowercase roman numerals (i, ii, iii, ...). The use of computers has removed the need for separate page numbers, but tradition is strong, and we continue to number the front matter with roman numerals; a report that does not do so is usually considered to be incorrect.

The title page is always page i, but it is artistically inappropriate for the number to appear there, so usually the first visible page number is on page ii or later.

The components of the front matter are listed in their usual order in the following paragraphs, although minor placement variations are possible.

Table 8.1   Physical components of a formal report. The body contains numbered sections, usually more than the four shown, with subsections as necessary. The body is preceded by the front matter, which is page-numbered with roman numerals, and followed by the back matter. The order of components may vary according to institutional format rules. Components marked with an asterisk (∗) are included when necessary; recommendations are often required.

Front matter		Front cover
		Title page
	∗	Letter of transmittal
	∗	Abstract
	∗	Key words
		Contents
	∗	List of figures
	∗	List of tables
	∗	List of symbols
	∗	Preface
	∗	Acknowledgments
		Summary (or Executive Summary)
Body	1.	Introduction
	2.	Analysis
	3.	Conclusions
	∗ 4.	Recommendations
Back matter		References
	∗	Bibliography
	∗	Appendices

Cover   Your company or department may have standard binding and cover requirements such as illustrated in Figure 8.2. The title of the report must be clearly visible on the external cover or binder; the reader should not have to open the report to see the title. In the absence of a special design, a clear plastic front cover that allows the title page to be read without opening the report is usually acceptable.

The standard components for the cover are

- title,
- author's name,
- date of publication,
- author's affiliation: company and department name.

Other items, such as library cataloguing information, the name of the recipient, or the author's student identification number and department, may be included on the cover as required.

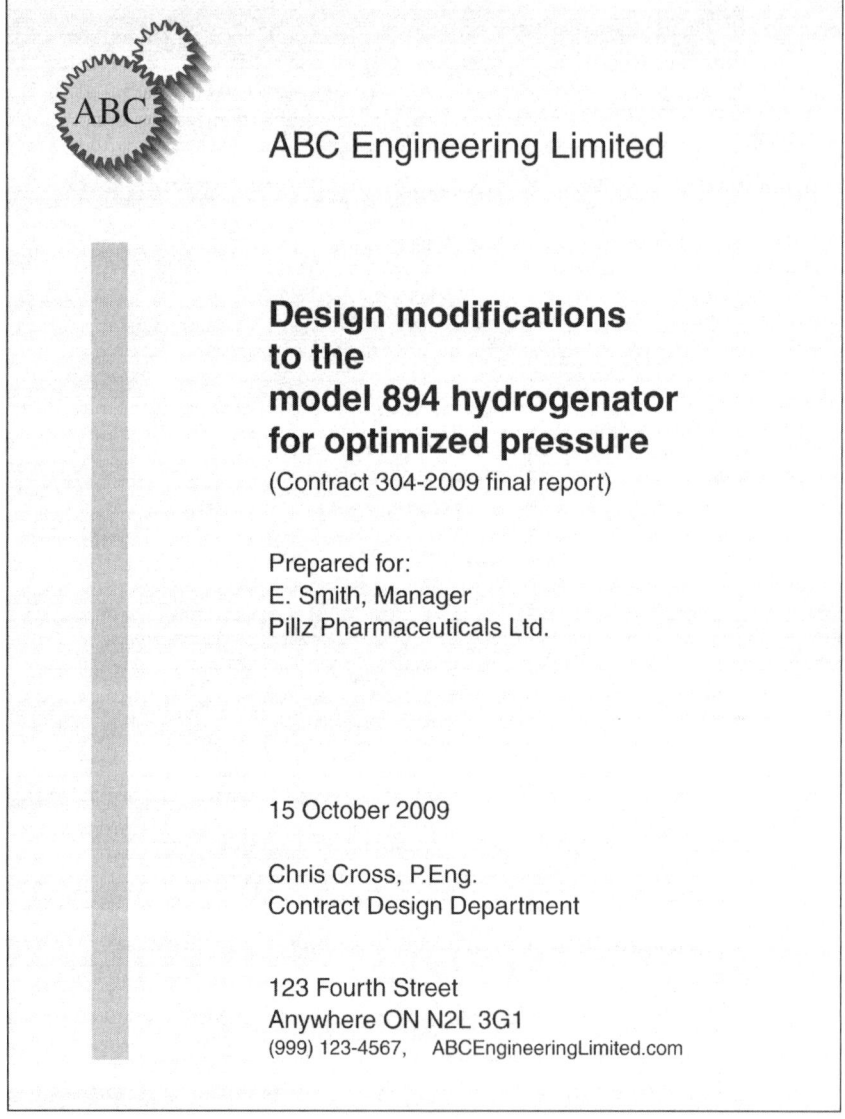

Figure 8.2   A typical company report cover. The key components are shown: title, "Prepared for" information, date, author, and author's affiliation.

Title page   The title page is the ultimate condensation of the report and must contain the following:

- the report title,
- author's name and address,
- date of report completion,
- the name and affiliation of the person for whom the report was prepared.

A title should not be too short for the intended reader to understand the report purpose. For example, consider a report investigating the cause of fires in a particular model of electrical control panel. Here are some example titles:

1. Control Panel Fires
2. The Cause of Six Fires in the Model 24 Control Panel
3. The Cause of Six Fires in Paneltronix Model 24 Control Panels Installed in Municipal Pumping Stations

The first title is much too vague for any circumstance. The second is adequate for a reader who knows what a Model 24 panel is but still too vague for a more general readership. The third probably is sufficiently clear and complete.

Choose precise, concrete words for the title, and avoid superfluous words such as *Report on . . .* or *Study of . . .* Do not merely name an object or machine; say something about it. For example, *Number Six Forging Press* identifies the machine but is vague. *Motor Replacement on the Number Six Forging Press* identifies the modification and the machine and might be much better, depending on the intended reader. A report title page may show the same basic information as the report cover, as shown in Figure 8.2, but must be adapted for specific contexts.

Letter of transmittal Normally a courteous, formal letter to the recipient accompanies a formal report when it is delivered. When the letter is meant to be seen by all the readers, a copy is bound into each report, usually immediately after the title page. The letter has a normal business format and is addressed to the official client or recipient of the report. It must

- refer to the report by title,
- identify the circumstances under which the report was prepared, for example, "in response to your letter of . . . ," "in fulfillment of contract number . . . dated . . . ," or "for academic credit following my work-term at XYZ Company,"
- identify the subject and purpose of the report.

In addition, the letter should

- direct attention to parts of the report that should receive particular attention,
- mention recommendations that require action,
- state how the recipient can ask for further information,
- acknowledge assistance received, unless there is explicit mention of assistance in the report,
- direct attention to the engineer's invoice for services, if it is attached.

Abstract Abstracts are concise, comprehensive summaries, typically limited to 50 to 200 words, suitable for storage in library indexes or other information-retrieval systems. A person searching a library index for material related to the report topic should be able to determine the gist of the report from the abstract, before obtaining the complete document. Since it briefly summarizes the whole report, the abstract typically contains the report purpose, methods used, results, and conclusions. Abstracts are required more often in published research reports and articles than in privately commissioned reports.

Some documents contain the heading "Abstract" but no "Summary" heading, in which case *Abstract* is a synonym for *Summary* as described on page 117.

Example 8.1
Two drafts of an
abstract

An inexperienced report writer made the mistake of attempting to write an abstract before the rest of the report. The result is below. However, after completing the report body and appendices, the writer realized that the draft abstract was not an effective summary but resembled an introduction; furthermore, it did not summarize all the report contents and conclusions as required. The writer then redrafted the abstract and concluded that it would have been better to compose the abstract last. The second draft is also shown.

Before

The Custom Products Group of ABC Engineering was contacted by Frico Refrigeration for advice on the design of the electronic controller for a new fan-ventilator product. Over six months, two proposed designs were developed: a custom circuit board and a commercially available microcontroller.

After

Two feasible designs for the electronic controller for the Frico ventilator are described. The specifications for a custom circuit board and interface are first given, and then a general-purpose microcontroller is analyzed. Cost and feature comparisons are made. Both designs are shown to be feasible, and investigation of the effect of features on sales rates is recommended.

Key words

If the report is to be indexed in an information-retrieval system, a list of descriptive words may be required so that a reader can identify documents of potential interest by searching the key word data. A list of key words for this book might contain *professional engineering, introduction to engineering, technical documents, technical measurements,* and *engineering practice,* for example. These are words or phrases that define the subject area of the report. They appear in published technical reports but are more common in technical research articles.

Table of contents

A table of contents is essential, even in a brief report. The numbered report section and subsection headings must be listed, with their page numbers. Usually the preface, summary, and lists of figures and tables appear in the table of contents, with their roman numeral page numbers.

List of figures, tables, symbols, and definitions

In all but small reports, the figures should be listed in the front matter to enable the reader to find them easily. The figure number, the caption or an abbreviated caption, and the page number are given for each figure. A separate list of tables serves a similar purpose, but a report containing only a few figures and tables might combine them into one list.

A list of symbols serves a different purpose: it identifies and defines acronyms, symbols used in formulas, or other special notation. Such a list should be included if symbols have been used with a meaning that the intended readers are not certain to know. The page of first use of each symbol should also be included.

Some reports include a glossary containing word definitions as well as symbol definitions. A list of symbols or glossary that is extensive can be put in the back matter as suggested in reference [6].

Preface A preface will normally be included when the letter of transmittal is not bound with the report and you want to give a message to the reader, such as advice for knowledgeable readers who need not consult all report sections. The acknowledgment of assistance may also be given in the preface or in a separate section as described below.

Acknowledgments Ethical behaviour requires an author to acknowledge any help given in producing the report, in obtaining data or analyzing it, for the loan of equipment, for permission to use copyrighted material in the report, or for other assistance. A separate "Acknowledgments" section is appropriate if there is no preface or bound transmittal letter or if the acknowledgments are extensive.

Summary (or executive summary) A technical report is not a novel in which the conclusion is cleverly concealed for 300 pages until the final chapter. The "Summary" section serves to outline the complete report in advance.

The summary is a complete, independent précis of the report, and it must summarize the introduction, body, conclusions, and recommendations in a few paragraphs, usually less than a full page. For many readers, the summary is the most important section. It is usually written last. Some standards list the summary as part of the report body.

Avoid describing the report in the summary. Do not say, "This report describes the modifications to . . . "; say what the modifications are, directly. Similarly, the statement "Recommendations are given" is vague; say what they are.

The title "Executive Summary" should be reserved for a summary that is written in non-technical language for persons—often company executives—who are not the primary readers of the report but who typically oversee or finance the project being described. An executive summary usually includes more background description than a normal summary and therefore may be longer, typically up to a few pages.

8.1.2   The report body

The body contains the essential report material, beginning with an introduction and ending with conclusions and often recommendations. These logical components are organized into numbered, titled sections. Some companies require the conclusions and recommendations sections to appear directly after the introduction section, for easy location and to convey the essential message early, leaving the details for later. The body pages are numbered using Arabic numerals, and the first page of the introduction is page 1.

In the body, the reader is led along a logical path through the sections and subsections that support the conclusions and recommendations. Headings in the body must be customized for the topic. The headings depend on whether the report describes a laboratory test, a product design, a field test, a failure analysis, a production efficiency problem, or other topic.

The usual numbering convention for report sections is illustrated below. The title of the introduction is usually "Introduction," but the other headings shown must be changed to suit the subject:

1. Introduction

1.1 Purpose of the Report

1.2 Background Information

2. Second Section

2.1 First Subsection of Section 2

2.1.1 Third-Level Heading

2.2 Second Subsection of Section 2

. . .

Subheadings beyond the third level should not be used. The section and subsection titles should be displayed in a distinctive font.

Introduction The introduction defines the purpose, scope, and methods of the investigation. It must also include sufficient background or history for the intended reader to understand its context. The introduction contains answers to the following:

- What are the purpose and scope of the report; that is, what questions are being investigated?

- Why is the report being written; that is, what is the motivation for the work?

- Who or what category of persons is intended to read the report?

- How was the work performed? The process by which the work was done and the report written is outlined.

Sufficient background must be included for the intended readers to understand the body, and the relationship of the report to existing documents cited as references must be described.

The introduction should conclude with a brief outline of the rest of the report, listing what is contained in each of the succeeding sections and the appendices if any.

Internal sections The subject matter determines the number of sections and their titles. The first of the following alternative ways of ordering sections is preferred when possible:

- in sequence from most important to least important,

- in problem, method, solution sequence,

- in cause-and-effect sequence,

- in chronological (time) sequence,

- in spatial (or location) pattern,

- by classification: group ideas and objects into similar classes,

- by partition: separate ideas and objects into component parts,

- by comparison: show similarities between ideas and objects,

- by contrast: show differences between ideas and objects,

- in order from general to specific.

A large report will not be read completely in one interrupted session; sections of it will be read independently at different times. Include an introductory paragraph or at least an introductory sentence at the beginning of each section, introducing its purpose and outlining its main results.

Diagrams and other artwork   Diagrams, photographs, charts, and tables are often essential, because complex structures and relationships are explained much more clearly by artwork than by written description. In fact, it is not unusual for a single diagram to convey the main message of a report. Illustrations are discussed in more detail in Chapter 9, but two rules are worth repeating here: each illustration, with an appropriate number and caption, should be inserted in the document just after its first mention in the text if possible; and every illustration must be mentioned in the text.

Citations   The words *citation* and *reference* are sometimes used as synonyms, but usually in the context of document production, a citation is a mention in the report text of an item in the list of references. Each reference in the list, which is in the back matter, provides the title, authorship, and other details of a document that supports or supplements the report. The purposes of a citation are

- to place the report in the context of existing documents;

- to mention background material that the reader is expected to know, or at least to know about, in order to read and understand the report;

- to give credit for material that has been quoted. Quotations from other documents are permissible if they are clearly distinguished from normal text, and if credit is given to their authors by proper citation.

- to give credit for material that has been rewritten or paraphrased. You are ethically required to acknowledge your sources.

- to mention related or similar material that is not included in the report;

- to allow a conclusion that depends, in part, on work contained in the reference documents;

- to add authority to a conclusion that is confirmed by work in an independent document.

As discussed in more detail in Section 8.1.3, each reference item has a label, such as "[3]" or "[Jones, 1999]." The label appears in a citation either

- as a noun, for example, "see [3], page 246," or "reference [3] concludes that ...";
  or

- as if it were parenthetic, for example, "the symbol % simply means the number 0.01 when used strictly [3]" or "In 1998, Gupta and Jones [3] concluded that ..."

Every cited document must be included in the References section (see Section 8.1.3), and every reference must be cited at least once in the document text.

Avoiding plagiarism As discussed in Section 3.5, plagiarism is the act of presenting the words or work of others as your own. Plagiarism is highly unethical. By using proper citations, you can avoid plagiarism while including brief quotations from other documents in your report, and you can refer to the analyses and conclusions of others. The following guidelines will help you to avoid plagiarism:

- Enclose borrowed wording in quotation marks, and cite the publication from which the wording is taken; for example: Jones [3] states, in a similar context, that "stiction is greater than friction."

- Cite each borrowed item parenthetically, even though you do not use the exact wording; for example: Jones [3] showed that under conditions of similar surface preparation but higher temperature, stiction is the dominant force, and friction is less important.

- Ensure that material you summarize or paraphrase is expressed in your own words. Changing minor wording or sentence structure is insufficient; read what you want to paraphrase, wait a significant time, write your paraphrase without referring to the source, and check that your work does not resemble the source. Add a citation, such as "the following summary follows the reasoning of [3] ... "

- Acknowledge collaborative work. It is normal to discuss your work with others and to develop joint solutions to similar problems. Acknowledgment of assistance or collaboration normally appears in the "Acknowledgments" section, unless a publication has been used or developed at the same time as your report, in which case a citation is appropriate. For example, "The data for Figure 5 were developed jointly with ... "

Conclusions A conclusion is a conviction reached on the basis of evidence and analysis given in the body. The evidence can be of two types: the author's work as presented in the report, or evidence in documents cited in the report. Every conclusion must be supported by the report itself, by cited references, or both.

Every investigation must reach a conclusion, even if it is merely that further investigation is needed or that the project was a failure. It should be possible to compare the conclusions section with the introduction section and find that every conclusion has been introduced in the introduction, and the outcome of every question introduced is found in the conclusions.

Many engineering projects result in recommendations in addition to conclusions, and it may be appropriate to use the title "Conclusions and Recommendations." The following discussion applies when a full section of recommendations is merited.

Recommendations A recommendation answers the reader's question, "What should I do about the situation?" Give clear, specific suggestions. A recommendation, which is imperative, should not be written as a conclusion, which is declarative. The first of the following two examples is a recommendation; the second is a conclusion:

- "Improve the transfer cooling by modifying the radiation fins."
- "The transformer cooling may be improved by modifying the radiation fins."

Recommendations require decisions to be made by someone in authority; therefore, each recommendation must be specific and complete. The reader usually needs to know what is to be done, when, whether it will disrupt normal operations, how long it will take, who will do it, and the cost. If required information is incomplete, the reader needs to know what next step should be taken.

8.1.3   The back matter

The back matter supports the report body, but with minor exceptions, understanding the body should not necessitate reading the back matter. The exception is this: large or numerous figures, tables, program listings, or other material that would disrupt the flow of the body should be put into appendices.

The body page numbering extends through the back matter, normally including the appendices.

References The title of the back matter section containing the list of references is simply "References" or sometimes "Documents Cited." Some standards list this section as part of the report body.

Each document in the list of references must be uniquely identified, together with its date, authors, and sufficient information for obtaining the document. Detailed rules for presenting this information have been developed. Technical societies publish guidelines that are available in printed versions or by web search (see Section 8.4, Problem 2), and of these, the IEEE style [7, 8] is followed in much technical literature. Most of these guidelines are derived from styles developed by the American Psychological Association (APA) [9], the Modern Language Association (MLA) [10], the University of Chicago Press [11], and the Council of Science Editors (CSE) [12]. These styles are discussed in many handbooks on English writing.

The order of the entries in a reference list is normally either

- alphabetical, by last name of first author, or
- by order of citation in the text.

Each entry is given a label, which is used in the citation and is either

- numerical, in list order, or
- a combination of author name and year of publication.

Your style guide should be obeyed, but the following examples show how different document types are typically entered into a list of references:

Book [1] T. K. Landauer, *The Trouble with Computers*. Cambridge, MA: The MIT Press, 1995.

Technical journal paper [2] L. Peng and P.-Y. Woo, "Neural-fuzzy control system for robotic manipulators," *IEEE Control Systems Magazine*, vol. 22, no. 1, pp. 53–63, 2002.

Conference paper   [3] J. Fredriksson and B. Egardt, "Nonlinear control applied to gearshifting in auto-mated manual transmissions," in *Proceedings of the 39th IEEE Conference on Decision and Control*, vol. 1, pp. 444–449, Sydney: Institute of Electrical and Electronics Engineers, 2000.

Web document   [4] International Bureau of Weights and Measures, *The International System of Units (SI)*. Sèvres, France: Bureau International des Poids et Mesures (BIPM), sev-enth ed., 1998. <http://www1.bipm.org/utils/en/pdf/si-brochure.pdf> (February 2, 2008).

Web documents create particular problems: the author may be difficult to identify, the contents or web address may change over time, and the validity of the content may be suspect. Publications of recognized technical societies, major commercial publishers, governments, or other authorities are to be preferred over others, and printed versions are generally preferable to web documents. Therefore, show the latest date that you accessed the file, and if a printed version exists, include its particulars in preference to or in addition to the web version. There are several different guidelines for including web addresses in references; the MLA style, outlined in [13], has been followed in this book.

The use of labels in citations is described above in Section 8.1.2. When the labels are numerical, then the third document, for example, in a list of references, is cited as [3], (3), or by superscript[3], depending on the chosen style. Otherwise, the label consists of the author name and year of publication; [Higham, 1998], for example.

Composing or maintaining a list of references by hand is problem-plagued and tedious, and these difficulties have influenced the guidelines mentioned previously. How-ever, the computer allows a simplified process by which document descriptions are retrieved from a central database or a local one provided by the report author, and auto-matically converted to the format required for the list of references. The references in this book, for example, were retrieved from a data file and formatted automatically using an IEEE style definition for BIBTEX, a program that produces files compatible with LATEX [14], the formatting software used for this book. Determine which system you pre-fer or are required to follow, and employ it consistently. You may use the citations and references in this book as a model if you wish.

If you must enter all references by hand, then the easiest method is usually to list them in order of citation, with author–date labels, since adding or deleting a reference does not require changing the labels or the order of the other references.

Bibliography   A bibliography section is included when documents that are not cited in the text are to be listed for the reader's benefit as background or further reading. The section usually begins with an introductory paragraph that describes the purpose and scope of the list. There also may be annotations (comments) included with each list entry to explain its relevance for the reader. The bibliography must be mentioned in the report text.

The bibliography format is the same as for references. The list is ordered, as appro-priate, alphabetically by author, chronologically by date of publication, or by topic.

Appendices   Appendices contain material that supplements the report body but is not an essential part of the text. Each appendix is given a label and a title, for example, "Appendix A: Source Code for the Interrupt Handler." The label for a single appendix is simply "Appendix."

Large or numerous figures, tables, program listings, or other material that would interfere with the clarity of the report body may be put into appendices and referred to in the main text. However, it should not be necessary to read appendices to understand how the report conclusions were obtained; material that is essential to the conclusions should be placed or, at least, summarized in the main text. Other typical appendix materials are the original letter from the person requesting the report, original laboratory data, laboratory instruction sheets, engineering drawings, and lengthy calculations.

An appendix that is produced together with the report body may contain citations or cross-references to other report contents, like any other section of the report. Independently produced material may also be included in an appendix and might contain its own bibliography, for example. The page numbering of stand-alone material may be retained if necessary, but page numbers corresponding to the report itself should be added.

8.2   Steps in writing a technical report

Creating a good report document requires English writing skills, proficiency with computer writing aids, a sense of design, and attention to detail. The possibilities for your working environment range from the desktop word processors found in small companies to elaborate project management software used in major organizations.

Desktop word processors allow you to write, revise, edit, and check what you have written. You can create or import high-quality diagrams and graphs. Spell checkers detect misspelled words, although they cannot always detect incorrect word use. Grammar checkers also help, but they do not capture the subtleties of the English language.

In a company environment, you may have publishing software designed for producing large documents. These tools are much more robust and flexible than word processors. They can reliably assemble a document from many sources and formats and automatically do numbering, cross-referencing, tables of contents, and indexing. They provide sophisticated formatting, special characters, and facilities to create formulas and equations. However, they may also require considerable skill to master.

Some companies not only allow the use of sophisticated writing aids, but require it. Project management software, for example, automatically saves documents as they are written and controls access to them so that only one person can modify a file at any time.

Plan, execute, revise   Whatever your computing environment, producing a report makes repeated use of a *plan, execute, revise* cycle to create the complete document and its component parts. The following steps apply to the complete report, but, with slight revision, they also apply to report components.

1. Identify the reader and the purpose of the report.

2. Plan and outline the report.

3. Organize the information, and fill in the details.

4. Complete the supporting sections.

5. Revise the material until it meets the stated purpose.

6. Submit the report.

Some engineers prefer a preliminary hand-drawn list or plan for each report component; others work entirely with their computer. In either case, the computer document begins as a skeleton to which detail is added at each step, as described in more detail below. Do not rush to write the detailed report body; the more thought you put into planning your document, the more logical its contents will be and the more efficient its production.

Identify the reader and the main message

Clarity and the needs of the reader are the two main considerations in report writing. Identify the reader before you begin to write. What is clear to one reader may be hopelessly opaque to another. Ideally, you will receive a memo from your boss requiring you to write a report for her or him, to answer specific questions, as part of the formal record of the progress made in an engineering project.

Without specific instructions, you must define the detailed purpose yourself. A report for your boss or a colleague in the same department requires much less background material and fewer definitions than a report for someone outside the company. In the absence of other information, write for a reader with your own general background but no specific knowledge of the special terms used in your report or the circumstances in which it is written.

Plan and outline the report

Once you know the general purpose of the report and its recipient, open the template provided by your company or a generic template that you have prepared or identified in advance. The template should contain standard report headings, but it also defines the report format: the choice of margins, point size, numbering system, front-matter components, and other production details.

Edit the title page by drafting a suitable title and entering the name of the report recipient. This page, and everything you write, is subject to later revision.

In the introduction, list the questions to be answered in the report, in point form and order of priority. Use this order, not just in this section, but also when presenting the details in the body of the report and when writing the conclusions and the recommendations.

Write tentative section headings in the report body, using the plan suggested for internal sections on page 118.

Organize the information, and fill in the details

When the reader and report objectives have been identified clearly, the next step is to organize the required information. If the work to be presented in the report is not complete, then your outline can assist you in setting work priorities and focusing on the questions to be answered.

Sort and organize the information as suggested on page 118. Revise the order and contents of the outline of the body as necessary.

Fill in the details of the body. Include an introductory sentence or paragraph for each section, particularly in a long report that will not be read from beginning to end at one time.

As each section is completed, write the corresponding paragraph or subsection of the conclusions and recommendations sections, and collect any required appendix material.

Once the body has been completed, write the introduction, changing it from the list of points or rough draft that was initially written. Make sure that each conclusion in the conclusions section corresponds to a part of the introduction.

Complete the supporting sections Once the report body is complete, the front matter and back matter can be finished. Word processors can produce tables of contents automatically, and some are capable of formatting the documents listed in the references section automatically. Write the summary section and the abstract, as described on pages 115 and 117. Add the appendices, and number the pages consecutively with the report body.

Revise Carefully read the report from beginning to end, making sure that the ideas are presented clearly and logically. Check the following:

1. Clarity: For every section and paragraph of the report, pose the question, "What will my reader ask at this point?" and make sure that the question is answered.

2. Logic: At every point in the document, make sure your reader knows why the material is important and placed where it is.

3. Authority: For every conclusion, imagine that your reader asks, "Why should I trust you?" Make sure that the support for the conclusion has been clearly identified in the analysis or the citations of references. This question applies to statements made in the interior of the report as well as in the conclusions section.

4. Spelling and punctuation: Use a spell checker, but do not rely totally on it.

5. Layout and appearance: Pay special attention that diagrams are clear.

The checklist in Section 8.3 below should be used at the final stage.

Submit the report The final report is usually bound and accompanied by a letter of transmittal for submission to the person who requested the report. Sometimes the letter is bound into the report after the title page, but often it is simply sent with the correct number of report copies to the receiver.

The letter of transmittal is a standard business letter, such as the one shown in Figure 6.2. Near the beginning, mention the name of the report and the terms of reference or contract under which it was written. Briefly describe the purpose of the report. It may be appropriate to describe any particular points you wish to be considered as the report is read. Usually the final paragraph encourages the reader to contact the sender if further information is required.

8.3 A checklist for engineering reports

This is a basic checklist for use before final submittal of your report; you may wish to add to it.

1. Letter of transmittal or cover letter (page 115; Section 8.1):
   - ☐ State the full title of the report.
   - ☐ Refer to the original request and purpose.
   - ☐ Acknowledge all assistance, and explain the extent of the help.

       □ State what parts of the report are your own work, if required.

       □ Date and sign the letter.

2. Cover or binder (page 113):

       □ The title and name of the author must be visible without opening the cover.

       □ The report must be securely bound.

       □ There must be a neat, professional appearance.

3. Title page (page 114):

       □ The title should be descriptive, specific, and sufficiently detailed for the reader to understand the purpose of the report.

       □ Your name, organization, and address must be shown.

       □ The date on which the report is submitted must be shown.

       □ The name of the recipient of the report must be shown.

4. Table of contents (page 116):

       □ The headings and heading numbers must be the same as in the report body.

       □ The correct page numbers must be given.

       □ Front-matter headings are included: "Summary," "List of Figures," ...

       □ Full appendix titles are included (not just "Appendix A," "Appendix B," ...).

5. List or lists of figures and tables (page 116):

       □ List all figures and tables that appear in the body of the report.

       □ Figure captions or abbreviated captions are included (not just "Figure 1," "Figure 2," ...).

6. Summary (page 117):

       □ State the purpose or goal of the work.

       □ Briefly state the methods used.

       □ Briefly state the conclusions and recommendations.

7. Introduction (page 118):

       □ State the purpose of the work; say what questions the report answers.

       □ State the scope of the work; outline what was done to complete the work.

       □ Mention the intended readership.

       □ Review the history, background, or previous work.

       □ Page 1 of the report is the first page of the introduction.

8. Body (page 117):

       □ It must be logically organized, with numbered section and subsection headings.

       □ All figures and tables must be numbered and captioned.

       □ Every figure and table must be referred to in the text.

       □ Landscape figures must be readable from the right of the page.

       □ Each figure or table is placed after its first reference.

9. Conclusions (page 120):

☐ Concisely restate the conclusions from the discussion in the detailed body sections.

☐ Each conclusion should correspond to a question introduced in the introduction section.

☐ Do not assume that the conclusions are obvious to the reader.

☐ Every conclusion must be supported by the report data and analysis or by correctly cited reference documents.

☐ Do not introduce irrelevant comments.

10. References (page 121):

☐ Every reference must be cited at least once in the body of the report.

☐ Cite using a standard method, such as by numbers in brackets [3].

☐ List the documents in a standard, consistent format.

☐ The reference documents are listed in a standard order, such as in order of citation or alphabetically by last name of the first author.

11. Appendices (page 122):

☐ Each appendix must be mentioned in the report body.

☐ Appendices contain only relevant supporting material.

☐ Appendices must be named descriptively (not just Appendix A, B, . . .).

12. General (Section 6.1.7; Chapter 7):

☐ Use the active voice as much as possible.

☐ Use the present or past tense but do not mix tenses.

☐ Use the spell-checking feature of your word processor.

☐ Proofread the report thoroughly.

☐ Make sure all pages are numbered in the correct order before binding.

8.4   Further study

1.   Choose the best answer for each of the following questions.

(a) The abstract and introduction of a report are normally written first.

    i. false   ii. true

(b) The essential elements of the title page of a formal report are

    i. title, author's name and address, date.

    ii. title, author's name, report purpose, recipient.

    iii. title, author's name and address, recipient, date.

(c) The front-matter pages of a formal report are numbered

    i. using lowercase roman numerals, with the title page as page i.

    ii. together with the rest of the report pages, with the title page as page 1.

    iii. using uppercase roman numerals.

(d) The conclusions must summarize all of the recommendations made in the report.

    i. false   ii. true

(e) Every reference, figure, table, and appendix must be mentioned in the body of the report.

    i. true   ii. false

(f) The essential items in the front matter of a formal report are

    i. front cover, title page, preface, acknowledgments, table of contents.

    ii. front cover, title page, summary, table of contents.

    iii. front cover, title page, letter of transmittal, table of contents.

    iv. front cover, title page, preface, table of contents.

(g) Engineering reports are not classified as legal documents.

    i. false   ii. true

(h) The main physical components of a formal report are

    i. title, letter of transmittal, summary, and table of contents.

    ii. front matter, body, back matter.

    iii. introduction, detailed analysis, conclusions.

(i) Normally the first page of the introduction section is page 1 of the report.

    i. false   ii. true

(j) Which statement is incorrect about the introduction of a formal report?

    i. It is usually placed in the front matter of the document.

    ii. It describes the questions that are answered in the conclusions.

    iii. It states what questions are addressed in the report, why it was written, for whom, and how.

    iv. It describes the context of the work performed.

2. Use your web browser to find and bookmark style guidelines published by

  (a) The American Society of Mechanical Engineers (ASME),

  (b) The American Society of Civil Engineers (ASCE),

  (c) The American Institute of Chemical Engineers (AIChE),

  (d) The Institute of Electrical and Electronics Engineers (IEEE),

  (e) The American Institute of Physics (AIP),

  (f) The American Mathematical Society (AMS),

  (g) The Association for Computing Machinery (ACM),

  (h) The Society for Industrial and Applied Mathematics (SIAM).

3.   Investigate the word-processing software on your computer or the computer system provided by your institution to find the default formats or templates for reports and other documents. How closely do they agree with the formats suggested in this text?

4.   Many engineering programs define requirements for formal reports such as are written in connection with work-terms or internships. Compare the standards that your reports must satisfy with the NISO standard [5]. Which of the many aids listed in Appendix A of this standard are you likely to need?

5.   Your boss is doing preliminary research for the construction of a factory for which power interruptions would have severe financial consequences, and she has asked you to report on the reliability of the electricity supply in your province. You reflect on how to write the report and decide that several factors are possibly relevant. The first is the generation capacity within the province, changes in supply in recent years, and plans for the future. The second is the history of demand and predictions of demand for the next few years. The third is the margin by which generation capacity exceeds current and predicted demand. The trends and predictions of these quantities are best explained with graphics. Write a report for your boss using graphs as appropriate to show trends and predictions. You will probably have to do an Internet search to find relevant information. If you cannot find suitable information for your province, use material from elsewhere in Canada.

6.   You work for a company that employs several dozen engineers and wishes to expand Canada-wide and internationally. The company specializes in non-destructive examination and testing of structures using a combination of electromagnetic and ionizing radiation as well as ultrasound. Because of the potential hazards involved if the measurements are not conducted properly, they should be supervised by a licensed engineer. In addition, the results of the testing typically bear on public safety. Therefore, your company wishes either to obtain permission for several of its engineers to practise widely elsewhere or to open branches containing locally licensed engineering practitioners. You have been given the task of investigating the conditions for licensure in Canadian provinces and territories, the United States, and the European Community. You have also been asked about the possibility of licensed engineers in your company obtaining permission to work in other provinces, the United States, and Europe. Investigate the information provided by licensing organizations in these areas, and write a report containing your findings.

7.   A company is proposing to open an engineering office in your home province. Before making a decision, the manager would like information about engineering salaries in the region. Search for engineering salary information and write a report, in the format discussed in this chapter, summarizing your findings. If information is unavailable or difficult to find, assume that the engineering manager may wish to establish an office in Ontario, where engineering salary data such as in Example 13.10 on page 196 is regularly made available by either the PEO or the OSPE. You may be able to show the

effect on annual salaries of the following factors: years since graduation, highest degree earned, size of organization, and region. Draw suitable graphics to show the relationships and trends of your data.

8.5	**References**

[1] R. Blicq and L. Moretto, *Guidelines for Report Writing*. Scarborough, ON: Prentice Hall Canada, fourth ed., 2001.

[2] J. N. Borowick, *Technical Communication and its Applications*. Upper Saddle River, NJ: Prentice Hall, 2000.

[3] D. G. Riordan and S. E. Pauley, *Technical Report Writing Today*. New York: Houghton Mifflin Company, 1999.

[4] N. J. Higham, *Handbook of Writing for the Mathematical Sciences*. Philadelphia: Society for Industrial and Applied Mathematics, second ed., 1998.

[5] National Information Standards Organization, *Scientific and Technical Reports— Preparation, Presentation, and Preservation*. Bethesda, MD: NISO Press, 2005. ANSI/NISO Z39.18–2005, Available by site search at <http://www.niso.org/> (September 30, 2008).

[6] National Information Standards Organization, *Scientific and Technical Reports—Elements, Organization, and Design*. Bethesda, MD: NISO Press, 1995. ANSI/NISO Z39.18–1995, <http://www.niso.org/standards/resources/Z39-18-1995.pdf> (March 9, 2008).

[7] IEEE, *IEEE Editorial Style Manual*. New York: Institute of Electrical and Electronics Engineers, 2007. <http://www.ieee.org/portal/cms_docs_iportals/iportals/publications/authors/transjnl/stylemanual.pdf> (March 9, 2008).

[8] M. Northey and J. Jewinski, *Making Sense, A Student's Guide to Research and Writing: Engineering and the Technical Sciences*. New York: Oxford, 2005.

[9] American Psychological Association, *Publication Manual of the American Psychological Association*. Washington, DC: American Psychological Association, fifth ed., 2001.

[10] J. Gibaldi, *MLA Handbook for Writers of Research Papers*. New York: Modern Language Association of America, fourth ed., 1995.

[11] University of Chicago Press, *Chicago Manual of Style*. Chicago: University of Chicago Press, fourteenth ed., 1993.

[12] Council of Science Editors, ed., *Scientific Style and Format: The CSI Manual for Authors, Editors, and Publishers*. Reston, VA: Council of Science Editors, seventh ed., 2006.

[13] J. E. Aaron and M. McArthur, *The Little Brown Compact Handbook*. Toronto, ON: Pearson Education, third Canadian ed., 1999.

[14] L. Lamport, *LaTeX, A Document Preparation System*. Reading, MA: Addison-Wesley Publishing Company, 1994.

Chapter

9 Report Graphics

Complex shapes and relationships are often best explained using graphics. Creating graphics is like writing: we must all understand the basic tools, but producing the best examples, such as the original on which Figure 9.1 is based, is an art.

In this chapter, you will learn

- some principles of good graphics,
- standard formats for technical graphs,
- the simplicity and utility of straight-line graphs,
- uses for logarithmic graph scales,
- a format for including engineering calculations in documents,
- the purpose and value of sketches.

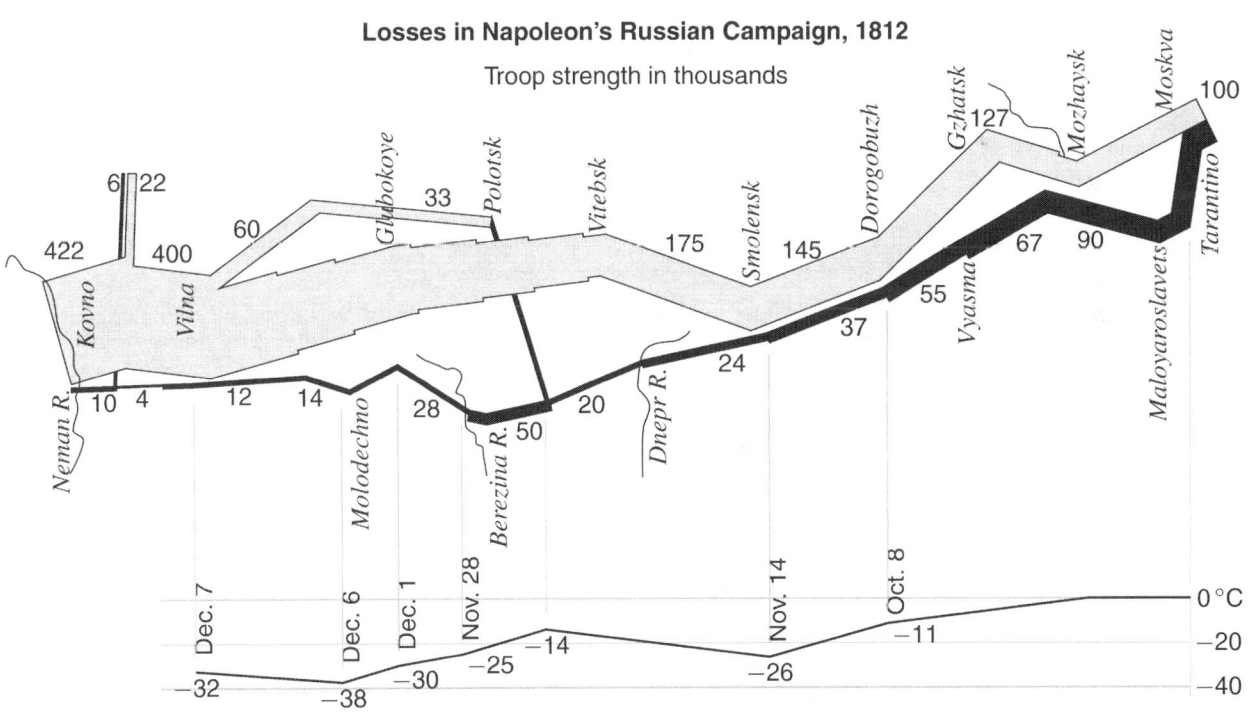

Figure 9.1 A graphic can explain a complex subject clearly. French engineer Charles Minard drew the first version of this chart in 1861, showing French army losses caused by the march, the main battles, disastrous river crossings, and wintry temperatures. The grey band shows the size and path of the advancing army, the black band traces the retreat, and temperatures are shown below. The army strength decreased from 422 000 troops to 10 000.

9.1 Graphics in engineering documents

This chapter focuses on graphics in engineering reports. In this context, the word *graphics* is synonymous with *figures*; the more general graphic design elements seen in magazines and advertising are absent. Specialized graphics such as mechanical, electrical, and chemical-process design drawings are not considered here in detail. Such material is often included in report appendices and should follow the standards of the applicable engineering discipline.

Numerous technical writing references contain advice on graphics [1, 2], but the design of graphics for reports is also a subject by itself; see references [3] and [4], for example. The authors of these publications recommend that graphics be designed according to the following principles:

- clarity: the graphic must display the correct message;

- efficiency: a significant amount of data is summarized, since small amounts might be better put in a table or the text;

- balance: the graphics and text complement each other, so that each graphic is discussed in the text and reinforces it.

Often the message to be delivered by a graphic is complicated; you must have clear insight before you can decide how to represent it. Several drafts may be required to make a satisfactory diagram. Consider yourself successful if the result replaces a difficult written explanation. As an example of complex data explained with minimal detail, Figure 9.1 accomplishes the feat of relating six variables: army strength, two position variables, direction of movement, date, and temperature. You might consider how long a written explanation would have to be to convey the same information, and whether it would have the same impact.

The simpler, the better, in graphics as in prose. Many computer programs are capable of producing good graphics, but beware of embellishing a diagram simply because the computer makes it easy.

Each figure in an engineering document must have a number and a caption. Tables normally have their own number sequence, separate from figures, although tables also require careful graphic design. Captions are placed below figures but above tables. The caption should make a point; it is not simply a title. The reader should be able to understand the principal message of the figure or table from its appearance and caption without reading the report text.

9.2 Standard formats for graphs

Graphs are employed to show trends and functional relationships in data. The standard format for line graphs is illustrated in Figure 9.2. Basic rules for this format and other graphical figures are given in the checklist below:

1. Fit each graphic within the typeset page margins. Use a landscape figure (having larger width than height) rotated 90° counterclockwise, to be viewed from the right,

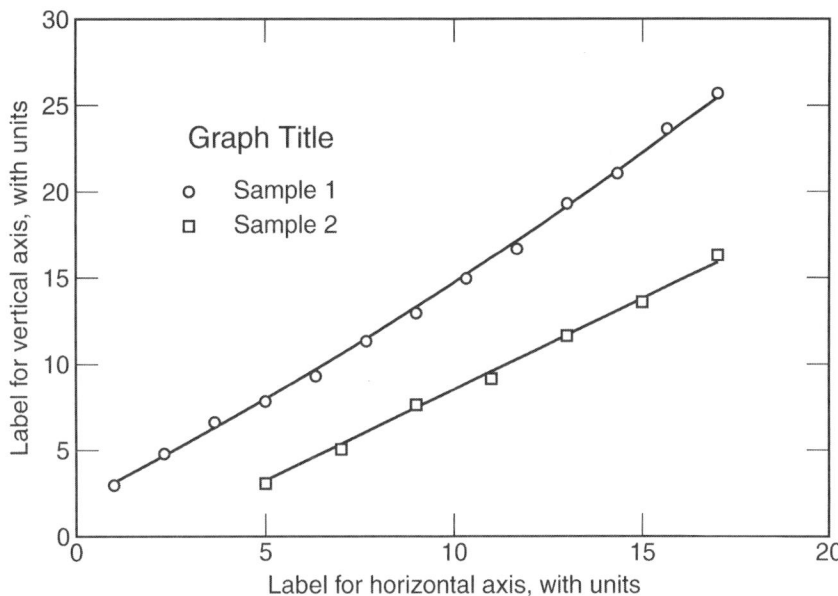

Figure 9.2   The standard format for a line graph is shown. Each axis must have a label with units. A legend is included to identify two or more curves or for required labelling that does not appear in the caption. The graph must accurately convey the desired message as simply and clearly as possible.

when more than the normal page width is required. For example, Figure 9.1 might be better drawn as a rotated landscape figure.

2. Plot the independent variable of a graph along the abscissa (horizontal axis) and the dependent (observed) variables along the ordinate (vertical axis).

3. Choose the scale of each axis to include sufficient detail on the graph for the correct conclusion to be deduced. Include suitable scale marks (tick marks) for numerical scales. Figure 9.2 shows short lines at the numbers. Insert smaller scale marks between the numbered marks when necessary. If the reader is expected to estimate the coordinates of points on the curves, then put grid lines on the graph background as in Figure 9.7 on page 138; omit them for clarity when the main message is the curve shape.

4. Label each axis to identify the quantity associated with it. The units of each quantity must be indicated using correct unit symbols as discussed in Chapter 10.

5. Make the graph labels readable without rotating the page. When this is impossible, make them readable when the page is rotated 90° clockwise. This rule applies to left-hand pages as well as to right-hand pages when the pages are printed on both sides of the paper.

6. Mark experimental data points with distinguishing symbols when two or more sets of data are plotted. The symbols overlay any lines that are drawn to show the trend in the data. Sometimes a vertical bar is added to each plotted point to show measurement uncertainty, as discussed in Chapters 11 and 12.

7. Include the zero value for each axis scale when the absolute value of quantities is important.

8. If the data points represent a continuous function, then fit a smooth curve through or near them. Do not draw a line if the data points do not represent a continuous function. A method for fitting a straight line to a set of points is discussed in Section 14.4.2.

9.2.1 Bar charts and others

Figure 9.3(a) contains a bar chart that compares the relative proportions of six quantities, arranged in order of size. Pie charts are often used for this purpose, perhaps because software for drawing them is readily available. However, comparing relative sizes of items is more difficult for pie charts than for bar charts, as shown in Figure 9.3(b); the best rule is to avoid pie charts when careful comparison of items is required.

There are many other kinds of charts, some of which are found in this book, as follows:

- Cumulative proportions, along with a bar graph, are shown in Figure 1.2 on page 7.

- A flowchart, which shows a series of steps, is shown in Figure II.1, page 79.

- Organizational charts show relationships between objects or people, as in Figure 4.2 on page 59.

- A line drawing shows essential features without extraneous detail, as in Figure 17.2 on page 264.

- A geographical chart relates quantities to map position, as in Figure 9.1 on page 131.

- Many other specialized drawing styles exist, such as those in Chapters 18 and 20.

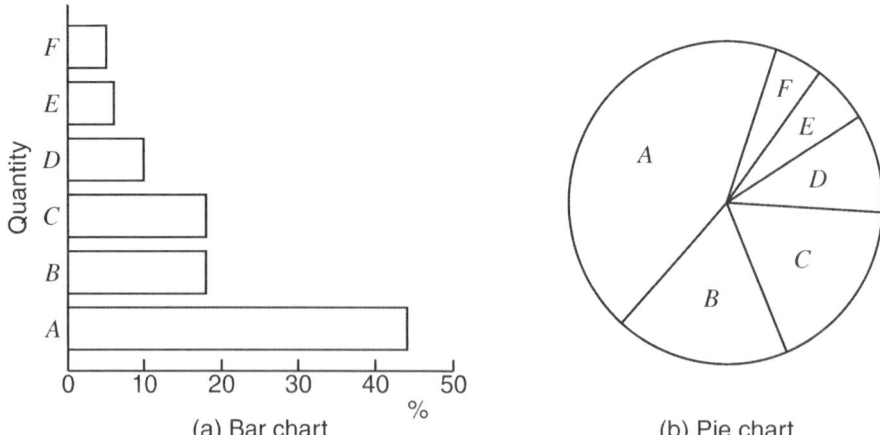

(a) Bar chart   (b) Pie chart

Figure 9.3 A bar chart (a) and pie chart (b) comparing six quantities. The proportions of B compared with C and of E compared with F are difficult to distinguish on the pie chart. Bar charts are preferred.

9.2.2   Straight-line graphs

Many physical phenomena are modelled by functions that correspond to straight-line graphs over significant ranges of variables. Straight-line functions are easy to under-stand, so they are often used in reports.

If x and y are variables, and m and b are constants, then the function

$$y(x) = mx + b \qquad (9.1)$$

results in a straight-line graph, as shown in Figure 9.4, with $y(x)$ as ordinate and x as abscissa. The constant m is the slope of the line, and b is the y-value at the intersection with the vertical axis.

Straight-line graphs such as Figure 9.4 and the lower graph in Figure 9.2 are some-times loosely described as "linear." Avoid this term when describing a straight line; the transformation from x to y defined in (9.1) is linear only if $b = 0$ (see Question 3 on page 145). However, along the bottom and left-hand scales of Figure 9.2, the distance from each scale mark to the origin is proportional to the numerical value at the scale mark, and these scales are therefore linear.

There are several ways, in addition to the formula given by (9.1), to describe the same straight-line function. For example, if a is the x-value at the intersection with the horizontal axis, then

$$y(x) = m\,(x - a), \qquad (9.2)$$

and if (x_1, y_1) is a point on the line, then, for all points (x, y) on the line,

$$y - y_1 = m\,(x - x_1). \qquad (9.3)$$

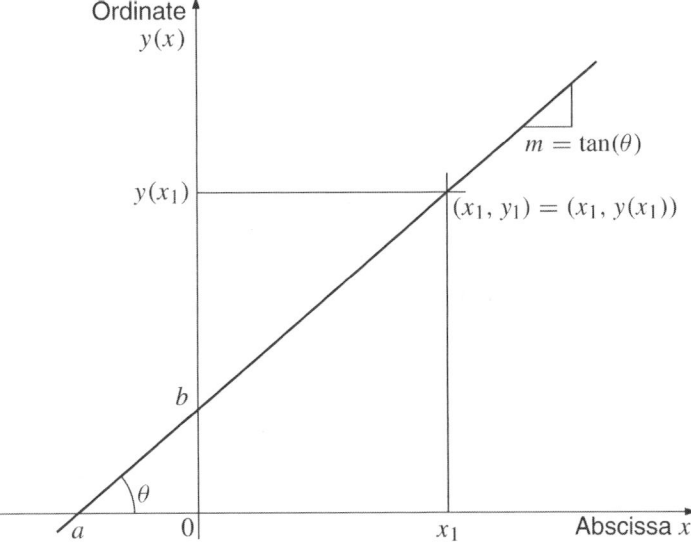

Figure 9.4   The graph of a straight-line function is defined by two independent numbers. Two ways of defining the function are $y(x) = mx + b$ and $y(x) = m\,(x - a)$.

The essential fact to note is that a straight-line function is defined by two independent quantities, such as m and b. Conversely, given the straight line, only two independent numerical quantities may be derived from it.

Given n measured data pairs (x_i, y_i), $i = 1, 2, \ldots n$, it may be desired to find two quantities, such as m and b, that define a straight line passing through or near the data points in the "best" way. The lower line of Figure 9.2 is an example. There are several ways of defining the best line, but the most common method is described in Chapter 14.

9.2.3 Logarithmic scales

Sometimes the domain or range of a function contains both very large and very small values, and a change of variables may be desirable in order to satisfy item 3 of the checklist in Section 9.2. "Moore's law," shown in Figure 9.5, illustrates such a change of variables. Gordon Moore, co-founder of the Intel corporation, predicted in 1965 [5] that the number of elements in integrated circuits would double every one to two years. Although there are indications that it may no longer be applicable, this prediction held true for more than three decades, perhaps partly because major business decisions were based on the assumption that it would, and engineers made it happen. A linear vertical scale would make the heights of all data points in the figure indistinguishably small, except for the largest few.

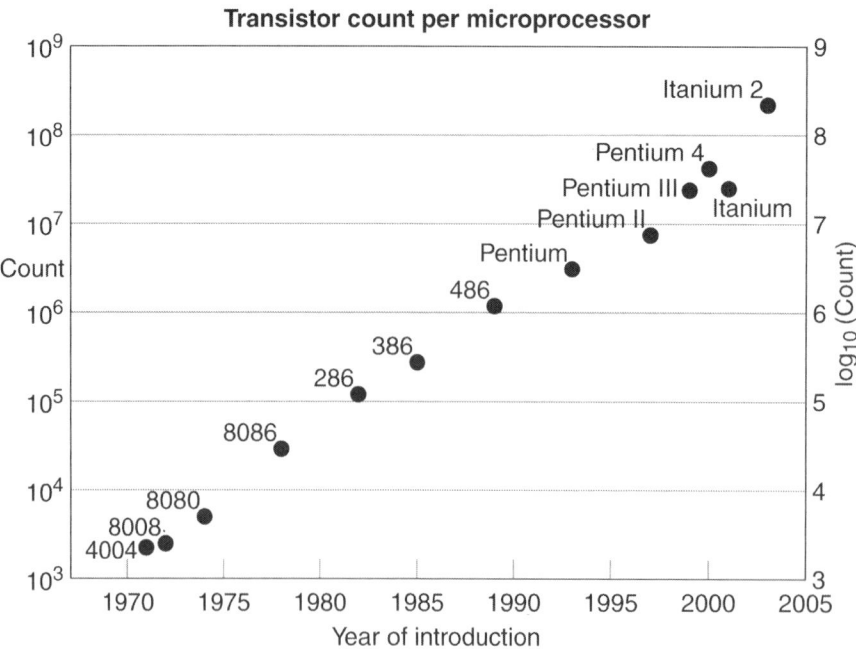

Figure 9.5 Moore's law states that the number of elements in integrated circuits doubles every one to two years. An approximately straight line results when the vertical scale is the logarithm of the dependent variable (data from [6]).

The right-hand scale in Figure 9.5 illustrates the fact that if

$$y = f(x), \tag{9.4}$$

and y takes on values from 10^3 to 10^9, then the function

$$v = \log_{10}(y) = \log_{10}(f(x)) \tag{9.5}$$

has a compressed range from 3 to 9. The scale marks along the left-hand scale are identical to those on the right-hand scale, but the count values, rather than their logarithms, have been written, as is normally preferable. The left-hand scale is described as "logarithmic." A graph with a logarithmic vertical scale and a linear horizontal scale is said to be "log-linear." Logarithmic scales cannot show a zero value since the logarithm of 0 does not exist.

A logarithmic change of variables may expand a scale rather than compress it. For example, Equation (9.5) expands the range of small values from 10^{-6} to 10^{-1} to the interval -6 to -1.

The decaying exponential function   The function shown in Figure 9.6 describes many physical phenomena such as decaying radioactivity, the cooling of a hot object, decaying chemical concentration, or decaying voltage across a charged object. In these cases, the physical variable $v(t)$, say, may be described by

$$v(t) = v_0\, e^{-t/T}, \tag{9.6}$$

where v_0 is the value of v at $t = 0$, and T is called the time constant of the function. Taking the logarithm gives

$$\log_{10} v(t) = \log_{10} v_0 + \log_{10}(10^{(\log_{10} e)(-1/T)\,t})$$
$$= \log_{10} v_0 + (-(\log_{10} e)/T)\, t, \tag{9.7}$$

which is a straight line on a log-linear graph, with vertical intercept $\log_{10} v_0$ (corresponding to v_0 on the logarithmic scale) and slope $-(\log_{10} e)/T$.

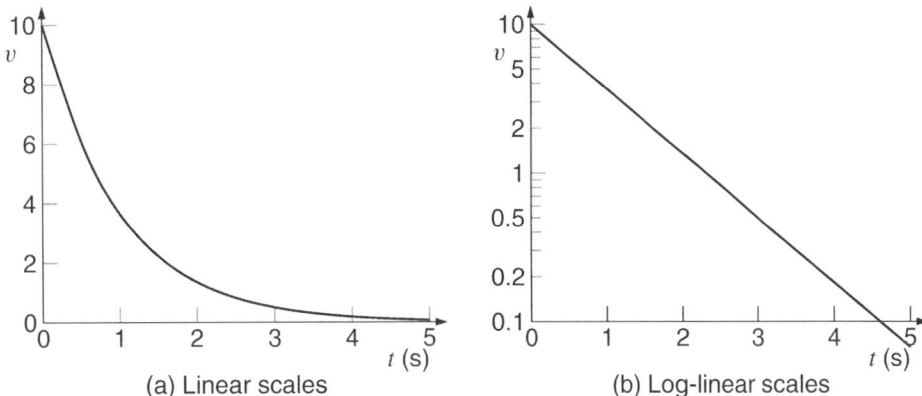

(a) Linear scales   (b) Log-linear scales

Figure 9.6   The linear scales in (a) display the rapid approach of the function to zero better than the log-linear plot in (b).

Figure 9.6 shows $v(t)$, plotted using linear scales and log-linear scales, for $v_0 = 10$ and $T = 1$. Notice that Figure 9.6(b) may give the wrong impression; the straight line appears to the eye to decrease by half its maximum value in half of the time and to have constant slope, whereas the untransformed graph (a) on the left shows a much faster decay and varying slope. A straight-line graph, or any other, should only be used when it gives the correct visual message.

Determining the initial value and time constant

Figure 9.6(b) may be the correct figure to draw in some circumstances. Suppose, for example, that we wish to obtain the initial value v_0 and the time constant T in Equation (9.6) from two or more measurements $v(t_1)$ $v(t_2), \dots$, and that these values have been used to draw Figure 9.6(b) with slope m and vertical intercept b. Then since $b = \log_{10}(v_0)$ in (9.7), the value v_0 can be read from the vertical intercept on the logarithmic scale. The time constant T is given by

$$T = \frac{\log_{10}(1/e)}{m} \tag{9.8}$$

or, since the change in the function in T seconds is $mT = \log_{10}(1/e)$, the constant T can also be obtained from the figure as the time in which the function decreases from any given value to $1/e = 36.8\,\%$ of the given value.

Other decaying exponential functions

More general exponential functions can be drawn as straight-line graphs in a way that is similar to the above example. Consider the function

$$y(t-t_0) = y_0\, \alpha^{\beta(t-t_0)}, \tag{9.9}$$

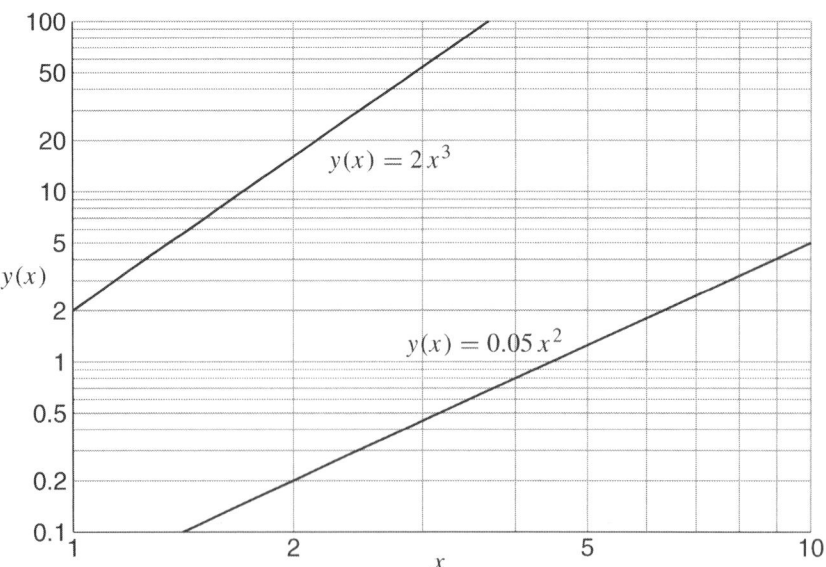

Figure 9.7 A power function $y(x) = \alpha\, x^\beta$ becomes a straight line on a log-log plot. Two examples are shown. Logarithmic grid lines have been superimposed on the graph to allow reading the scale values of points on the line. Only scale marks should be drawn if the reader does not require such detail.

where y_0, α, β, and t_0 are constants. With the substitutions $\alpha = 10^{\log_{10}\alpha}$ and $x = t - t_0$, and by taking the logarithm, we get

$$\log_{10} y(x) = \log_{10} y_0 + \beta(\log_{10}\alpha)\, x. \tag{9.10}$$

The result is a straight line on a log-linear graph, with vertical intercept y_0 and slope $\beta(\log_{10}\alpha)$.

The power function　The function shown in Figure 9.7 is a second common form for physical models. This function is described by the equation

$$y(x) = \alpha\, x^{\beta}, \tag{9.11}$$

where α and β are constants. Taking the logarithm, we get

$$\log_{10} y(x) = \log_{10}\alpha + \beta \log_{10} x. \tag{9.12}$$

The function produces a straight line if $\log_{10} y(x)$ is plotted versus $\log_{10} x$ on linear scales or, alternatively, if $y(x)$ is plotted versus x on a "log-log" plot, in which both the vertical and horizontal scales are logarithmic, as in Figure 9.7.

9.3　Engineering calculations

Numerical and symbolic calculations are key components of engineering practice and sometimes are crucial factors in major decisions. Therefore, calculations must be done with care and recorded in reports or other documents with great clarity and sufficient detail to allow verification of the results. Critical calculations are frequently checked by others under several circumstances, such as the following:

- For important decisions involving large expenditures of money or risk to human life, calculations are typically double-checked by a second engineer. Responsibility remains with the first engineer, but this checking provides protection against incorrect assumptions or incomplete analysis.

- Minor changes to the calculations for a project may make them applicable to a later project. The calculations and their range of validity should be easily understood by competent engineers who have not been part of the initial working group.

- In legal cases, such as inquests, civil suits, and disciplinary hearings, the engineering records of a project may be required as evidence, for the scrutiny of the court and expert witnesses. Clear assumptions and unambiguous conclusions are required to protect the calculations from challenge.

A standard format for calculations helps to ensure clarity and correctness and to guide the engineer's thought process. The format must be clear and logical, whether the calculations concern the depth of I-beams in a bridge structure, the diameter of pipes in a heat exchanger, the pitch diameter of a transmission gear, or the parameters in an electronic design. Some people have a natural talent for writing technical calculations neatly and logically; others must practise to achieve an acceptable standard.

In development laboratories, calculations are traditionally kept in bound logbooks as a permanent record; however, electronic records are now possible and increasingly used. Companies have standard formats for such files. At the end of a project, they are collected, indexed, and stored as a permanent record of the project. If project-management software is used, this process may be performed automatically.

As part of your professional education, university assignments are expected to conform to a standard format for engineering calculations. Figure 9.8 illustrates a typical

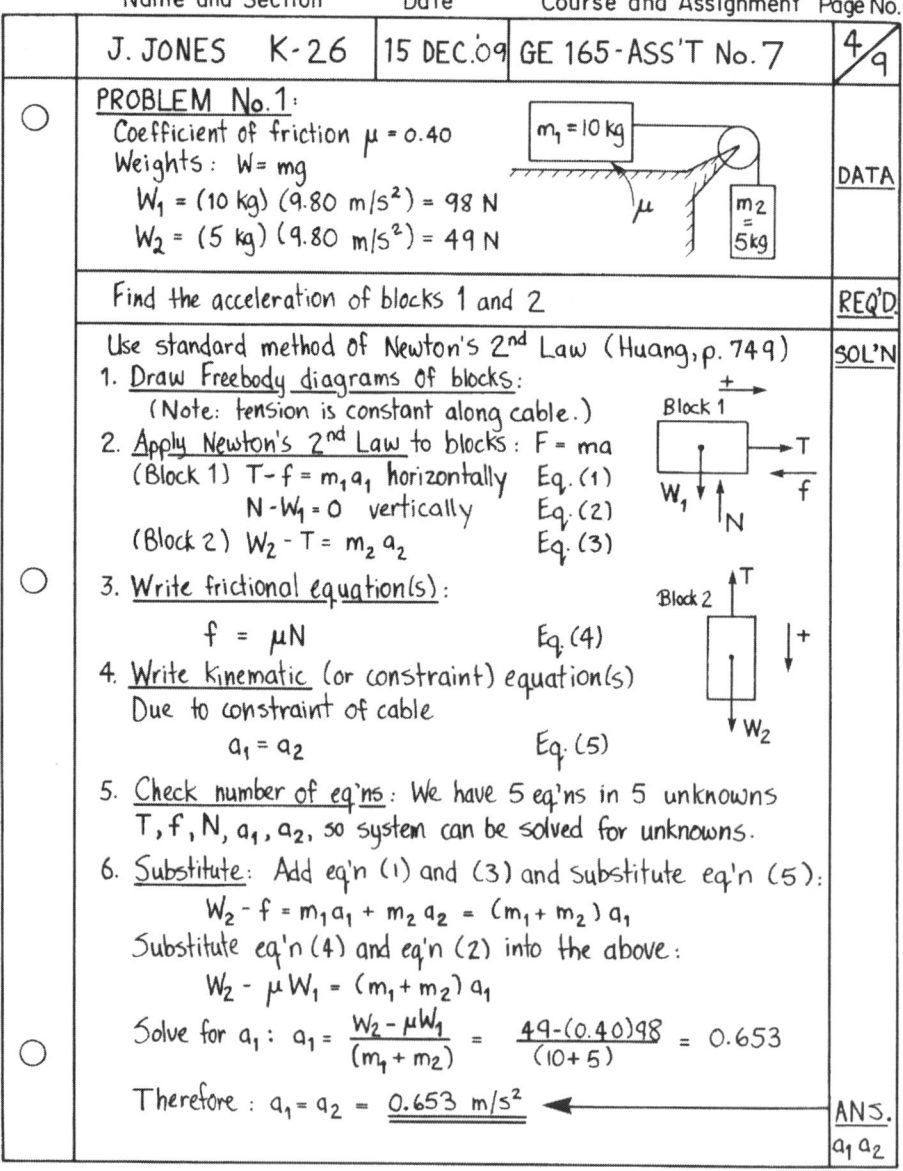

Figure 9.8 A standard format should be used for engineering calculations. This example analyzes a problem requiring elementary dynamics.

problem in elementary dynamics prepared by hand in a standard format. Similar pages can be produced using software that will store the written material and do the calculations. However, the computer does not eliminate the need for clarity of thought and exposition. As the figure shows, the following are standard requirements.

1. A space is reserved across the top for identification, including:
   - author's name,
   - date prepared,
   - project name (or course number),
   - page number and number of pages in the document.
2. The left margin is sufficiently wide for binding.
3. The right margin should be wide enough to be able to flag important items.
4. A statement of the problem is required, including the given data.
5. The calculations are presented clearly. Answers must include units and should be highlighted.

Engineers use a great variety of software for tasks such as modelling, analysis, and prediction. However, as mentioned in Section 3.5, the professional engineer is legally responsible for the calculations and conclusions and cannot blame the computer if the output is incorrect or is misunderstood. A careful record of the data, computations, and results must be kept, containing at least the following information:

- the name of the producer of the file,
- the date and time,
- the name and version of the software producing the result, and
- sufficient information to uniquely identify the input data.

9.4   Sketches

Freehand sketches are an essential aid for developing and explaining new ideas. A sketch such as Figure 9.9, for example, can effectively communicate ideas that would require long written explanation. Because of the speed and ease with which they can illustrate ideas and relationships, sketching techniques have not been superseded by computer software. New or evolving ideas are often sketched first, then later redrawn as detailed engineering drawings. However, well-prepared sketches may be used in all but the most formal of reports.

In industrial design offices, the first sketch of a new idea is dated and filed; it becomes a formal record of intellectual property and may be very valuable for patent or copyright purposes. Sketching can also quickly record project decisions and details that might be lost prior to the production of complete project documents.

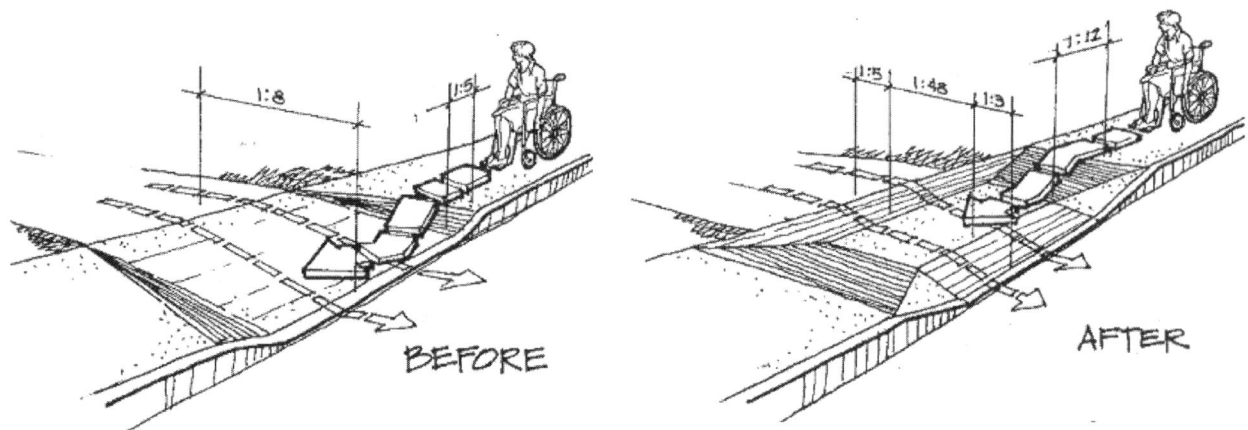

Figure 9.9   A sketch conveys essential details informally and complements or replaces long written explanation (from [7]).

   One sketching technique that engineers have traditionally cultivated is the ability to print alphabetic letters precisely. Figure 9.10 shows a form of "standard gothic" that is suitable for freehand work and similar to lettering that was required for production drawings before the widespread use of computers. Considerable effort and discipline is required to achieve professional quality, but both precision and speed can be developed with practice. One of the most important techniques, aside from drawing and spacing let-

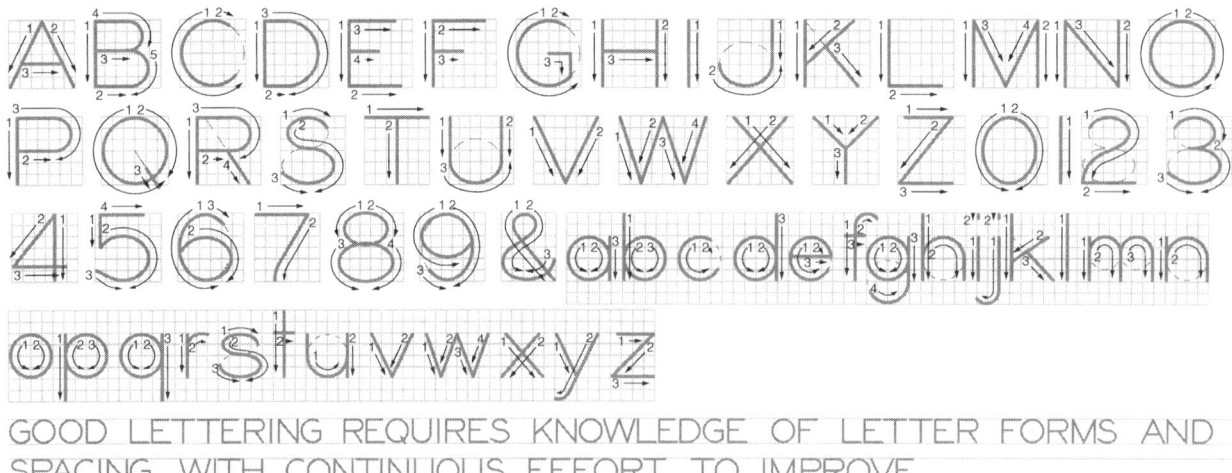

Figure 9.10   Standard letters for sketches and draft diagrams are illustrated, with arrows showing the order of the drawing strokes. All uppercase letters are six units high, and all are six units wide except for those in the name TOM Q. VAXY, which are five units wide, and the letters I and W. Always draw guidelines as shown in the bottom example. Letters should be closely spaced with their placement adjusted to minimize differences between the visually perceived interletter spaces. Interword spaces are approximately the size of the letter O (stroke data from [8]).

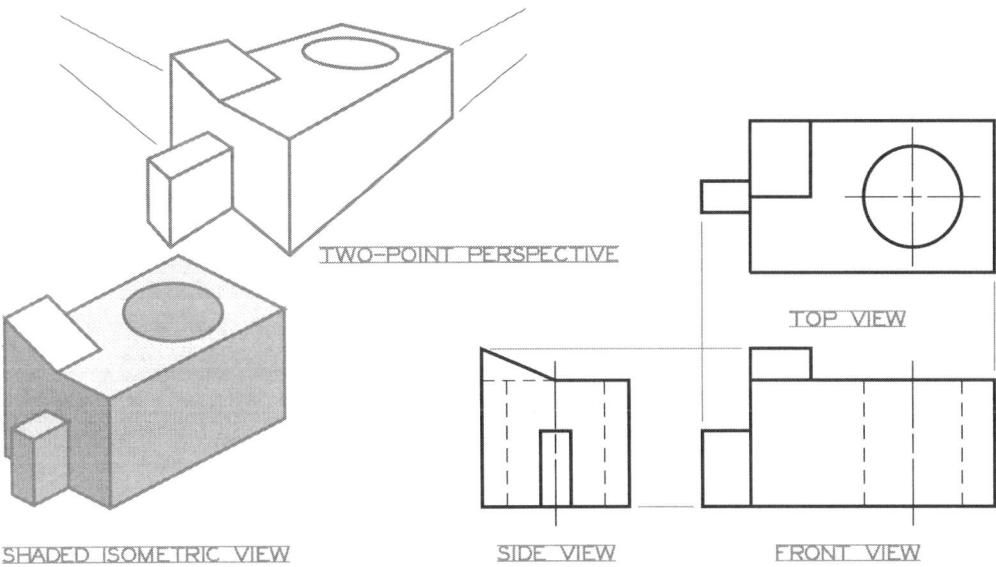

Figure 9.11   Typical views used in sketches and other drawings are shown. The two pictorial views display more than one face of the object. An isometric view shows the height, width, and depth distances to the same scale. In two-point perspective, the parallel horizontal lines of the principal faces of the object converge. The top, side, and front orthographic projections show any line in a plane parallel to the paper (viewplane) to true length.

ters correctly, is to always draw guidelines for the lettering as illustrated in Figure 9.10, since uneven heights or baselines are very visible. Formal illustrations should be lettered by computer.

Figure 9.11 illustrates some of the many views that may be chosen to describe a simple object. Computer-aided technical drawing is taught in many engineering courses, and requires artistry combined with knowledge of views, projections, sections, perspective, and the graphical symbols and fabrication processes of specific industries. For general advice, consult a textbook [9] or the standard published by the Canadian Standards Association (CSA) on general drawing principles [10].

Standard line weights and patterns are employed for both sketches and mechanical drawings; Figure 9.12 illustrates the basic types and their typical uses.

Some of the manual drafting skills that once were required of all engineers have been replaced by proficiency with software drawing programs, which remove the manual

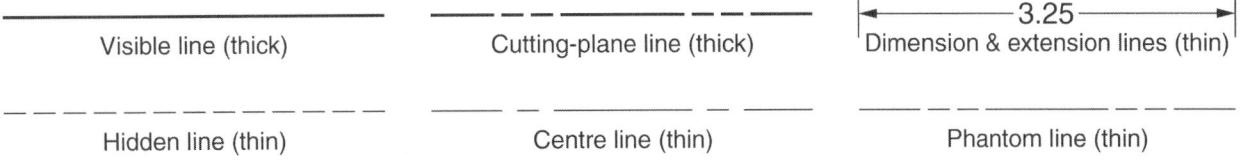

Figure 9.12   The lines used in sketches and other drawings.

dexterity and part of the tedium formerly associated with drawing. However, the ability to visualize and compose a drawing is a skill to be practised and valued, and is aided but not replaced by the use of computers.

9.5 Further study

1. Choose the best answer for each of the following questions.

(a) Engineering calculations should always include

 i. the author, date, project, and page numbers.

 ii. the date, project, page numbers, and conclusions.

 iii. the author, page numbers, and conclusion.

(b) Detailed grid lines should be included on a graph

 i. when the main message of the graph is the curve shape.

 ii. normally.

 iii. when the reader is expected to estimate the coordinates of points on the curves.

(c) Standard line graphs preferably show

 i. only one vertical scale.

 ii. the independent variable along the vertical axis, and the dependent variable along the horizontal axis.

 iii. the independent variable along the horizontal axis, and the dependent variable along the vertical axis.

 iv. only one horizontal scale.

(d) Sketches and handwritten calculations are normally considered to be legal documents only if they are included as part of a report.

 i. true   ii. false

(e) It is incorrect to state that a logarithmic graph scale

 i. can be used to graph a power function as a straight line.

 ii. should always be used if a straight-line function results, since this is the simplest function to understand.

 iii. can be used to graph an exponential function as a straight line.

 iv. cannot show the zero value.

(f) Referring to a graphic in the text of an engineering report is optional.

 i. true   ii. false

(g) On a logarithmic scale

 i. the scale numbers indicate the range of the logarithm of a value, but they are spaced in proportion to the value.

 ii. the scale numbers are placed linearly.

 iii. the scale numbers indicate the range of a value, but they are spaced in proportion to the logarithm of the value.

(h) Which of the following is not listed as one of the principles of graphic design?

    i. parsimony: graphics should not take up too much space in a document

    ii. efficiency: a significant amount of data is summarized

    iii. clarity: the graphic must display the correct message

    iv. balance: the graphics and text complement each other

(i) The zero value is included on the scale of the axis of a graph

    i. when the absolute value of quantities is important.

    ii. when deviations of the graph or graphs from a nonzero value are important.

    iii. when the relative sizes of quantities are important.

    iv. when the graph is a straight line.

(j) A straight-line graph is defined by

    i. an arbitrary number of independent quantities.

    ii. three independent quantities.

    iii. two independent quantities.

---

2. Two lines have been drawn through the data points in Figure 9.2. The lower line is a straight-line graph.

  (a) By reading from the lower graph, find constants m and b such that Equation (9.1) is a model for the function,

  (b) Similarly, find constants m and a such that Equation (9.2) is a model for the lower graph.

  (c) The upper line is not straight. Can you estimate constants b, m_1, and m_2 such that the equation $y(x) = b + m_1 x + m_2 x^2$ models the data?

---

3. The word *linear* means *line-like* or *straight line-like* in some contexts, but in mathematics, the axioms defining a linear transformation are: $y(x_1 + x_2) = y(x_1) + y(x_2)$ for all x_1 and x_2, and $y(\alpha x) = \alpha y(x)$ for all x and constant α. For the transformation defined by (9.1), write the formulas for $y(x_1 + x_2)$ and for $y(x_1) + y(x_2)$. Compare the two formulas to see why they are equal for all x_1 and x_2 only if $b = 0$.

---

4. Suppose that the angle θ and horizontal intercept a are known in Figure 9.4. Write the formulas for the slope m and intercept b.

---

5. As described in Section 9.2.3, Moore's law states that the number of elements per integrated circuit doubles every T years. For the data in Figure 9.5, this law is modelled by the formula

$$n(t) = n(1971) \times 2^{(t-1971)/T}.$$

By fitting a straight line approximately to the data in the figure, determine the doubling period T that applies to the figure.

6. Draw the graph of the functions (a) $y(x) = 0.3\,x^{2.5}$ and (b) $y(x) = 60\,x^{-3}$ on Figure 9.7.

7. You are given a function $y(x) = f(x)$, where $f(x)$ is known, but is neither an exponential function such as in Equation (9.9) nor a power function such as Equation (9.11). Describe how you could choose the scales of a graph so that the function is a straight line when plotted on the graph.

8. Figure 9.8 illustrates a page of hand calculations in standard format. Suppose that you would like to reproduce the same page, but entirely by software, such that if you were to change any parameter, m_1, m_2, or μ, say, then the correct answer would be calculated and appear automatically on the page. Find at least one software product or method of performing such calculations.

9.6 References

[1] D. Beer, ed., *Writing and Speaking in the Technology Professions*. Hoboken, NJ: Wiley-Interscience, second ed., 2003.

[2] J. S. VanAlstyne, *Professional and Technical Writing Strategies*. Upper Saddle River, NJ: Prentice Hall, 1999.

[3] W. S. Cleveland, *The Elements of Graphing Data*. Monterey, CA: Wadsworth Advanced Books and Software, 1985.

[4] E. R. Tufte, *The Visual Display of Quantitative Information*. Cheshire, CT: Graphics Press, 1983.

[5] G. E. Moore, "Cramming more components onto integrated circuits," *Electronics*, vol. 38, no. 8, 1965.

[6] Intel Corporation, "Moore's law," 2008. <http://www.intel.com/technology/mooreslaw/index.htm> (March 9, 2008).

[7] J. Barlow, P. Cannon, D. Dawson, H. D'Lil, L. Kozisek, G. Hilberry, E. Hilton, J. Markesino, C. Nagel, M. O'Connor, D. Ryan, E. Vanderslice, F. Wai, and M. Wales, *Public Rights-of-Way Access Advisory Committee Final Report*. Washington, DC: U.S. Government, Architectural and Transportation Barriers Compliance Board, 2001. <http://www.access-board.gov/prowac/commrept> (March 9, 2008).

[8] F. E. Giesecke, A. Mitchell, and H. C. Spencer, *Technical Drawing*. New York: The Macmillan Company, fourth ed., 1958.

[9] F. E. Giesecke, A. Mitchell, H. C. Spencer, I. L. Hill, J. T. Dygdon, and J. E. Novak, *Technical Drawing*. New York: Prentice Hall, twelfth ed., 2003.

[10] Canadian Standards Association, *Technical Drawings—General Principles*. Rexdale: Canadian Standards Association, 1983. CAN3-B78.1-M83, reaffirmed 1990.

Part III Engineering Measurements

5″ Railroad Transit

Catalog No. 50

Figure III.1   A railroad transit from the Bausch & Lomb 1908 *Catalog of Engineering Instruments*. Much of North America was surveyed with such instruments, and the surveys continue to be used today. (Photo courtesy of American Artifacts.)

Engineering decisions, whether in design, planning, scheduling, or fabrication, are often based on measured information. However, measurements are inherently inexact. The skill, science, and mathematics required to make precise measurements and interpret the effect of measurement uncertainty are a basic part of your engineering education. Part III of this book introduces you to the following topics.

Chapter 10   **Measurements and units:** This chapter reviews the unit systems used in Canadian engineering practice. The SI (Système Internationale or International System) units are emphasized, but other systems are also described.

Chapter 11   **Measurement error:** Measurement, as distinct from counting, is always a form of estimation and is therefore inexact. This chapter describes the sources of measurement uncertainty and gives rules for representing inexact quantities.

Chapter 12   **Error in computed quantities:** How does measurement uncertainty affect a quantity derived from measurements? This chapter shows how to calculate or estimate the uncertainty in a derived quantity as a function of measurement uncertainties.

Chapter 13   **Basic statistics:** Engineering is based on science as well as human factors; science is based on measurement, and modern measurement relies heavily on statistics. This brief introduction to basic statistics enables you to describe and compare sets of data using measures of central tendency (mean, mode, and median) and measures of dispersion (standard deviation and variance).

Chapter 14   **Gaussian law of errors:** Random errors, which are a fundamental component of measurement uncertainty, may often be described by the Gaussian probability distribution. This chapter defines the Gaussian distribution and shows how to use it to estimate errors and interpret data.

Chapter 10

Measurements and Units

Prior to the systematic definition of the unit systems that are taken for granted today, many regions and towns had their own distinct standards for measuring commodities (see Figure 10.1). This chaos has been simplified gradually by international agreement, starting almost two centuries ago, although the need for uniform measurement standards was recognized much earlier.

This chapter describes the engineering unit systems in current use, explains the difference between fundamental and derived units, and outlines a useful method for converting units. You will learn

- the definition, components, and uncertainty of measurements;

- the elements of the unit systems used by engineers in this and other countries;

- rules for correctly writing quantities with units;

- elementary descriptions of the most commonly used unit quantities in the SI and FPS unit systems;

- the usefulness of unit algebra.

The analysis of existing measurement methods and the invention of new techniques is an active technical art, with a community of specialists and regular journal publications and conferences [1], but the fundamentals are firmly fixed, except for occasional slight redefinition of the basic units.

Figure 10.1 Before the introduction of SI units, measurements were specific to particular purposes, trade goods, or locations. The iron reference standards in this digitally enhanced photograph are attached to the gateway wall of the pre-revolutionary Hôtel de Ville in Laon, France. The T shape provides reference lengths for measuring barrels, the rectangles for bricks and roof tiles, and the rightmost bar is for measuring cloth.

10.1 Measurements

Engineers frequently conduct or supervise tests and experiments that require physical measurements, such as tests of material qualities, soil and rock properties, manufacturing quality control, and experimental verification of design prototypes. A measurement that is adequate for the job at hand must be sufficiently accurate and repeatable, two properties that will be discussed in Chapter 11. This chapter concentrates on unit systems.

Measurements and units

A *measurement* is a physical quantity that has been observed and compared to a standard quantity, called a *unit*. The written representation of the measurement consists of two parts: a numerical value and a name, symbol, or combination of symbols that define the reference standard. For example, a distance may be measured in metres, millimetres, inches, feet, or other units, and the numerical value will depend on the unit chosen. The unit is a physical quantity that has been accepted according to experience and often international agreement as the standard by which certain measurements will be made. For example, the kilogram is defined to be the mass of a metallic object that is kept in Paris, France.

Base and derived units

Some units of measurement are defined in terms of others; for example, pressure may be measured in pascals. One pascal is defined as one newton per square metre, and since its definition depends on other units, the pascal is called a *derived unit*. The units from which all others are derived are called fundamental units or *base* units. How many fundamental units are required? For all normally measured physical quantities, the answer is seven. The several unit systems in common use provide definitions for fundamental units and a list of other units derived from them.

Dimensions

The words *units* and *dimensions* are sometimes used as synonyms; however, in the context of measurements, it is better to say that the dimensions of speed, say, are "distance divided by time," whereas the units of speed may be "kilometres per second" or "millimetres per year." Thus the fundamental dimensions include mass, distance, time, and other quantities corresponding to the fundamental quantities in a unit system.

10.2   Unit systems for engineering

The set of definitions and rules called the International System of Units, the Système International d'Unités, or simply SI [2], has been adopted by almost all countries, with the notable exception of the United States where, nevertheless, the system is used in specific industries.

Introduction of the metric system

The story of the engineering project to define the standard metre has been the subject of a best-selling historical account [3] and at least one historical novel. In the late 1700s, when new scientific thought was flourishing and Europe was in political ferment, a measurement system "for all people, for all time" was proposed, to be derived from nature rather than from a fabricated object. The reference metre was chosen to be one ten-millionth of the distance at sea level from the north pole to the equator. Determining this length required a survey of a significant distance along a meridian of longitude and extrapolation to the quarter meridian of Earth.

The French Académie des Sciences chose the meridian from Dunkerque through Paris, France, to Barcelona in Spain, and charged two groups led by savants of the highest reputation with the survey mission. A major engineering project, the survey lasted seven years and required specially designed instruments, perseverance in the face of accident, revolution, and war, and written reports. However, the nature and inevitability of measurement error (see Chapter 11) were not yet fully understood. An unexplained systematic discrepancy led the leader of one of the two survey parties to delay his report and to alter data by hand. Then the extrapolation of the survey to the quarter meridian was hindered by new knowledge of the irregularity of the shape of Earth. The delay caused by surveying difficulties and the withheld data had, in the meantime, forced a provisional length of the metre to be chosen. Although its precision has been refined several times since the original definition, this unintended length is still in use. Modern measurements show the quarter meridian to be approximately 10 002 000 m.

The survey project was conducted primarily in France, but the results were of international interest. The new system was introduced in France and ignored by most people, then rejected by Napoleon in 1812, and finally reintroduced in 1837 after several changes of government. The British were little inclined to adopt a system from a country with which they had been at or near war for decades, and which had produced the antimonarchist excesses of the French Revolution. The need for uniform standards was acknowledged in the United States, but relations with France were strained at times and, in some quarters, the adoption question was reduced to a simplified choice: "Shall we mold our citizens to the law, or the law to our citizens?" The latter was preferred. Acceptance was often associated with political upheaval. The system was adopted as a symbol of nationhood in Germany and Italy during political unification, in the Spanish and French colonies by decree, in Russia and Eastern Europe after the revolution of 1917, in Japan in 1945 at the end of war, in India at independence from Britain in 1947, and in China after the Communist accession in 1949. Britain, Australia, and Canada finally began a systematic switch to SI in the 1970s. The United States also began adoption at that time under President Ford, but the initiative was cancelled by President Reagan.

There are several conclusions that can be drawn from this early engineering project. One is that a technical decision may have to be taken under conflicting arguments about its validity. Indeed, although almost all scientific measurements use SI units, the system is still the subject of emotional debate. A second conclusion is that decisions affecting broad society may be accepted only very slowly and can easily become political issues.

One might ask why other measurements, such as time, are not based on a decimal system. Such units were proposed and a scheme in which there were three 10-day intervals (called decades) in a month, 10 hours of 100 minutes in a day, and 400 degrees in a circle was used for a while, but soundly rejected by most people. It appears that the base-60 scale used now for time and angle originated in Sumeria about 4000 years ago.

Absolute and gravitational systems

The SI base units and some derived units are listed in Table 10.1. The SI unit system is called an *absolute* system because mass is a fundamental unit and Newton's second law (force = mass × acceleration) is invoked to derive the force due to gravity. *Gravitational* systems adopt force as a fundamental unit and determine mass from Newton's second law. Unit systems are classified according to whether they are gravitational or absolute and whether they use metric or English units.

In Canada and the United States, the traditional unit system is an English gravitational system, called the FPS gravitational system, in which distance, force, and time are

Table 10.1   Base and derived SI units

Symbol	Unit Name	Quantity	Definition
m	metre, meter	length	base unit
kg	kilogram	mass	base unit
s	second	time	base unit
K	kelvin	temperature	base unit
°C	degree Celsius	temperature	(kelvin temperature) $- 273.15$
N	newton	force	$\mathrm{m} \cdot \mathrm{kg} \cdot \mathrm{s}^{-2}$
J	joule	energy	$\mathrm{N} \cdot \mathrm{m} = \mathrm{m}^2 \cdot \mathrm{kg} \cdot \mathrm{s}^{-2}$
W	watt	power	$\mathrm{J/s} = \mathrm{m}^2 \cdot \mathrm{kg} \cdot \mathrm{s}^{-3}$
Pa	pascal	pressure	$\mathrm{N/m}^2 = \mathrm{m}^{-1} \cdot \mathrm{kg} \cdot \mathrm{s}^{-2}$
Hz	hertz	frequency	s^{-1}
	**Electrical and Electromagnetic Units**		
A	ampere	current	base unit
C	coulomb	charge	$\mathrm{A} \cdot \mathrm{s}$
V	volt	potential	$\mathrm{J/C} = \mathrm{m}^2 \cdot \mathrm{kg} \cdot \mathrm{s}^{-3} \cdot \mathrm{A}^{-1}$
Ω	ohm	resistance	$\mathrm{V/A} = \mathrm{m}^2 \cdot \mathrm{kg} \cdot \mathrm{s}^{-3} \cdot \mathrm{A}^{-2}$
S	siemens	conductance	$1/\Omega = \mathrm{m}^{-2} \cdot \mathrm{kg}^{-1} \cdot \mathrm{s}^3 \cdot \mathrm{A}^2$
F	farad	capacitance	$\mathrm{C/V} = \mathrm{m}^{-2} \cdot \mathrm{kg}^{-1} \cdot \mathrm{s}^4 \cdot \mathrm{A}^2$
Wb	weber	magnetic flux	$\mathrm{V} \cdot \mathrm{s} = \mathrm{m}^2 \cdot \mathrm{kg} \cdot \mathrm{s}^{-2} \cdot \mathrm{A}^{-1}$
T	tesla	flux density	$\mathrm{Wb/m}^2 = \mathrm{kg} \cdot \mathrm{s}^{-2} \cdot \mathrm{A}^{-1}$
H	henry	inductance	$\mathrm{Wb/A} = \mathrm{m}^2 \cdot \mathrm{kg} \cdot \mathrm{s}^{-2} \cdot \mathrm{A}^{-2}$
	**Dimensionless Quantities**		
rad	radian	plane angle	$\mathrm{m/m} = 1;\ 1/(2\pi)$ of a circle
sr	steradian	solid angle	$\mathrm{m}^2/\mathrm{m}^2 = 1;\ 1/(4\pi)$ of a sphere
mol	mole	particle count	base unit ($\simeq 6.02 \times 10^{23}$)
	**Light**		
cd	candela	intensity	base unit
lm	lumen	flux	$\mathrm{cd} \cdot \mathrm{sr}$
lx	lux	illuminance	$\mathrm{lm/m}^2$

fundamental quantities, measured in units of feet, pounds, and seconds. Both FPS and SI systems are in use in Canada, and students must be familiar with both.

The SI and the FPS systems, as well as other unit systems sometimes found in references, are compared in Table 10.2 and discussed briefly below.

Absolute systems

- The SI system is preferred in Canada, although quantities such as land survey measurements performed using other systems can be expected to be encountered indefinitely. Where exports, particularly to the United States, are significant, many industries continue to use versions of English or U.S. unit systems.

Table 10.2   Comparison of unit systems

	Dimensions	Absolute		Gravitational	Hybrid
		SI	CGS	FPS	American
Fundamental	Force [F]	–	–	lb	lbf
	Length [L]	m	cm	ft	ft
	Time [T]	s	sec	sec	sec
	Mass [M]	kg	g	–	lbm
Derived	Force [F]	newton (kg · m/s^2)	dyn (g · cm/sec^2)	–	–
	Mass [M]	–	–	slug (lb · sec^2/ft)	–
	Energy [LF]	joule (N · m)	erg (cm · dyne)	ft · lb	ft · lbf
	Power [LF/T]	watt (N · m/s)	erg/sec	ft · lb/sec	ft · lbf/sec
	Pressure [F/L^2]	pascal (N/m^2)	dyne/cm^2	lb/ft^2	lbf/ft^2

- The CGS system uses centimetres, grams, and seconds and was previously used extensively in science.

- The FPS absolute system is very rarely used and is not listed in the table.

Gravitational systems

- The FPS gravitational system is the traditional system used in North America. Students must be familiar with it for a few more decades.

- The metric MKS gravitational system is rarely used. This system is not listed in Table 10.2.

Hybrid systems   The American engineering system and the European engineering system are occasionally still encountered. These are called hybrid systems, since both mass and force are fundamental units; the European hybrid units are not shown in the table.

Hybrid systems were popular for a time because units on both sides of Newton's second law equation could be arbitrarily defined; that is, the weight of an object (which is a force) could also be used as its mass. However, the use of both units, such as pound mass (lbm) and pound force (lbf), leads to confusion. To use hybrid systems, Newton's second law must be rewritten as

$$F = k_g ma, \qquad (10.1)$$

where the constant k_g equals $1/g$, and where g is the acceleration of a falling object (32.2 ft/sec^2 or 9.80 m/s^2). This system should be avoided, as should the European metric hybrid system mentioned above.

| 10.3 | **Writing quantities with units** |

Standardized rules for correctly and unambiguously representing quantities with units should be followed whenever possible. There are sometimes very slight differences between published standards, usually because of the time required for approval of modifications. This book follows reference [4]. Figure 10.2 shows an example of a quantity with units to illustrate the following items.

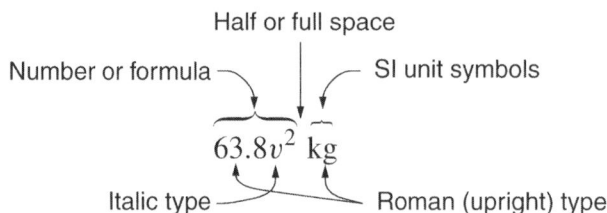

Figure 10.2   A physical quantity is written as a number or a formula representing a number, followed by unit symbols.

1. The symbols denoting a physical quantity consist of a number or formula that represents a number, which may be negative, followed directly by its unit symbol or symbols. The number is separated from the unit symbols by a narrow space or, if unavailable, by a normal space, except that no space is used with the symbols $°, ', ''$ when they designate plane angle as degrees, minutes, and seconds, respectively:

    $$10.25\,\text{km} \quad \text{not} \quad 10.25\text{km}$$
    $$15\,°\text{C} \quad\quad \text{not} \quad 15°\text{C}$$
    $$37.5° \quad\quad\quad \text{not} \quad 37.5\,°$$

2. The unit symbols must be in roman (upright) type regardless of the surrounding text. The number is normally in roman type and may be in decimal, scientific, or other notation as discussed in the next chapter. If the number is represented by a formula, then the symbols in the formula are normally in italic or "math italic" type with superscripts and subscripts in italic or roman type as appropriate:

    $$10.25\,\text{km} \quad \text{not} \quad 10.25\,km$$
    $$ab^2c\,\text{N} \quad\quad \text{not} \quad \text{ab}^2\text{c}\,\text{N}$$

3. SI unit symbols are lowercase in general, except when they are derived from the proper name of a person, in which event they normally begin with a capital letter, regardless of whether there is a prefix:

    $$15\,\text{kN} \quad \text{not} \quad 15\,\text{kn}$$

    However, when the full name of a unit, such as a newton, is used in a sentence but is not the first word, the name is not capitalized. The name "degree Celsius" is a special case.

4. The unit may be preceded by a magnitude prefix listed in Table 10.3. The prefix is regarded as an inseparable part of the symbol; thus $3\,\text{km}^2 = 3 \times (\text{km})^2 = 3 \times 10^6\,\text{m}^2$,

Table 10.3   Magnitude prefixes in the SI unit system by symbol, name, and numerical value.

Y	yotta	10^{24}	M	mega	10^{6}	f	femto	10^{-15}
Z	zetta	10^{21}	k	kilo	10^{3}	a	atto	10^{-18}
E	exa	10^{18}	m	milli	10^{-3}	z	zepto	10^{-21}
P	peta	10^{15}	μ	micro	10^{-6}	y	yocto	10^{-24}
T	tera	10^{12}	n	nano	10^{-9}			
G	giga	10^{9}	p	pico	10^{-12}			

not $3 \times 1000 \times (\text{m})^2 = 3000\,\text{m}^2$. An international standard for computer-related magnitude prefixes that are powers of 2 has been proposed but is not in widespread use. Other computer-related magnitude prefixes are typically not separated by a space: the M in 128MB stands for 1024×1024, so 128MB $= 134\,217\,728$B.

5. Unit symbols, together with prefixes if any, are treated like algebraic factors that can be multiplied or divided with numbers or other unit symbols. Multiplication between units is indicated by a half-height (centred) dot or a half-space, division by a slash (solidus) or a negative superscript:

$$27\,\text{kg} \cdot \text{m/s}^2 \quad \text{or} \quad 27\,\text{kg}\,\text{m/s}^2 \quad \text{or} \quad 27\,\text{kg} \cdot \text{m} \cdot \text{s}^{-2}$$

6. Unit symbols are not modified with subscripts:

$$V_{\text{max}} = 200\,\text{V} \quad \text{not} \quad V = 200\,\text{V}_{\text{max}}$$

7. It must be clear which numerical quantity and unit symbols are related:

$$3.7\,\text{km} \times 2.8\,\text{km} \qquad \text{not} \quad 3.7 \times 2.8\,\text{km}$$
$$87.2\,\text{g} \pm 0.4\,\text{g} \text{ or } (87.2 \pm 0.4)\,\text{g} \quad \text{not} \quad 87.2 \pm 0.4\,\text{g}$$

8. Unit symbols are not mixed with unit names in expressions, and mathematical symbols are not applied to unit names. Thus, in a sentence, use kg/m^2 or kilogram per square metre, but not kilogram/m^2.

9. When long numbers are written in groups of three, the groups should be separated by a non-breaking half space rather than the comma often used in North America or the period used in some European countries:

$$93\,000\,000 \text{ miles} \quad \text{not} \quad 93{,}000{,}000 \text{ miles}$$
$$21\,298.046\,83\,\text{m} \quad \text{not} \quad 21298.04683\,\text{m}$$

Grouping is strongly preferred for numbers containing five or more adjacent digits. Four-digit numbers are not split except for uniformity when they are in a table with longer numbers or compared with longer numbers, as in the third column of Table 1.1 on page 10, for example. If a number is split on one side of the decimal point, it is usually split on the other side as well.

| 10.4 | **Basic and common units** |

Unit definitions in both SI and FPS systems are described below and compared in Table 10.2. Handbooks [5, 6] and textbooks [7] list others. When working with different unit systems, you must be aware of the fundamental difference between absolute and gravitational systems of units and the importance of Newton's second law in defining these systems.

Force *Force* is a push or a pull; it causes acceleration of objects that have mass. In the FPS gravitational system, force is a fundamental unit, expressed in pounds (lb). In SI, the unit of force is the newton (N), which is the force required to give a mass of one kilogram an acceleration of one metre per second squared, derived using Newton's second law. A newton is about one-fifth of a pound or, more accurately, $1\,\text{lb} = 4.448\,222\,\text{N}$.

Length *Length* or distance is a fundamental unit in all systems. The FPS unit is the foot (ft) and the SI unit is the metre (m). To convert from one system to the other, by definition, $1\,\text{ft} = 0.3048\,\text{m}$. The metre is defined to be the distance travelled by light in a vacuum in $1/299\,792\,458$ of a second. The United States disagrees with most other countries on the spelling, but either *meter* or *metre* is accepted when used consistently. In Canada, it is generally accepted that a *metre* is a unit of measurement, and a *meter* is a measuring instrument.

Mass *Mass* is a measure of quantity of matter. Mass should not be confused with weight, which is the force of gravitational acceleration acting on the object; thus a change in gravitational acceleration causes a change of weight but not of mass. The kilogram is the fundamental unit of mass in SI and is defined to equal the mass of a specific object kept in Paris. In the FPS gravitational system, the unit of mass is the slug, derived using Newton's second law. One slug is the mass that accelerates at one foot per second squared under a force of one pound: $1\,\text{slug} = 14.593\,90\,\text{kg}$.

Time *Time* is a fundamental unit in all systems and is measured in seconds, abbreviated "sec" in the FPS gravitational system, but simply "s" in SI. The second is defined as $9\,192\,631\,770$ periods of the radiation corresponding to the transition between the two hyperfine levels of the ground state of the cesium 133 atom.

Pressure *Pressure* has the derived units of force per unit area. In the FPS gravitational system, pressure is measured in pounds per square inch (psi) or pounds per square foot (lb/ft^2). In SI units, the unit of pressure is the pascal (Pa): $1\,\text{Pa} = 1\,\text{N/m}^2$, and $1\,\text{psi} = 6.894\,757\,\text{kPa}$.

Work and energy *Work* is defined as the product of a force and the distance through which the force acts in the direction of motion. *Energy* has the same units and is the capacity to do work. Both are derived units in the SI and FPS systems. In the FPS gravitational system, work is measured in foot pounds (ft lb), inch pounds (in lb), horsepower hours (HP hr), British Thermal Units (BTU), and other units; the SI unit is the joule (J), and $1\,\text{ft}\cdot\text{lb} = 1.355\,818\,\text{J}$.

Power *Power* is the rate of doing work. In the FPS gravitational system, power is measured in foot pounds per second (ft lb/sec) or horsepower (HP) ($1\,\text{HP} = 550\,\text{ft}\cdot\text{lb/sec}$). In SI units, power is measured in watts (W): $1\,\text{W} = 1\,\text{J/s} = 1\,\text{N}\cdot\text{m/s}$, and $1\,\text{ft}\cdot\text{lb/sec} = 1.355\,818\,\text{W}$.

Temperature *Temperature* is an indirect measure of the amount of heat energy in an object. There are four common temperature scales: Fahrenheit, Rankine, Celsius, and kelvin.

On the Fahrenheit scale used in the FPS system, water freezes at $32\,^\circ\text{F}$ and boils at $212\,^\circ\text{F}$. Fahrenheit temperatures can be converted to Rankine by adding a constant, since the degree size is equal; $t\,^\circ\text{F}$ corresponds to $(459.67+t)\,^\circ\text{R}$. The Rankine scale measures temperature from absolute zero and is still in common use in engineering thermodynamics in North America.

On the Celsius scale (which is a slight modification of the older centigrade scale), water freezes at $0\,^\circ\text{C}$ and boils at $100\,^\circ\text{C}$. In fundamental SI units, water freezes at $273.15\,\text{K}$. The two scales have identical unit difference, so that $t\,^\circ\text{C}$ corresponds to $(273.15+t)\,\text{K}$. Absolute zero, where molecular motion stops, is the same in both Rankine and SI units ($0\,^\circ\text{R} = 0\,\text{K}$). Note that the degree symbol ($^\circ$) is not used with the kelvin symbol in SI notation.

Dimensionless quantities Some measured quantities are defined as ratios of two quantities of the same kind, and thus the SI units cancel, leaving a derived unit, which is the number 1. Such measured quantities are described as being dimensionless, or of dimension 1. Many coefficients used in physical modelling and design are dimensionless; for example, the coefficient of friction is the ratio of two forces. Another example is the measurement of angle, but in this context the names "radians" and "degrees" are used for the dimensionless units. Recall that an angle in radians is the ratio of arc length over radius, and in degrees, it is the ratio of arc length divided by circumference, multiplied by the number 360, a scale factor.

The litre The litre, a special name for one cubic decimetre, is not included in the SI units but is accepted for use with them and is often given the symbol L to avoid confusion with the number 1.

U.S. and imperial units There are differences between the U.S. and imperial units of identical name that deserve attention, particularly the units for volume. One Canadian or imperial gallon equals $4.546\,09 \times 10^{-3}\,\text{m}^3$, whereas the U.S. gallon equals $3.785\,412 \times 10^{-3}\,\text{m}^3$, or about $83.3\,\%$ of the imperial gallon. There are 160 imperial fluid ounces per imperial gallon but 128 U.S. fluid ounces per U.S. gallon. A summary such as reference [6], Appendix B of [4], or Chapter 5 of [7] should be consulted for lists of conversion factors.

10.5   Unit algebra

As seen in Section 10.3, a measured quantity is written as a number, or more generally as a formula for a number, together with the associated unit symbol or symbols. Ordinary algebra can be performed using such quantities, with the restriction that only quantities with the same units can be equated, added, or subtracted.

Consider the conversion of quantities between unit systems. Suppose, for example, that a measured power has been found to be $20\,000\,\text{ft}\cdot\text{lb/min}$, and this is to be converted to horsepower, which is defined as $550 \, \text{ft} \cdot \text{lb/sec}$. Then the written measured quantity is modified as shown:

$$20\,000 \frac{\text{ft} \cdot \text{lb}}{\text{min}} \times \frac{1 \, \text{min}}{60 \, \text{sec}} \times \frac{1 \, \text{HP}}{550 \, \text{ft} \cdot \text{lb/sec}} = \frac{20\,000}{60 \times 550} \, \text{HP} = 0.606 \, \text{HP}, \qquad (10.2)$$

where the original quantity has been multiplied by the number $1 = \frac{1 \, \text{min}}{60 \, \text{sec}}$ and again by $1 = \frac{1 \, \text{HP}}{550 \, \text{ft} \cdot \text{lb/sec}}$ in order to cancel the unwanted unit symbols. Sometimes, particularly when the numerical values are given by formulas rather than by pure numbers, the unit symbols are enclosed in square brackets, [ ], to avoid confusion.

Unit algebra may also be performed without specifying the unit system, as in the dimensional analysis [8] of models. Let [T] represent an arbitrary unit of time, and similarly let mass be represented by [M] and length by [L]. Consider the equation from dynamics for distance s travelled by an object in time t, as the result of constant acceleration a, with initial velocity v:

$$s = vt + \frac{1}{2}at^2. \qquad (10.3)$$

To derive the dimension of s, rewrite Equation (10.3) using only the dimension symbols for each of the quantities:

$$\text{dimension of } s = \frac{[L]}{[T]} \times [T] + \frac{[L]}{[T]^2} \times [T]^2, \qquad (10.4)$$

and then perform cancellations to reduce the right-hand side to $[L] + [L]$. These two terms can be added since they have the same dimension. Therefore the dimension of s is [L], or length, as required.

The simple concept that units obey the rules of algebra as if they were numbers is a powerful check on calculations. Checking by unit conversion and cancellation that all terms of an equation have the same units, or that the dimensions are equal as in the above example, gives confidence that a blunder has not been committed in developing the equation.

10.6 Further study

1. Choose the best answer for each of the following questions.

(a) The written representation of a measurement contains

    i. a numerical value and symbols indicating the unit or units of the measurement.

    ii. a numerical value.

    iii. an exact count of the number of units of a quantity that are present.

(b) The unit of energy in the SI system is the

    i. pascal.   ii. newton.   iii. watt.   iv. metre/second.   v. joule.

(c) In an absolute system of units, force is

      i. a fundamental or base unit.   ii. a derived unit.

(d) An inductor is an electrical component that transfers energy to and from a magnetic field. The inductance L of a linear inductor is given by $L = \lambda/i$, where λ is the magnetic flux and i is the current passing through the element. The unit of flux is the weber (Wb), which in base SI units is $m^2 \cdot kg \cdot s^{-2} \cdot A^{-1}$, and the unit of current is the ampere (A). Therefore, the base SI units of inductance are

      i. $m^2 \cdot kg \cdot s^{-2} \cdot A^{-2}$ ii. $m^2 \cdot kg \cdot s^{-2}$ iii. $V \cdot s \cdot A^{-1}$

(e) The correctly written quantity in the following list is

      i. (2 to 18) MHz   ii. 2 to 18 MHz   iii. 2 MHz − 18 MHz

(f) What must the units of α be if the following equation is correct?

$$\left(\alpha \frac{km}{m^3} \right)^{-1} = \frac{8.83\,L}{100\,km}$$

      i. m   ii. $\frac{ft^3}{L}$ iii. $\frac{mi}{gal}$ iv. $\frac{m^3}{L}$ v. $\frac{km}{m^3}$

(g) The incorrectly written quantity in the following list is

      i. $60\,kg \pm 5\,kg$ ii. $(60 \pm 5)\,kg$ iii. $60 \pm 5\,kg$ iv. 60 ± 3.5

(h) In the FPS gravitational system, the unit of mass is the slug.

      i. true   ii. false

(i) The incorrectly written quantity with SI units in the following list is

      i. $78.3 \times 10^{-7}\,C\,m^{-2}$ ii. $78.3 \times 10^{-7}\,C/m^2$

      iii. $78.3 \times 10^{-7}\,C \cdot m^{-2}$ iv. $78.3 \times 10^{-7}C/m^2$

(j) Inspect the following quantities carefully. The correctly written measurement is

      i. $28.45\pi r^3 \rho\,kg$ ii. $28.45\pi r^3 \rho\,kg$ iii. $28.45\pi r^3 \rho\,kg$

      iv. $28.45\pi r^3 \rho\,kg$ v. $28.45\pi r^3 \rho kg$

---

2.   The equation $\sigma = My/I$ gives the stress σ at a distance y from the neutral axis of a beam subject to bending moment M. The dimensions are $[F]/[L]^2$ for stress, $[F][L]$ for moment, and $[L]$ for distance. Determine the dimensions of variable I. Can you guess what I represents, from its dimensions?

---

3.   An object weighs 100 lb on Earth. What would it weigh on the Moon, where the acceleration of gravity is $1.62\,m/s^2$? What is its mass on the Moon?

---

4.   The fuel consumption of vehicles is given as miles (1 mile = 1.609 344 km) per gallon in the United States, but in litres ($1\,L = 10^{-3}\,m^3$) consumed per 100 km in Canada. Find the formula for converting the first quantity to the second. What is 32 miles per gallon expressed in litres consumed per 100 km?

5.   A direct current $i = 10$ A passes through a wire of resistance $R = 200\,\Omega$, producing heat energy at the rate $i^2 R$. How much heat is generated in three minutes? Convert the result to FPS units.

6.   This is a question that uses dimensional algebra. A pendulum can be constructed from a string with length ℓ and an object with mass m. Suppose that we suspect by observation that the period T of the pendulum swing depends on length ℓ, acceleration of gravity g, and mass m. Because such relationships are generally algebraic, we suppose that

$$\text{(dimensions of } T) = \text{(dimensions of } \ell)^a \times \text{(dimensions of } g)^b$$
$$\times \text{(dimensions of } m)^c,$$

for some unknown constants a, b, and c. Rewrite this equation using only generalized dimension symbols as in Equation (10.4). By examining this result, determine these constants and, hence, the general relationship between the period of a pendulum and the parameters ℓ, g, and m.

7.   A railway locomotive with a mass of approximately 100 tonnes is travelling at 100 kilometres per hour. In order to stop the locomotive, the kinetic energy must be dissipated as heat. Approximately how many litres of water at $100\,°\mathrm{C}$ would this energy convert to steam? You may need to search to find the heat of vaporization of water.

10.7   References

[1] Institute of Physics, *Measurement Science and Technology*. IOP Publishing Limited, 2008. <http://www.iop.org/EJ/journal/MST> (March 9, 2008).

[2] International Bureau of Weights and Measures, *The International System of Units (SI)*. Sèvres, France: Bureau International des Poids et Mesures (BIPM), eighth ed., 2006. <http://www.bipm.org/utils/common/pdf/si_brochure_8_en.pdf> (June 23, 2008).

[3] K. Alder, *The Measure of All Things, the Seven-Year Odyssey and Hidden Error That Transformed the World*. New York: The Free Press, Simon & Schuster, Inc., 2002.

[4] B. N. Taylor, *Guide for the Use of the International System of Units (SI)*. Gaithersburg, MD: National Institute of Standards and Technology, 1995. <http://physics.nist.gov/cuu/pdf/sp811.pdf> (March 9, 2008).

[5] F. Cardarelli, *Encyclopaedia of Scientific Units, Weights and Measures*. New York: Springer-Verlag, 2003.

[6] T. Wildi, *Units and Conversion Charts*. New York: IEEE Press, 1991.

[7] B. S. Massey, *Measures in Science and Engineering: Their Expression, Relation and Interpretation*. Chichester: Ellis Horwood Limited, 1986.

[8] E. S. Taylor, *Dimensional Analysis for Engineers*. Oxford: Clarendon Press, 1974.

Chapter

11

Measurement Error

A special vocabulary and special techniques have been developed for describing the inevitable deviations of physical measurements from true values. Computations containing inexact quantities are also possible. In this chapter, you will learn

- the definition of a traceable measurement;
- a classification of uncertainties into systematic and random effects;
- correct use of the words *accuracy, precision,* and *bias;*
- how to write inexact quantities using engineering and other notation;
- the correct use of significant digits.

11.1 Measurements, uncertainty, and calibration

What is a measurement? One possible answer might be "A property of a physical object that can be represented using a real number." However, we have to distinguish between counting and measuring using an instrument such as in Figure 11.1. Counting is exact if the set of objects to be counted can be defined and it does not contain too many members. Thus we can count the people in an aircraft, the welds in a pipeline, and with difficulty the number of ants in a colony. However, the grains of sand on a beach are probably too many to count. Therefore, let us distinguish between counting, which is exact, and measuring, which is not. The quantities that are measured are those for which base and derived units described in Chapter 10 exist.

Measured physical quantities may be inexact because the property being measured is not precisely defined. For example, to speak of the thickness of an object, such as the

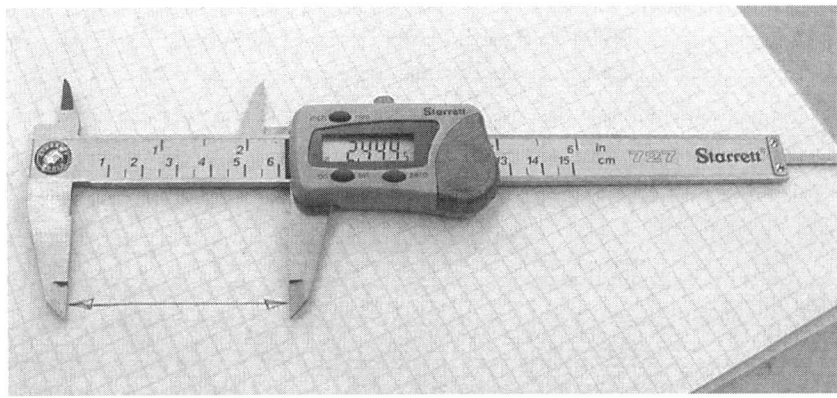

Figure 11.1 A modern digital-readout vernier caliper, with a resolution of 0.0005 inch, or 0.01 mm.

asphalt of a highway, is to assume two parallel plane boundaries, which cease to exist when the object is looked at closely enough.

Range of uncertainty

Even when it is possible to define a "true" value of a physical quantity, the instrument used to obtain this value cannot be constructed perfectly. The true value is conceptually a point on the line representing the real numbers as illustrated in Figure 11.2, but because of imperfect measurement, generally only an estimated true value can be obtained, together with an interval in which the true value lies. The difference between the true value and the measured value is called *measurement error* or *observation error.* In this context, a measurement error does not mean a mistake or blunder. The interval determines the unknown error range, which is synonymously called the *range of uncertainty,* or simply *uncertainty.*

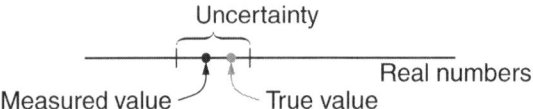

Figure 11.2   Illustrating a true (exact) value, which is a point on the line of real numbers, and its measured estimate, together with an interval of uncertainty.

Traceable measurements

A measurement is fundamentally a comparison of a physical object to a standard physical object. Let us suppose that the vernier caliper of Figure 11.1 has been employed to measure the thickness of an object. If the instrument has been calibrated using a set of gauge blocks manufactured for the purpose, and the gauge block manufacturer certifies their accuracy by comparison with a national standard that has been calibrated to the SI standard, then the measurement is said to be *traceable* to the international standard. In Canada, the Institute for National Measurement Standards [1] of the National Research Council provides calibration standards for factories and laboratories.

11.2   Systematic and random errors

Measurement errors are classified into two categories, systematic errors and random errors, although it is not always easy to identify the extent to which each is present in specific cases. The categories are defined and illustrated below.

11.2.1   Systematic errors

A systematic error is a consistent deviation, also called a *bias* or an *offset,* from the true value. The error has the same magnitude and sign when repeated measurements are made under the same conditions. Systematic errors can sometimes be detected by careful analysis of the method of measurement, although usually they are found by calibration or by comparing measurements with results obtained independently. Three types of systematic errors are usually encountered: natural, instrument, and personal error, as follows.

Natural error   *Natural error* arises from environmental effects. For example, temperature changes affect electronic components and measuring instruments. It may be possible to identify the effects of these phenomena and to apply a correction factor to the measurements. For example, the air buoyancy of a mass weighed by a high-precision equal-arm balance must be calculated, in order to remove its effect from the reading.

Instrument error   The second type of systematic error is *instrument error,* or *offset,* and is caused by imperfections in the adjustment or construction of the instrument. Some examples are misaligned optics, meter zero-offset, and worn bearings. In precise work, instruments are usually calibrated (checked) at several points within their range of use, and a calibration curve is included with the instrument. The calibration data permits the engineer to correct the readings for instrument errors. For example, steel surveying tapes are occasionally stretched a few millimetres during use. A stretched tape gives distance measurements that are slightly low. However, if the tape is calibrated, distance measurements can be corrected for the stretch.

Personal error   The third type of systematic error is *personal error.* Such error results from habits of the observer; for example, one person may have a tendency to estimate scale values that are slightly high; another may read scale values that are low. Personal error can be reduced by proper training.

11.2.2   Random errors

Random errors are the result of small variations in measurements that inevitably occur even when careful readings are taken. For example, in repeated measurements, an observer may exert slightly different pressures on a micrometer or connect voltmeter leads in slightly different locations. Random errors from the true value do not bias the measurement, but produce both positive and negative errors with zero mean value. If several measurements are made, there is no reason to favour one observation over any other, but their arithmetic mean is normally a better estimate of the true value than is any single measurement. The estimate improves as the number of observations increases, as discussed in Chapter 14. This technique yields an effective countermeasure against random errors: take repeated measurements and compute the mean of the results.

11.3   Precision, accuracy, and bias

The terms *accuracy* and *precision* are often misused when measurements are being described. These two words are not synonyms.

The precision of a measurement refers to its repeatability. A *precise* measurement has small random error and, hence, the discrepancies between repeated measurements taken under the same conditions are small.

A measurement that is close to the true or correct value is said to be *accurate.* However, *accuracy* has more than one meaning; some authors use this word to mean low total error, but others use it to mean low systematic error. When this term is used, its meaning should be made clear; otherwise use the word *bias.* To say that a measurement is unbiased is to mean unambiguously that there is no discernible systematic error. The target analogy shown in Figure 11.3 illustrates these words and their relationship to systematic and random errors. The spread (also called scatter or dispersion) of the hits is an example of random error caused by the random motion of the shooter. The average displacement between the hits and the target centre (bull's-eye) is an example of systematic error, such as might be caused by misalignment of the gun sights. Therefore, a small spread of hits located far from the centre is precise but inaccurate shooting, while a large spread around the centre is imprecise but unbiased shooting.

If unbiased measurements are repeated, their mean value approaches the true value as the number of observations increases. That is, if there is no systematic error, the effect of random error can be reduced by averaging.

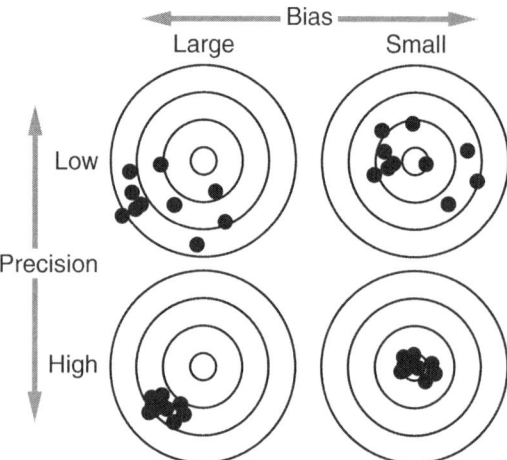

Figure 11.3   Precision, bias, and accuracy: The two right patterns have small bias (small systematic error); the two bottom patterns have high precision (small random error). The lower right pattern is accurate; less commonly the two right patterns would be said to be accurate.

11.4   Estimating measurement error

When systematic errors are known and quantified, the numerical observations can be corrected to remove bias. The offset is subtracted from the readings, or the instrument is recalibrated to eliminate the offset. However, the random error remains and must be estimated and quoted with the measurement. In all cases, the technical specifications of an instrument are the first source of error estimates, but reading errors must also be considered, as follows.

Estimating the position of the needle of a gauge or meter requires interpolation between scale markings, and the precision of the result is limited by the ability of the human eye to resolve small distances. Then the precision of the readings will be a fraction of the amount between scale markings, typically one-tenth, and often this will be an adequate estimate of the measurement precision.

Readings taken from instruments with digital display must also be interpreted carefully, since the specifications might indicate less precision than the number of digits displayed. For example, a typical uncertainty for moderately priced multipurpose instruments is 0.1 % of the reading plus one digit.

The precision of a measurement may become evident when the observation is repeated; any digit that changes is suspect. Include only one suspect digit, and record it as the least significant digit in the numerical value of the measurement.

11.5 How to write inexact quantities

A recorded measurement is complete only when a statement about its uncertainty is included. The uncertainty is normally an estimated range and may be stated explicitly, which is always correct but sometimes inconvenient and repetitive, or implicitly, in which case care must be used both in writing and in interpreting the numbers. Showing the uncertainty requires a clear and unambiguous notation.

A quantity written using an ordinary decimal number, for example, 30 140.0, is said to be in *fixed notation*. If it is written as a decimal number with one non-zero digit to the left of the decimal and a power of 10 is appended as a scale factor, for example, $3.014\,00 \times 10^4$, it is in *scientific* notation. If the exponent of 10 is a multiple of 3 to correspond to the SI prefixes, for example, 30.1400×10^3, the number is in *engineering* notation.

11.5.1 Explicit uncertainty notation

Sets of measured values are often listed together in tables. When the values have the same uncertainty, it may be specified in a caption or table heading. Otherwise, the uncertainty must be given with each measurement. The measurement may be expressed in fixed, scientific, or engineering notation, and the uncertainty may be given in either absolute or relative form, as will be described.

In fixed notation, the numerical value of the measurement is written together with an estimate of the half-range, when the range is symmetric about the measured value as is often true. For example, a certain measurement of the speed of sound v_s in air at 293 K gives

$$v_s = 343.5\,\text{m/s} \pm 0.9\,\text{m/s} = (343.5 \pm 0.9)\,\text{m/s} \tag{11.1}$$

where the estimated maximum error is 0.9 m/s. The unit symbol "m/s" is treated as an algebraic factor multiplying the contents of the parentheses. Rather than indicating the *absolute uncertainty* as above, the *relative uncertainty* may be written as a percentage of the magnitude. In this case,

$$\frac{0.9\,\text{m/s}}{343.5\,\text{m/s}} = 0.26 \times 10^{-2} = 0.26\,\%, \tag{11.2}$$

so that the measured speed can be written as

$$v_s = 343.5\,(1 \pm 0.26\,\%)\,\text{m/s}. \tag{11.3}$$

The unit symbol "%" simply means the number 0.01 when used strictly (see Chapter 11 of [2]), but the ambiguous notation $343.5 \, \text{m/s} \pm 0.26 \, \%$ is sometimes accepted.

Using scientific notation, the above measurement is

$$v_s = (3.435 \pm 0.009) \times 10^2 \, \text{m/s} \tag{11.4}$$

which means that v_s lies between $3.426 \times 10^2 \, \text{m/s}$ and $3.444 \times 10^2 \, \text{m/s}$. The uncertainty is written with the same scale multiplier as the value of the measurement. In engineering notation, the above speed is

$$v_s = (343.5 \pm 0.9) \times 10^0 \, \text{m/s}. \tag{11.5}$$

11.5.2 Implicit uncertainty notation

If a quantity is written without an explicit uncertainty, the uncertainty is taken to be ± 5 in the digit immediately to the right of the least significant digit. For example, the statement $t = 302.8 \, \text{K}$ is equivalent to $t = (302.8 \pm 0.05) \, \text{K}$. This convention establishes the uncertainty only to within a factor of 10. Moreover, in fixed notation, ambiguity may result, as shown in the next section.

11.6 Significant digits

The significant digits of a written quantity are those that determine the precision of the number. It is generally a serious mistake to record a quantity to more significant digits than justified by the measurement, with the exception that one or more suspect digits is sometimes retained when the effect of random errors will be reduced by averaging as discussed in Section 11.6.2.

Relative precision is unaffected by the power-of-10 scale factors in scientific or engineering notation. Significant digits are identified as follows:

- All non-zeros are significant, and all zeros between significant digits are significant.
- Leading zeros are not significant.
- Trailing zeros are significant in the fractional part of a number.

In fixed notation, trailing (right-hand) zeros in a number without a fractional part may cause ambiguity. Therefore, if right-hand zeros should be considered to be significant, explicit mention of the fact must be made, or the uncertainty must be included. The following are some examples of numbers in fixed notation:

0.0350 oz	3 significant digits
90 000 000 miles	1 significant digit, ambiguous
92 900 000 miles	3 significant digits, ambiguous
(92 900 000 ± 500) miles	5 significant digits

In scientific notation, the integer part is normally one non-zero digit, and the number of significant digits is one more than the number of digits after the decimal point. To illustrate, the distance from Earth to the Sun is written as shown:

9×10^7 miles	1 significant digit
9.3×10^7 miles	2 significant digits
9.290×10^7 miles	4 significant digits

In engineering notation, all of the digits except leading zeros are considered to be significant, and the exponent of 10 in the scale factor is a multiple of 3. In engineering notation, the above distances would be written:

0.09×10^9 miles	1 significant digit
93×10^6 miles	2 significant digits
92.90×10^6 miles	4 significant digits

11.6.1   Rounding numbers

When a number contains digits that are not significant, rounding is used to reduce the number to the appropriate number of significant digits. Rounding should not be confused with truncation, which means simply dropping the digits to the right of a certain point. Rounding drops digits but may increase the rightmost retained digit by 1, typically according to the following rules.

When the leftmost discarded digit is less than 5, then the retained digits are unchanged; for example, rounding to three or two digits,

$$3.234 \quad \to \quad 3.23 \quad \text{or} \quad 3.2$$
$$9.842 \quad \to \quad 9.84 \quad \text{or} \quad 9.8. \tag{11.6}$$

When the leftmost discarded digit is greater than 5, or is a 5 with at least one non-zero digit to its right, then the rightmost retained digit is increased by 1; for example, rounding to three or two digits,

$$3.256 \quad \to \quad 3.26 \quad \text{or} \quad 3.3$$
$$9.747 \quad \to \quad 9.75 \quad \text{or} \quad 9.7 \text{ but } not \text{ } 9.8. \tag{11.7}$$

When the leftmost discarded digit is a 5 and all following digits are 0, then the rightmost retained digit is unchanged if it is even; otherwise it is increased by 1. Compared to rounding to the next higher digit, this rule avoids introducing a bias in the mean by rounding a set of numbers. The resulting final digit is always even; for example, rounding to two digits,

$$3.25 \quad \to \quad 3.2$$
$$3.250 \quad \to \quad 3.2$$
$$3.350 \quad \to \quad 3.4. \tag{11.8}$$

11.6.2   The effect of algebraic operations

When computations are performed using inexact measured quantities, the results are also inexact. The quantities are entered to the number of digits justified by experimental error, and computations are performed to the full precision of the computer. However, the final computed value must not be written using more significant digits than are justified; it must be rounded to imply the correct interval of uncertainty. Basic rules will be given for writing the result of simple operations, anticipating the more detailed analyses in Chapters 12 to 14.

Addition Let x and y be two measured values, with (unknown) measurement errors Δx, Δy, respectively, in the ranges given by implicit notation. Then the sum of the true values is

$$(x + \Delta x) + (y + \Delta y) = (x + y) + (\Delta x + \Delta y) \tag{11.9}$$

where $x + y$ is the calculated sum and $\Delta x + \Delta y$ the resulting error. The magnitude of the error in the sum is $|\Delta x + \Delta y|$, corresponding to an absolute uncertainty that is at most twice the uncertainty of the least precise operand. Therefore, for a small number of additions or subtractions, the result is often rounded to the absolute uncertainty of the least precise operand.

Subtraction The analysis for addition shows that the absolute uncertainty can at most double with each addition. This conclusion is similarly true for subtraction, but the *relative* uncertainty, and hence the number of significant digits, may change drastically. For example, $5.75 - 5.73 = 0.02$ with an implied error of ± 0.01, so the result can only be written to one significant digit at most, rather than the three digits of the operands. The following examples illustrate addition and subtraction:

4.16	0.123	6.162	25.4
-12.3214	-178	-12.3214	3.1416
91.2	0.002164	6.150	0.3183
83.0	-178	-0.009	28.9

Multiplication and division The relative uncertainty in a product or ratio is typically at most the sum of the relative uncertainty of the factors. This result is shown for the product as follows; a similar analysis holds for division. Given two quantities $x + \Delta x = x(1 + \Delta x/x)$ and $y + \Delta y = y(1 + \Delta y/y)$ as before, the relative errors are $\Delta x/x$ and $\Delta y/y$, often expressed in percent. Then the computed product of the two measured values x and y is simply xy, but including the errors in the computation gives

$$(x + \Delta x)(y + \Delta y) = xy + y\,\Delta x + x\,\Delta y + \Delta x\,\Delta y$$
$$= xy\left(1 + \frac{\Delta x}{x} + \frac{\Delta y}{y} + \frac{\Delta x}{x}\frac{\Delta y}{y}\right). \tag{11.10}$$

Typically, the relative errors $\Delta x/x$ and $\Delta y/y$ are small, so the rightmost term in the parentheses can be ignored, and the relative error of the product is at most approximately twice the largest relative uncertainty of the factors. Consequently, the result of a multiplication is often rounded to the number of significant digits of the factor with the fewest significant digits. The following examples illustrate this rule:

$$
\begin{aligned}
2.6857 \times 3.1 &= 8.3 \\
(489.5)^2 &= 239\,600 \\
236.52/1.57 &= 151 \\
25.4 \times 0.866\,025 &= 22.0
\end{aligned}
\tag{11.11}
$$

Caution   These simple rules must be used with care, since implicit notation specifies precision only to within a factor of 10. Furthermore, the rules for single operations are not always applicable to complex calculations. Consider the product of four measured values, each given to two significant digits, implying an uncertainty of ± 0.05:

$$p = 1.1 \times 1.2 \times 1.3 \times 2.4 = 4.1184. \tag{11.12}$$

Rounded to two digits according to the precision of the individual factors, the result would be written as $p = 4.1$, which implies that $4.05 \le p \le 4.15$. However, in the worst case, when the errors have the same sign and maximum magnitude, the actual range minimum and maximum are given by the calculations

$$\begin{aligned} 1.05 \times 1.15 \times 1.25 \times 2.35 &= 3.5470, \\ 1.15 \times 1.25 \times 1.35 \times 2.45 &= 4.7545; \end{aligned} \tag{11.13}$$

that is, $3.5470 \le p \le 4.7545$. If the worst case is required, the result should be written as $p = 4$, implying that $3.5 \le p \le 4.5$, which approximates the calculated range. However, judgment may be needed about whether a worst-case analysis is appropriate; Chapter 12 discusses alternative range calculations.

This example illustrates a key method for performing complex calculations: specify the precision of the arguments; perform intermediate calculations to the full precision available; determine the precision of the result; and round the result to show the correct precision.

Arithmetic mean   The arithmetic mean of a set of unbiased measurements of a single quantity is more precise than individual measurements in the set and so may be written with additional significant digits. Anticipating the discussion in Section 14.2.1, the uncertainty in the mean of n observations is $1/\sqrt{n}$ of the uncertainty in one observation. Each reduction of uncertainty by a factor of 10 allows an additional significant digit, so appending m digits to the mean implies that

$$\frac{1}{\sqrt{n}} = \left(\frac{1}{10}\right)^m. \tag{11.14}$$

Inverting and squaring both sides gives

$$n = 10^{2m}, \tag{11.15}$$

from which

$$m = 0.5 \log_{10} n. \tag{11.16}$$

Converting m to integer values produces Table 11.1, which shows that no digits should be added for a mean of fewer than 10 measurements, one for 10 to 999, and so on, or perhaps more conservatively, depending on the uncertainty and independence of individual measurements.

Table 11.1 Significant digits appended to the mean of n measurements to indicate increased precision compared to a single measurement.

Measurements	Appended Digits
n	m
1 to 9	0
10 to 999	1
1000 to 99 999	2
$\vdots$	$\vdots$
10^k (k odd) to $10^{k+2}-1$	$(k+1)/2$

11.7 Further study

1. Choose the best answer for each of the following questions.

(a) A measurement of a constant pressure is repeated 15 times using a meter that has a precision of 0.5 %. The arithmetic mean of these readings is calculated to be 40.405 kPa. In a report, this value should be written as

 i. 40.40 kPa ii. 40 kPa iii. 40.405 kPa iv. 40.4 kPa

(b) A set of measurements having insignificant systematic error but significant random error is

 i. imprecise and biased. ii. imprecise and unbiased.
 iii. precise and unbiased. iv. precise and biased.

(c) The number 0.745×10^4 is in scientific notation.

 i. false ii. true

(d) Four measured values have been written with their precision expressed implicitly and combined in the formula $(1.32 \times 40.31) - (0.09738 \times 501.80)$. Expressed to the correct number of significant digits, the result is

 i. 4.3×10^0 ii. 4.344 iii. 4.3 iv. 4.340

(e) Identify the incorrect statement.

 i. A set of precise measurements may contain significant error.

 ii. Systematic and random errors can simultaneously affect a measurement.

 iii. Left-hand zeros are significant in fixed notation.

 iv. All digits to the right of the decimal point are significant in scientific notation.

 v. The significant digits of a number give a measure of the precision of the number.

(f) The following measurements have been made: 84.52, 3.0, 41.081. Their correctly written sum is

 i. 128.6010 ii. 129 iii. 128. iv. 128.6 v. 128.60

(g) Compared to the precision of individual measurements, the arithmetic mean of 40 measurements subject to random error can be written using
   i. two additional significant digits.
   ii. one fewer significant digit.
   iii. one additional significant digit.
   iv. as many significant digits as for the individual measurements.

(h) The arithmetic mean of a set of unbiased measurements of the same quantity
   i. has unknown precision compared to individual measurements.
   ii. is more precise than a single measurement in the set.
   iii. is as precise as a single measurement in the set.
   iv. is less precise than a single measurement in the set.

(i) Several numbers have been rounded to three significant digits as shown. The incorrectly computed operation is
   i. $27.550 \rightarrow 27.6$ ii. $27.500 \rightarrow 27.5$ iii. $27.450 \rightarrow 27.5$
   iv. $27.350 \rightarrow 27.4$ v. $27.455 \rightarrow 27.5$

(j) Written in engineering notation, the quantity $13\,824\,\mathrm{s}$ is
   i. $13.824 \times 10^3\,\mathrm{s}$ ii. $139.32 \times 10^3\,\mathrm{s}$ iii. $1.3824 \times 10^4\,\mathrm{s}$
   iv. $1.3824\,\mathrm{ks}$ v. $13\,824\,\mathrm{s}$

---

2. A dinosaur skeleton discovered in 1990 was estimated to be $90\,000\,000$ years old. Does this mean that the skeleton was $90\,000\,010$ years old in the year 2000? (Adapted from reference [3])

---

3. The original observations by Dr. Wunderlich of the temperature of the human body were averaged and rounded off to $37\,°\mathrm{C}$, which was the nearest Celsius degree. Therefore, the precision of the measurement was presumably $\pm 0.5\,°\mathrm{C}$. When this measurement was converted to the Fahrenheit scale, it became $98.6\,°\mathrm{F}$. What is the correct interpretation of the tolerance on the Fahrenheit figure?

Recently, extensive measurements of body temperature gave a mean of only $98.2\,°\mathrm{F}$. Convert this to an equivalent kelvin temperature with the proper tolerance. (Adapted from reference [3])

---

4. The notation $343.5\,\mathrm{m/s} \pm 0.26\,\%$ was described as ambiguous on page 165. What rule about units does it break?

---

5. Write the answers to the following exercises according to the discussion in this chapter.
   (a) $xy/z =?$, where $x = 405\,\mathrm{V}$, $y = 53.92\,\mathrm{V}$, and $z = 16.02$
   (b) $a + b + c =?$, where $a = 28.1$, $b = 97$, and $c = 43.567$
   (c) $\alpha t =?$, where $\alpha = 0.0143\,\mathrm{s}^{-1}$ and $t = 30\,\mathrm{s}$
   (d) $p - q - r =?$, where $p = 33.6$, $q = 18.1$, and $r = 3.53$

6. Perform the calculation shown, and express the result to the correct number of significant digits. Which of the three operands is the most precise? How much does the most precise operand contribute to the final answer?

$$28\,402 + 1.30 \times 10^4 - 3100 + 32.897\,34$$

11.8 References

[1] National Research Council Canada, "Institute for national measurement standards," 2002. <http://inms-ienm.nrc-cnrc.gc.ca/main_e.html> (March 9, 2008).

[2] B. N. Taylor, *Guide for the Use of the International System of Units (SI)*. Gaithersburg, MD: National Institute of Standards and Technology, 1995. <http://physics.nist.gov/cuu/pdf/sp811.pdf> (March 9, 2008).

[3] J. A. Paulos, *A Mathematician Reads the Newspaper*. New York: Harper Collins, 1995.

Chapter

Error in Computed Quantities

Calculations performed with inexact data give inexact results as we saw in the previous chapter. Some measured variables may contribute more to the uncertainty in the final result than others, and it might be useful to improve the measurement procedures for those variables. Hence, the key questions are the following: what is the uncertainty in the calculated result, and which measured data items are most significant in contributing to the uncertainty in the result? Furthermore, the uncertainties of the arguments may not be given as absolute values or ranges, but as relative quantities. The requirement may then be to find the relative uncertainty in the calculated result and the measured items that contribute most to it.

Using elementary mathematical notation, suppose that x, y, z, ... have inexact values and that the function $f(x, y, z, ...)$ is to be calculated. Let the measured values of the arguments be x_0, y_0, z_0, ..., respectively, so that the nominal computed value is $f(x_0, y_0, z_0, ...)$. Then we must calculate the range of f given the ranges of its arguments and the sensitivity of f to deviations of its arguments from their measured values. In this chapter, you will learn

- the direct method of calculating the exact range of a function when the measurement error range is known,

- how to approximate the worst-case range for more complex calculations, using the differentiation techniques of elementary calculus,

- how to approximate the range when estimates of typical measurement errors, rather than their range, are known.

12.1 Method 1: Exact range of a calculated result

The first method calculates uncertainty by calculating the range of the final result, using the range of the input variables and the following assumptions:

Assumptions
1. The exact range of the computed quantity $f(\cdots)$ is required.
2. The exact ranges of the arguments x, y, z, ... are known.

These assumptions are not always strictly true, since the range of a measured quantity is often an estimate rather than an exact quantity. However, this basic method requires no approximations as the other methods do.

Example 12.1
Measurement of resistance
The resistance in ohms (Ω) of an element in an electric circuit is to be calculated, given the measured values $V = 117\,\text{V}$ of the voltage across the element and $I = 2.23\,\text{A}$ of the current through the element. The resistance of the element is defined to be $R(V, I) = \dfrac{V}{I}$. Assume that the ranges of I and V are

$$I_{\min} \leq I \leq I_{\max} \quad \text{and} \quad V_{\min} \leq V \leq V_{\max}$$

where I_{min}, I_{max}, V_{min}, and V_{max} are known positive values. Then the nominal resistance is the value

$$R = \frac{V}{I} = \frac{117}{2.23}\,\Omega = 52.5\,\Omega.$$

For this simple function the range of the calculated resistance is seen to be

$$\frac{V_{min}}{I_{max}} \le R \le \frac{V_{max}}{I_{min}}.$$

Using the extreme values for V and I implied by their significant digits,

$$\frac{V_{min}}{I_{max}} = \frac{116.5}{2.235}\,\Omega = 52.13\,\Omega \le R \le 52.81\,\Omega = \frac{117.5}{2.225}\,\Omega = \frac{V_{max}}{I_{min}}.$$

Notice that the nominal value of R is not at the midpoint of its calculated range.

Example 12.2
Mass of a conical volume of crushed glass

A quantity of fine crushed glass for recycling has been piled in an approximate cone of diameter d and height h, and its mass is to be calculated from its density ρ. The volume of a cone is one-third the height times the area of the base, so the mass is given by the formula $m(\rho, h, d) = \rho \times \frac{h}{3} \times \pi \left(\frac{d}{2}\right)^2 = \frac{\pi}{12}\rho h d^2$. The measured values are $d = 1.73\,\text{m} \pm 0.05\,\text{m}$ and $h = 1.03\,\text{m} \pm 0.05\,\text{m}$, where the uncertainty arises since the pile is not perfectly circular or conical. The density of fine crushed glass is given as $(1120 \pm 30)\,\text{kg/m}^3$.

The nominal calculated mass is

$$m = \frac{\pi}{12} \times 1120 \times 1.03 \times 1.73^2\,\text{kg} = 904\,\text{kg}.$$

The numerical ranges of the arguments of the function $m(\rho, h, d)$ are

$$1090 \le \rho \le 1150, \quad 0.98 \le h \le 1.08, \quad 1.68 \le d \le 1.78,$$

and since m is proportional to ρ, h, and d^2, its minimum and maximum occur at the respective minimum and maximum of its arguments, so that

$$m \ge \frac{\pi}{12} \times 1090 \times 0.98 \times 1.68^2\,\text{kg} = 789\,\text{kg} \quad \text{and}$$

$$m \le \frac{\pi}{12} \times 1150 \times 1.08 \times 1.78^2\,\text{kg} = 1030\,\text{kg}.$$

Again, the nominal calculated value does not occur at the midpoint of the range.

Extreme values

It should not be concluded from the above two examples that the extreme values of the range of a function always occur at the extreme values of the ranges of its arguments. Figure 12.1 illustrates a function $f_1(x)$ of one argument x for which the range is not obtained by calculating the function at maximum or minimum values of x. However if, over the range of x, the function is bounded and monotonic, that is, always non-decreasing or always non-increasing, as illustrated by $f_2(x)$ in the figure, then its extreme values are reached at the extreme values of x. If a function of several variables has this property with respect to each of its arguments, as is true for differentiable functions and sufficiently small argument ranges, then the range of the function is found relatively simply by searching at the extreme values of the arguments.

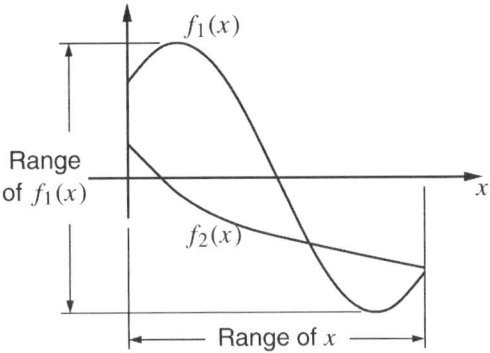

Figure 12.1   Computing $f_1(x)$ at the upper and lower boundaries of x does not give the range of $f_1(x)$ over the range of x. The extremes of $f_2(x)$, which is monotonic and more typical for small ranges of x, are reached at the extremes of x.

12.2 | Method 2: Linear estimate of the error range

The second method for calculating the range of a function of inexact variables gives an approximation of the result of the first method. An approximation is often sufficient since the range of measured quantities is typically estimated rather than known exactly, and further approximation does not change the character of conclusions based on the measurements. It turns out that the magnitude of the uncertainty in the calculated value is the sum of the effects of the errors in the variables. This method provides an essential result that will also be used in the third method. The assumptions are as follows:

Assumptions

1. The range calculated in the previous method is to be *approximated.*

2. The function $f(x, y, z, \ldots)$ is differentiable at the measured values $x_0, y_0, z_0, \ldots$ of its arguments.

3. The measurement errors in the measured values are independent of each other and may be either positive or negative.

4. The ranges of the measured values are known.

For simplicity, consider first a function $f(x)$ of one variable, as illustrated in Figure 12.2. At nominal (i.e., measured) value $x = x_0$, $f(x)$ has the value $f(x_0)$ and slope $\left.\dfrac{df}{dx}\right|_{x=x_0}$, assuming that the derivative exists at x_0. Then from the definition of the derivative, for small changes Δx in the measured value, the function changes by an amount Δf given approximately by

$$\Delta f \simeq \left.\frac{df}{dx}\right|_{x=x_0} \Delta x. \tag{12.1}$$

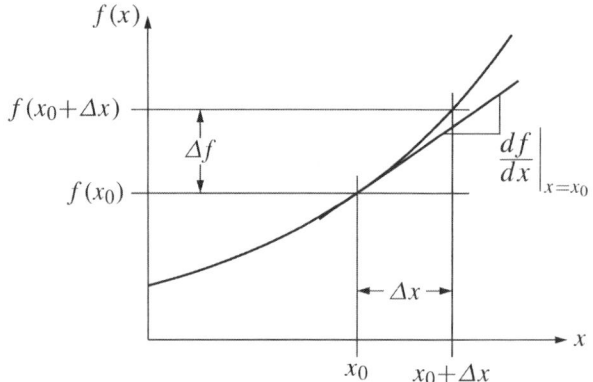

Figure 12.2   The change in $f(x)$ is approximately the slope of f at x_0 multiplied by the deviation Δx.

Example 12.3
Uncertainty
calculation

Suppose that the quantity $f(x) = 1 + x^2$ is to be calculated and that the measured value of x is $x_0 = 0.64$, implying an uncertainty of ± 0.005 and a nominal value $f(x_0) = 1 + 0.64^2 = 1.41$. Differentiating the formula for $f(x)$ gives the formula

$$\frac{df}{dx} = 2x,$$

but when this formula is evaluated at the measured value, we get

$$\left.\frac{df}{dx}\right|_{x=x_0} = 2x|_{x=0.64} = 1.28$$

which is a constant. Then the uncertainty in the calculated quantity is approximately

$$\Delta f \simeq 1.28 \times 0.005 = 0.0064.$$

12.2.1   Sensitivities

Expressing the result of the last section in words, the approximation illustrated in Figure 12.2 for the change in $f(x)$ near the point of tangency is

$$\text{change in } f(x) \simeq (\text{rate of change with respect to } x) \times \Delta x. \tag{12.2}$$

To extend this result to a function of two variables, consider Figure 12.3, which shows a surface $f(x, y)$ as a function of x and y, with a plane tangent to the surface at the point $x_0, y_0, f(x_0, y_0)$. In moving from the tangent point to a nearby point on the surface, the f-distance moved is approximately the slope of the plane measured in the x-direction times the x-distance moved, plus the slope in the y-direction times the y-movement. A similar argument holds for functions of several variables:

$$\text{change in } f(x, y, z, \ldots) \simeq$$

$$(\text{rate of change with respect to } x) \times \Delta x$$

$$+ (\text{rate of change with respect to } y) \times \Delta y$$

$$+ (\text{rate of change with respect to } z) \times \Delta z$$

$$+ \cdots \tag{12.3}$$

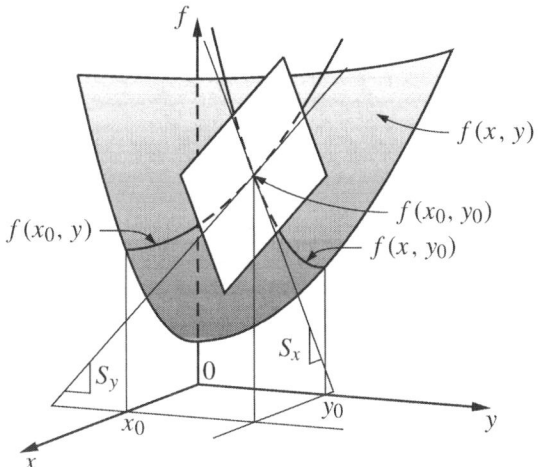

Figure 12.3   The plane is tangent to the surface $f(x, y)$ at $f(x_0, y_0)$. Moving a small distance Δy in the y-direction from (x_0, y_0) changes f to $f(x_0, y_0 + \Delta y)$ on the $f(x_0, y)$ curve. The resulting change in f is approximately $S_y \Delta y$, the height change obtained by moving in the plane along the tangent line. Similarly, a change Δx from x_0 changes f by approximately $S_x \Delta x$, and x-movement Δx together with y-movement Δy changes f by approximately $S_x \Delta x + S_y \Delta y$.

for small Δx, Δy, Δz, ..., provided that all the required quantities exist and are unique. Rewriting this approximation more symbolically, we have

$$\Delta f \simeq S_x \Delta x + S_y \Delta y + S_z \Delta z + \cdots \tag{12.4}$$

where the rates of change of f with respect to its arguments are *sensitivity constants, sensitivity coefficients,* or simply *sensitivities,* calculated as follows:

The sensitivity constant S_x is obtained by calculating the formula for the derivative of f with respect to x, treating y, z, ... as constants, and evaluating the formula at the measured values by substituting $x = x_0$, $y = y_0$, $z = z_0$, ...

The sensitivity constant S_y is obtained by calculating the formula for the derivative of f with respect to y, treating x, z, ... as constants, and evaluating the formula at the measured values by substituting $x = x_0$, $y = y_0$, $z = z_0$, ...

Further constants are calculated similarly. The derivatives calculated above are called *partial derivatives* of f, and are written $\frac{\partial f}{\partial x}$, $\frac{\partial f}{\partial y}$, ... instead of $\frac{df}{dx}$, $\frac{df}{dy}$, ... to indicate that differentiation is performed with respect to the indicated variable while keeping the others constant, but in all other respects the derivatives are the same as in elementary calculus. These calculations are related to Figure 12.3, for example, by recalling the definition of a derivative as a limit and observing that, in the figure, the slopes S_x and S_y are given by

$$S_x = \lim_{\Delta x \to 0} \frac{f(x_0 + \Delta x, y_0) - f(x_0, y_0)}{\Delta x}, \tag{12.5}$$

and

$$S_y = \lim_{\Delta y \to 0} \frac{f(x_0, y_0 + \Delta y) - f(x_0, y_0)}{\Delta y}. \tag{12.6}$$

Example 12.4
Sensitivity constants
for Example 12.1,
resistance

In Example 12.1, the formula for the computed result was

$$R(V, I) = \frac{V}{I},$$

from which the required derivatives and associated sensitivities are

$$\frac{\partial R}{\partial V} = \frac{1}{I}, \quad S_V = \frac{1}{I}\bigg|_{\substack{V=117 \\ I=2.23}} = \left(\frac{1}{2.23}\right) \Omega/V = 0.448 \ \Omega/V,$$

$$\frac{\partial R}{\partial I} = \frac{-V}{I^2}, \quad S_I = \frac{-V}{I^2}\bigg|_{\substack{V=117 \\ I=2.23}} = \left(\frac{-117}{2.23^2}\right) \Omega/A = -23.5 \ \Omega/A.$$

Thus, near the computed resistance, the rate of change of resistance is $0.448 \ \Omega/V$ with respect to voltage changes and $-23.5 \ \Omega/A$ with respect to current changes.

Example 12.5
Sensitivity constants
for Example 12.2,
mass of conical
volume

In Example 12.2, the formula for the computed result was

$$m(\rho, h, d) = \frac{\pi}{12}\rho h d^2,$$

so that the sensitivities are calculated as follows:

$$\frac{\partial m}{\partial \rho} = \frac{\pi}{12}hd^2, \quad S_\rho = \frac{\pi}{12}hd^2\bigg|_{\substack{\rho=1120 \\ h=1.03 \\ d=1.73}} = 0.807 \ \text{m}^3,$$

$$\frac{\partial m}{\partial h} = \frac{\pi}{12}\rho d^2, \quad S_h = \frac{\pi}{12}\rho d^2\bigg|_{\substack{\rho=1120 \\ h=1.03 \\ d=1.73}} = 878 \ \text{kg}/m,$$

$$\frac{\partial m}{\partial d} = \frac{\pi}{6}\rho hd, \quad S_d = \frac{\pi}{6}\rho hd\bigg|_{\substack{\rho=1120 \\ h=1.03 \\ d=1.73}} = 1045 \ \text{kg}/m.$$

12.2.2 Relative sensitivities

The sensitivity constants defined in Section 12.2.1 allow an answer to the first part of the question posed at the beginning of this chapter: for small measurement errors, the measurements that potentially contribute most to changes in the calculated result are those that have the largest sensitivity constants. However, a modified question may be posed: what relative change in f results from small relative changes in its arguments, and for relative changes, which variables may have the largest effect on the computed result?

For a simplified notation, write f_0 in place of $f(x_0, y_0, z_0, \ldots)$. Then divide Equation (12.4) by f_0 so that the left side $\frac{\Delta f}{f_0}$ is the relative change in the computed result f_0, assuming that f_0 is non-zero. The result can be written

$$\frac{\Delta f}{f_0} \simeq \frac{S_x x_0}{f_0}\left(\frac{\Delta x}{x_0}\right) + \frac{S_y y_0}{f_0}\left(\frac{\Delta y}{y_0}\right) + \frac{S_z z_0}{f_0}\left(\frac{\Delta z}{z_0}\right) + \cdots$$

$$= S_x^r \frac{\Delta x}{x_0} + S_y^r \frac{\Delta y}{y_0} + S_z^r \frac{\Delta z}{z_0} + \cdots \tag{12.7}$$

where $\frac{\Delta x}{x_0}$ is the relative change in x, and so on for the other variables. The *relative sensitivity coefficients*, written $S_x^r = \frac{S_x x_0}{f_0}$ and similarly for the other variables, have been given a superscript r to distinguish them from previously defined quantities. These coefficients are purely numeric (dimensionless) since they are ratios of quantities with identical units.

We can now answer the second part of the question posed at the beginning of the chapter: the argument for which a relative change produces the largest relative deviation of f from f_0 is the argument with the largest relative sensitivity coefficient.

Example 12.6
Relative sensitivities for Example 12.4, resistance

Example 12.4 investigated the influence of changes in measured values on changes in calculated resistance. In terms of relative changes, the result looks somewhat different. For the formula for $R(V, I)$, the relative sensitivity constants are

$$S_V^r = \frac{S_V V_0}{R(V_0, I_0)} = \frac{0.448 \times 117}{52.5} = 1.00,$$

$$S_I^r = \frac{S_I I_0}{R(V_0, I_0)} = \frac{-23.5 \times 2.23}{52.5} = -1.00.$$

Thus, a change of one percent in either V or I will result in a change of approximately one percent in the calculated resistance, and the two measured values V and I are equally important in terms of relative changes. The computed coefficients have a suspicious simplicity that will be explained in Section 12.2.4.

12.2.3   Approximate error range

The approximation for the change in the computed function $f(x, y, z, \ldots)$ was given in Equation (12.4), where it was assumed that the deviations Δx, Δy, Δz, $\ldots$ are independent and may be of either sign. The calculated constants S_x, S_y, S_z, $\ldots$ also may be of either sign. To approximate the worst-case deviation $|\Delta f|_{\max}$, take absolute values to get the formula

$$|\Delta f|_{\max} \simeq |S_x||\Delta x| + |S_y||\Delta y| + |S_z||\Delta z| + \cdots, \tag{12.8}$$

which is an approximation to the exact result discussed in Section 12.1. This formula is particularly simple to use if the sensitivity constants can be obtained easily.

Example 12.7
Computed range for
Example 12.4,
resistance

From the values given in Examples 12.1 and 12.4, the approximate range of the computed result is

$$|\Delta R|_{\max} \simeq |S_V||\Delta V| + |S_I||\Delta I|$$
$$= (|0.448| \times |0.5| + |-23.5| \times |0.005|)\,\Omega = 0.3\,\Omega,$$

giving the calculated resistance as

$$R = (52.5 \pm 0.3)\,\Omega,$$

which may be compared with the exact result given in Example 12.1.

Example 12.8
Computed range for
Example 12.5, mass
of conical volume

The numerical values from Examples 12.2 and 12.5 give

$$|\Delta m|_{\max} \simeq |S_\rho||\Delta\rho| + |S_h||\Delta h| + |S_d||\Delta d|$$
$$= (|0.807| \times |30| + |878| \times |0.05| + |1045| \times |0.05|)\,\mathrm{kg}$$
$$= 120\,\mathrm{kg},$$

giving the calculated mass as $m = (904 \pm 120)\,\mathrm{kg}$, which may be compared with the exact result of Example 12.2.

12.2.4 Application of Method 2 to algebraic functions

The quantities computed from measured data are often simple algebraic functions of their arguments. Examples of such simple functions will be investigated here.

Linear combinations Simple addition and subtraction are special cases of functions that are linear combinations of variables with constant coefficients, for example,

$$f(x, y, z) = ax + by - cz, \tag{12.9}$$

with constants a, b, c.

For this simple function, the exact range of Method 1 is given by

$$|\Delta f|_{\max} = |a||\Delta x| + |b||\Delta y| + |-c||\Delta z|, \tag{12.10}$$

that is, absolute values of all terms are added.

Method 2 requires sensitivity constants, but in this case these are simply the constant coefficients of the variables in the expression, so that

$$\Delta f = S_x\Delta x + S_y\Delta y + S_z\Delta z = a\Delta x + b\Delta y - c\Delta z, \tag{12.11}$$

giving exactly the same result as Method 1:

$$|\Delta f|_{\max} = |a||\Delta x| + |b||\Delta y| + |-c||\Delta z|. \tag{12.12}$$

In the special cases of addition and subtraction, simply add the absolute error magnitudes as in Section 11.6.2; that is, if

$$f(x, y) = x + y \tag{12.13}$$

then

$$|\Delta f|_{\max} = |1||\Delta x| + |1||\Delta y| = |\Delta x| + |\Delta y|. \tag{12.14}$$

Example 12.9
Addition and
subtraction

$$\left.\begin{array}{l} x = 1.5 \pm 0.1 \\ y = 2.3 \pm 0.2 \end{array}\right\} \rightarrow x + y = 3.8 \pm 0.3 \text{ and } x - y = -0.8 \pm 0.3$$

Multiplication or division

For these two operations, the rule is to add relative uncertainties, as in Section 11.6.2, rather than absolute uncertainties. If a product is written as

$$f(x, y) = xy, \tag{12.15}$$

then its sensitivity with respect to x is the value of y, and vice versa, so that

$$|\Delta f|_{\max} \simeq |y||\Delta x| + |x||\Delta y|. \tag{12.16}$$

Dividing both sides by $|f(x, y)| = |x||y|$ to obtain the relative uncertainty of the result gives

$$\frac{|\Delta f|_{\max}}{|f|} \simeq \frac{|y||\Delta x|}{|x||y|} + \frac{|x||\Delta y|}{|x||y|} = \frac{|\Delta x|}{|x|} + \frac{|\Delta y|}{|y|}, \tag{12.17}$$

which is the sum of relative uncertainties. A similar derivation for simple division produces the same right-hand side.

Example 12.10
Multiplication

$$\left.\begin{array}{l} x = 1\,(1 \pm 1\,\%) \\ y = 2.3\,(1 \pm 2\,\%) \end{array}\right\} \rightarrow xy = 2.3\,(1 \pm 3\,\%) \text{ and } x/y = 0.43\,(1 \pm 3\,\%).$$

Algebraic terms

The relative sensitivity coefficient for the term $f(x) = x^a$ is a. This very useful result is shown as follows:

$$S_x^r = \frac{S_x x_0}{f_0} = \frac{(dx^a/dx)|_{x=x_0}\, x_0}{f_0} = \frac{ax_0^{a-1} x_0}{x_0^a} = a. \tag{12.18}$$

Now, consider a product of terms of the above form, for example,

$$f(x, y, z) = A\frac{x^a y^b}{z^c} = Ax^a y^b z^{-c}. \tag{12.19}$$

Then the relative sensitivity coefficient for x is

$$S_x^r = \frac{(Aax_0^{a-1} y_0^b z_0^{-c})x_0}{f_0} = \frac{(Aax_0^{a-1} y_0^b z_0^{-c})x_0}{Ax_0^a y_0^b z_0^{-c}} = a \tag{12.20}$$

as before, and similarly for the coefficients S_y^r and S_z^r. Applying Equation (12.7), which is a sum of relative changes multiplied by their relative sensitivity coefficients, gives the approximate relative change for the term defined by (12.19):

$$\frac{\Delta f}{f_0} \simeq a\frac{\Delta x}{x_0} + b\frac{\Delta y}{y_0} - c\frac{\Delta z}{z_0}. \tag{12.21}$$

If the relative changes are maximum range values, then the maximum relative change in the term is

$$\left|\frac{\Delta f}{f_0}\right|_{\max} \simeq |a|\left|\frac{\Delta x}{x_0}\right| + |b|\left|\frac{\Delta y}{y_0}\right| + |-c|\left|\frac{\Delta z}{z_0}\right|. \tag{12.22}$$

Example 12.11
Algebraic term:
volume of a cylinder

Certain cylindrical containers manufactured for preserved fruit have a nominal diameter $d = 75 \, \text{mm}$ and height $h = 50 \, \text{mm}$, with manufacturing tolerances of 0.5 % for diameter and 0.8 % for height. The container volume is $(\pi/4) \, d^2 \, h$, and by looking at the exponents for d and h we can write immediately that the maximum relative change in volume will be approximately

$$2 \left| \frac{\Delta d}{d_0} \right| + 1 \left| \frac{\Delta h}{h_0} \right| = 2 \times (0.5\,\%) + 1 \times (0.8\,\%) = 1.8\,\%.$$

Example 12.12
Algebraic term:
Example 12.2, mass
of conical volume

In contrast to the sensitivity constants for the recycled-glass example of Example 12.5, the relative sensitivities are simply the powers of the variables in the expression

$$m(\rho, h, d) = \frac{\pi}{12} \rho h d^2,$$

giving

$$\left| \frac{\Delta m}{m} \right|_{\text{max}} \simeq |1| \left| \frac{\Delta \rho}{\rho} \right| + |1| \left| \frac{\Delta h}{h} \right| + |2| \left| \frac{\Delta d}{d} \right|.$$

Thus, for relative changes, the computed quantity is most sensitive to the measured value of the diameter d, which is squared.

12.3   Method 3: Estimated uncertainty

The third method for calculating the uncertainty of f assumes a context that differs slightly from that of the first two methods. In particular, an estimate of the typical magnitude of random measurement errors is assumed to be available rather than their exact range as before. For a more detailed discussion of this estimate, consult a textbook such as references [1–3], for example, or the international standard that applies [4, 5].

Let $f(x, y, z, \ldots)$ be the function to be calculated from inexact values $x = x_0 + \Delta x$, $y = y_0 + \Delta y$, $z = z_0 + \Delta z$, $\ldots$.

Random errors in $x, y, z, \ldots$ will produce a random error in f. A positive change in f caused by error in one variable may be partially cancelled by a negative change caused by independent error in another, and the more variables there are, the less likely that the effects of their errors will combine in the same direction, as assumed in the worst-case analysis of the first two methods. Rather, the method of this section is used to estimate the typical magnitude of the random error in f instead of its worst-case magnitude.

For the typical value of $|\Delta x|$, we shall use the standard deviation of x, also called the standard error, a statistical term [2, 3] to be discussed further in Chapter 13, defined in the current context as the square root of the mean value of $(\Delta x)^2$, with a similar definition for other variables. We are required to find a typical $|\Delta f|$, which will be done by finding the standard deviation of f, as follows:

Assumptions

1. The standard deviation of the value of the function is required, rather than the maximum deviation.

2. The errors in the variables are independent random quantities symmetrically distributed about a mean of zero, and the standard deviations of the variables are known.

3. The function is differentiable at the measured values of its arguments.

Under the above assumptions, the approximate change in the calculated value of f for changes in its arguments is given by Equation (12.4), where the formulas for the partial derivatives of f with respect to its arguments have been evaluated at the measured values to calculate the sensitivity constants S_x, S_y, S_z .... .

A simple derivation will be deferred to Section 12.3.1; the required result will be simply stated here as follows:

$$\text{standard deviation of } f = \sqrt{S_x^2 |\Delta x|^2 + S_y^2 |\Delta y|^2 + S_z^2 |\Delta z|^2 + \cdots} \qquad (12.23)$$

where the values of $|\Delta x|$, $|\Delta y|$, ... inserted in the formula are the standard deviations of x, y, ..., respectively, and the positive square root is used. This quantity is smaller than the worst-case deviation given by Equation (12.8), for identical sensitivities and $|\Delta x|$, $|\Delta y|$, ..., and accounts for the partial cancellation of random error effects mentioned above.

A similar result also holds when Equation (12.23) is modified to contain relative quantities, as follows:

$$\text{standard deviation of } \frac{\Delta f}{f_0} = \sqrt{(S_x^r)^2 \left|\frac{\Delta x}{x_0}\right|^2 + (S_y^r)^2 \left|\frac{\Delta y}{y_0}\right|^2 + (S_z^r)^2 \left|\frac{\Delta z}{z_0}\right|^2 + \cdots}.$$
$$(12.24)$$

Example 12.13
Estimated error in
Example 12.1,
resistance

The measured voltage V in Example 12.1 will be assumed to have standard deviation equal to 0.5, rather than a range of exactly ± 0.5 as before. Similarly, the measured current will be assumed to have standard deviation of 0.005, so that, using the sensitivity constants previously calculated in Example 12.4, the error in the calculated resistance is

$$\text{standard deviation of } R = +\sqrt{S_V^2 |\Delta V|^2 + S_I^2 |\Delta I|^2}$$
$$= \left(+\sqrt{(0.448)^2 |0.5|^2 + (-23.5)^2 |0.005|^2}\right) \Omega$$
$$= 0.25 \, \Omega,$$

giving the result that the calculated resistance is $R = (52.5 \pm 0.25) \, \Omega$, where the uncertainty is a statistically estimated value rather than a maximum value as it was in Examples 12.1 and 12.7.

Example 12.14
Estimated error in
Example 12.11,
volume of a cylinder

For the cylindrical containers of Example 12.11, the relative change as a function of variations in diameter and height was found to be

$$2 \times (0.5\,\%) + 1 \times (0.8\,\%).$$

If, however, the 0.5 % variability in diameter and 0.8 % variability in height are assumed to be estimated random quantities (standard deviations) rather than maximum values, then the square root of the sum of squares is taken:

$$\sqrt{(2 \times (0.5\,\%))^2 + (1 \times (0.8\,\%))^2} = 1.3\,\%,$$

and the expected change in volume is less than the maximum change calculated in Example 12.11.

12.3.1 Derivation of the estimated value

The origins of Equation (12.23) are illustrated in Figure 12.4 and can be developed as follows. In Equation (12.4), repeated here,

$$\Delta f \simeq S_x \Delta x + S_y \Delta y + S_z \Delta z + \cdots, \tag{12.4}$$

each of the quantities Δx, Δy, Δz, ... is now assumed to be random, with zero mean value. To calculate the standard deviation of f, we shall square both sides and take the positive square root of the expected (mean) value. Squaring gives

$$(\Delta f)^2 \simeq S_x^2 (\Delta x)^2 + S_y^2 (\Delta y)^2 + S_z^2 (\Delta z)^2 + \cdots$$
$$+ 2 S_x S_y \Delta x \Delta y + 2 S_x S_z \Delta x \Delta z + 2 S_y S_z \Delta y \Delta z + \cdots. \tag{12.25}$$

We shall use the facts that the mean of a sum is the sum of the means, and that the mean of a random quantity multiplied by a constant is the constant times the mean of the random quantity.

In calculating the expected value, or mean, all products of the form $(\Delta x)^2$, $(\Delta y)^2$, ... on the right-hand side of (12.25) are non-negative and therefore have non-negative mean value. However, because Δx, Δy, ... , are random, independent, and symmetrically distributed about zero, products of the form $\Delta x \Delta y$, $\Delta x \Delta z$, ... may be either positive or negative, and will have an expected value of zero, as illustrated in Figure 12.4. From the above reasoning, we conclude that

expected value of $(\Delta f)^2 \simeq$

expected value of $\left(S_x^2 (\Delta x)^2 + S_y^2 (\Delta y)^2 + S_z^2 (\Delta z)^2 + \cdots \right)$

$= S_x^2 \times (\text{expected value of } (\Delta x)^2) + S_y^2 \times (\text{expected value of } (\Delta y)^2)$

$+ \cdots. \tag{12.26}$

This is a standard result [2, 3], often called the "propagation of error formula." By the definition of the standard deviation of f, the positive square root of both sides of Equation (12.26) is now taken, giving Equation (12.23), as required.

Equation (12.24) is derived in the same way, starting with Equation (12.7) instead of Equation (12.4).

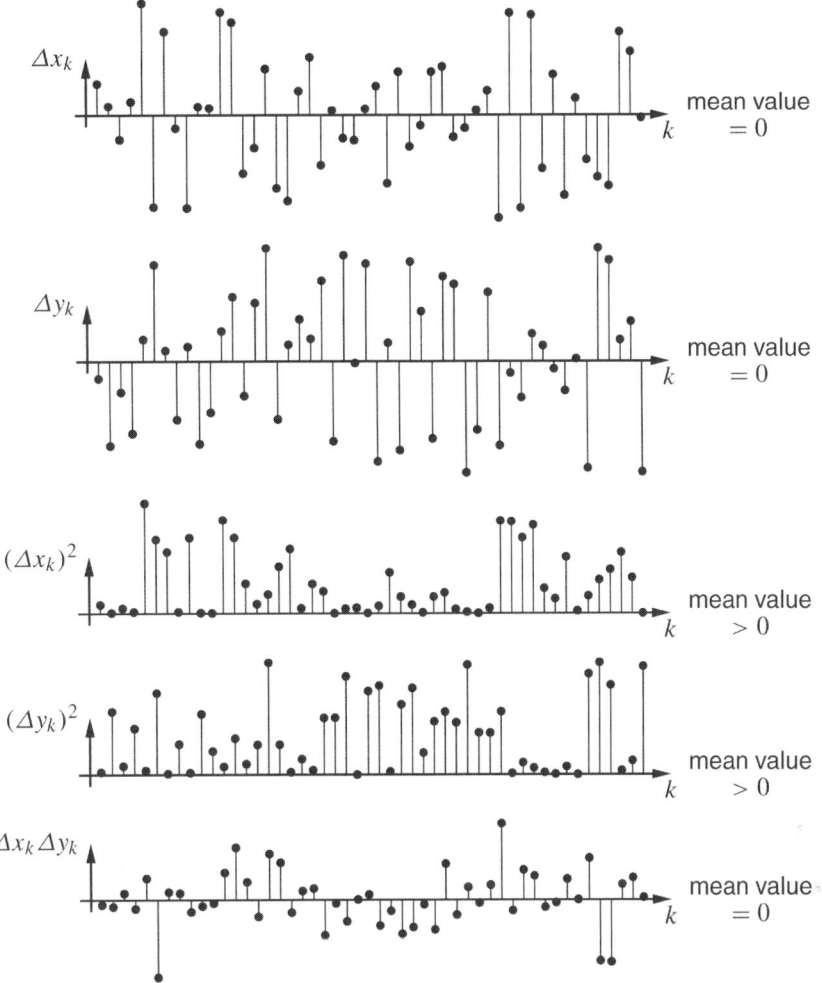

Figure 12.4 Graphs of samples of zero-mean independent random variables Δx_k and Δy_k, $k = 1, \ldots$, showing that the mean of the squared quantities is positive, but the products $\Delta x_k \Delta y_k$ may be positive or negative and have an expected (mean) value of zero.

12.4 Further study

1. Choose the best answer for each of the following questions.

(a) In the formula $q = mv^2/2$, with $m = 1.50\,\text{kg}$ and $v = 10\,\text{m/s}$, the estimated independent random errors in m and v are 3 % and 2 %, respectively. The resulting estimated random error in q is approximately

    i. 4   ii. infinite   iii. 81   iv. 0.5   v. 6

(b) The area of a circular object is to be calculated from its measured diameter. If the diameter value has a relative error range of ± 0.3 %, what is the approximate relative error range of the calculated area?

    i. ± 0.3 square units   ii. ± 0.4 %   iii. ± 0.6 %   iv. ± 0.3 %

(c) The area of a rectangular piece of ground is to be calculated by multiplying the measured length $a = 1287.1$ m by the measured width $b = 902.3$ m. These measurements are subject to random relative error estimated at 0.02%. The estimated relative random error in the area is approximately

    i. 0.03% ii. 0.04% iii. 0.02% iv. $350\,\text{m}^2$

(d) Three quantities measured to 0.2% accuracy are used to compute the function $f = \pi\dfrac{1.204^2 \times 1.010}{2.500^2}$. The maximum relative error $|\Delta f/f|$ of the result is approximately

    i. an unknown value   ii. 1% iii. 10% iv. 0.3% v. 3% vi. 20%

    vii. 0.2%

(e) The linear approximation of the maximum error in q in the formula $q = mv^2/2$ subject to independent errors Δm and Δv is

    i. $|\Delta q| \simeq |v/2||\Delta m| + |mv||\Delta v|$

    ii. $|\Delta q| \simeq (v^2/2)(\Delta m) + (mv)(\Delta v)$

    iii. $|\Delta q| \simeq \sqrt{(v^4/4)(\Delta m)^2 + (m^2v^2)(\Delta v)^2}$

    iv. $|\Delta q| \simeq |v^2/2||\Delta m| + |mv||\Delta v|$

(f) The centrifugal force corresponding to a mass m in circular motion with radius r and speed v is mv^2/r. The approximate change in this force for small changes Δm, Δv, and Δr is

    i. $(v^2/r)\Delta m + 2(m/r)v\Delta v - (mv^2/r^2)\Delta r$

    ii. $\left|(v^2/r)\Delta m\right| + |2(m/r)v\Delta v| + \left|(mv^2/r^2)\Delta r\right|$

    iii. $(v^2/r)\Delta m + 2(m/r)v\Delta v + (mv^2/r^2)\Delta r$

(g) In the formula $f = vr^2h^4$ the quantity v is known exactly, r is known to 1.5%, and h is known to 1%. Assuming that these errors are independent random quantities, the estimated error in the computed values is approximately

    i. 5% ii. 3% iii. 2% iv. 7%

(h) A quantity measured as 300.3 is to be squared. The implied error in the result is

    i. ± 0.1 ii. ± 30 iii. ± 15 iv. ± 300.3 v. ± 0.05

(i) For the three measurements $A = 0.55 \pm 0.03$, $B = 3.25 \pm 0.01$, $C = 0.32 \pm 0.02$, the linear approximation method gives the maximum error in the calculated quantity $g = (A/C) + 3BC$ as approximately

    i. 0.4 ii. 4×10^{-4} iii. 0.2 iv. infinite   v. 12%

(j) In the formula $q = mv^2/2$, with $m = 1.50\,\text{kg}$ and $v = 10\,\text{m/s}$, the maximum error in m is 3% and the maximum error in v is 2%. The resulting maximum error in q is approximately

    i. 4 ii. 81 iii. infinite   iv. 5 v. 0.5

2.   The formula for the volume of a cylinder is $v = \pi r^2 h$, where r is the radius and h is the height. A certain cylinder has radius $r = 0.020$ m and height $h = 0.15$ m.

   (a) Calculate the sensitivity coefficients for changes in v with respect to errors in r and h at the given values. Be sure to use correct units.

   (b) Calculate the relative sensitivity coefficients for errors in the given quantities. Errors in which quantity cause the largest relative change in the calculated value?

3.   The mass of a small copper disk is to be calculated from measurements obtained using a vernier caliper with a precision of ± 0.002 cm. Copper has a specific gravity (density compared to the density of water) of between 8.8 and 8.95. The measured diameter is $d = 11.04 \times 10^{-3}$ m, and the measured thickness is $t = 2.58 \times 10^{-3}$ m.

   (a) Calculate the sensitivity coefficients of the mass for changes in the measured values and the given density.

   (b) Calculate the relative sensitivity coefficients for errors in the given quantities. Errors in which quantity cause the largest relative change in the calculated value?

4.   Consider three measurements x, y, z, as listed below along with estimates of the error in each measurement:

$$x = 6.31 \pm 0.04, \quad y = 9.23 \pm 0.01, \quad z = 16.3 \pm 0.5,$$

and consider also the derived quantities d, e, f, g, shown below:

$$d = x + 2y + 3z, \quad e = xy/z, \quad f = x^2 yz^{1/2}, \quad g = (x/z) + 3y$$

   (a) Compute the nominal values of d, e, f, and g.

   (b) Using the exact-range method, compute the maximum and minimum values for d, e, f, and g.

   (c) Using the linear approximation method, compute the error in each derived quantity.

   (d) Using Method 3, and assuming that the measurement uncertainties given above are independent estimated errors, compute the estimated error in each derived quantity.

5.   A mass m hung by a string of length ℓ forms a pendulum that oscillates with a period T. The equation for the period of the pendulum can be derived easily and used to find the acceleration due to gravity. From such an analysis, the formula for the period is $T = 2\pi \sqrt{\ell/g}$; rearranging, $g = 4\pi^2 \ell T^{-2}$.

   The string length is measured to be $\ell = (2.000 \pm 0.003)$ m, and the time for the pendulum to make 10 complete oscillations is $10T = (28.5 \pm 0.5)$ s.

   (a) Calculate the acceleration g due to gravity.

   (b) Calculate the percent error in g using the three methods given in this chapter.

6. The length x and width y of a rectangular plot of land are to be measured, with estimated errors Δx and Δy, respectively.

(a) Sketch the rectangle, with measured dimensions x, y.

(b) On the sketch, superimpose the rectangle that corresponds to measurements $x + \Delta x$, $y + \Delta y$, assuming positive Δx, Δy.

(c) On the sketch, identify the change in area between the two measurements, as estimated by Method 2 of this chapter.

(d) On the sketch, identify the area that is the difference between the actual change in area and the estimated change in area.

12.5 References

[1] N. C. Barford, *Experimental Measurements: Precision, Error and Truth.* New York: John Wiley & Sons, second ed., 1985.

[2] R. V. Hogg and J. Ledolter, *Applied Statistics for Engineers and Physical Scientists.* New York: Macmillan, 1992.

[3] S. B. Vardeman, *Statistics for Engineering Problem Solving.* New York: IEEE Press, 1994.

[4] ISO, *Guide to the Expression of Uncertainty in Measurement.* Geneva: International Organization for Standardization, 1993. Corrected and reprinted 1995.

[5] B. N. Taylor and C. E. Kuyatt, *Guidelines for Evaluating and Expressing the Uncertainty of NIST Measurement Results.* Gaithersburg, MD: National Institute of Standards and Technology, 1994. Technical Note 1297, <http://physics.nist.gov/Document/tn1297.pdf> (March 9, 2008).

Chapter

13

Basic Statistics

A knowledge of statistics can guide the engineer in collecting and analyzing measured data and is particularly useful for evaluating random measurement errors [1]. Systematic measurement errors can be removed by calibration, after which statistics are used to reduce the effect of random errors and to establish a level of confidence in measured values.

Chapters 11 and 12 discussed inexact numbers with a prespecified range of uncertainty or inexactness. Often, however, the uncertainty is unknown in advance and must be determined through experimentation and observation.

In this chapter, you will learn

- methods for calculating the central value of a set of measured data,
- how to describe the spread of measured values,
- a method for calculating the relative standing of measurements, and
- how to present measured data using frequency distributions such as histograms.

This chapter defines and discusses three classes of statistics for describing data: measures of central value, spread, and relative standing. The central value of data is described by the median, mode, and mean; spread is described by the range, variance, and standard deviation; and relative standing is described in terms of quartiles, deciles, and percentiles. Basic probability is the subject of Chapter 14. Most engineering programs include further courses in probability, statistics, or both; these chapters introduce but do not replace such courses. Consult references such as [1–5] to probe further.

13.1 Definitions and examples

Some basic statistical terms to be used later are defined in this section.

Observation An *observation* is simply another term for a measured value. Depending on the circumstances, the value may be the result of counting or of comparison with a standard quantity as described in Chapter 10.

Parent population The *parent population* of a measurement variable is the complete set of all possible observations of the variable. Statistics that describe a parent population are usually written using Greek letters, such as μ and σ.

Sample A *sample* is a set of observations consisting of one or more measurements taken from the parent population. As a result, a sample is a subset of the parent population. If the parent population is infinitely large, then the number of different possible samples is also infinitely large. Statistics that describe a sample are usually written using English letters, such as m and s.

Sample distribution A *sample distribution* is a description of the frequencies with which the values of observations in a given sample occur. Distributions are typically described by functions representing the probability of occurrence of observations as in Example 13.1 below or by cumulative probability functions as in Example 13.3 below.

Parent distribution The *parent distribution* is a description of the frequencies with which values of observations from the parent population occur. If the set of all possible observations corresponds to an interval on a line representing the real numbers, then the function is continuous. If the set corresponds to a sequence of points on the line, the function is said to be discrete.

Error An *error* is the difference between a measured value and the true value of a variable; it is the total effect of both systematic errors (bias) and random errors (imprecision), as discussed in Chapter 11.

Statistic A *statistic* is a calculated value that characterizes data in the presence of random uncertainties. The arithmetic mean, for example, provides one method to characterize the central value of a measured quantity.

Descriptive and inferential statistics The discipline of *statistics* involves two broad goals: the techniques of *descriptive statistics* are used to describe complex data simply, and those of *inferential statistics* to estimate or predict the properties of a parent population from a sample. Other uses of statistical techniques include hypothesis testing, risk assessment, and decision-making. Table 13.1 contains examples of description and inference.

Table 13.1 Examples of description and inference. Samples are always described, but the properties of a parent population may be inferred from those of a sample.

Example	Information provided
A class is tested on their knowledge of thermodynamics.	The mean of the class grades provides a description of the entire class (the parent population). Inference is not necessary.
A hockey player is given a contract on the basis of ability to score goals.	Inference from past games to future performance is required. Past goals are the sample; past and future goals are the parent population.

Example 13.1
Student grades on a 20-point scale

An engineering school assigns test marks as integers from 0 to 20. Table 13.2 shows the marks obtained on all past tests. The bell-shaped histogram of Figure 13.1 represents the distribution of this sample and gives the best estimate of the parent distribution, which includes future observations. The relative frequency of any mark is calculated by dividing the number of occurrences by the total number of observations. The relative scale on the right of the figure allows the comparison of histograms for samples of similar tests with different observation counts, since the horizontal and vertical scales will then be identical.

Table 13.2   Mark occurrences n_i and relative frequency p_i for each mark $x_i = i, i = 0, 1, \ldots 20$ on past tests.

x_i	0	1	2	3	4	5	6	7	8	9	10
n_i	1	0	2	7	28	86	163	367	889	1845	3311
p_i	0.000	0.000	0.000	0.000	0.000	0.001	0.002	0.004	0.009	0.018	0.033

x_i	11	12	13	14	15	16	17	18	19	20
n_i	5392	7944	10 472	12 506	13 393	13 034	11 431	8869	6272	3856
p_i	0.054	0.079	0.105	0.125	0.134	0.130	0.114	0.089	0.063	0.039

Engineers look for methods to ensure the quality of their work by making independent checks of calculations. In this calculation, the sum of relative frequencies given in Table 13.2 equals 1 by definition, and summing the tabulated values provides an independent check on the calculations of these values. In Table 13.2 the sum equals 0.999; the slight difference from 1 is caused by rounding errors that were introduced in converting fractions to decimal numbers.

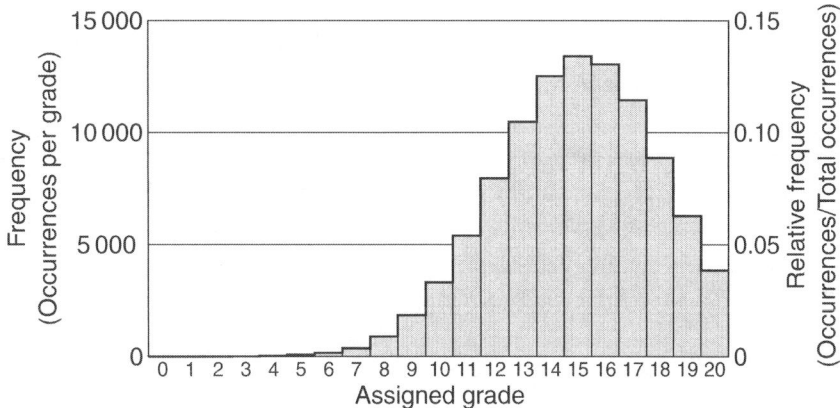

Figure 13.1   Language test mark distribution for all past tests as given in Table 13.2. The relative frequency scale on the right allows a comparison of this distribution function with histograms for samples with different occurrence count.

Example 13.2
True value of a single measurement: a construction stake

The elevation change between a reference point on a construction site and a construction stake elsewhere on the site is required. A true value exists, but an observation will inevitably be in error. The engineer records the measured value along with an estimate of the range in which the true value is expected to be found. The value is written as described in Chapter 11, for example, as (7.15 ± 0.02) m.

Example 13.3
Ball bearing diameter variations: cumulative distributions

High precision balls for ball bearings are graded according to surface roughness and deviation from a perfect sphere. The parent population of ball diameters is infinite assuming that balls can be produced and measured indefinitely. For a sample of 10 balls, the deviations from the desired diameter in microinches were measured as -1.31,

−1.12, −1.04, −0.41, 0.11, 0.21, 0.23, 0.45, 1.07, and 1.16. Figure 13.2(a) displays these values and is called the sample cumulative distribution function. As sample size is increased indefinitely, the graph is expected to approach a continuous function such as Figure 13.2(b), which requires a fractional vertical scale since the population is infinite, and which has a range from 0 to 1.

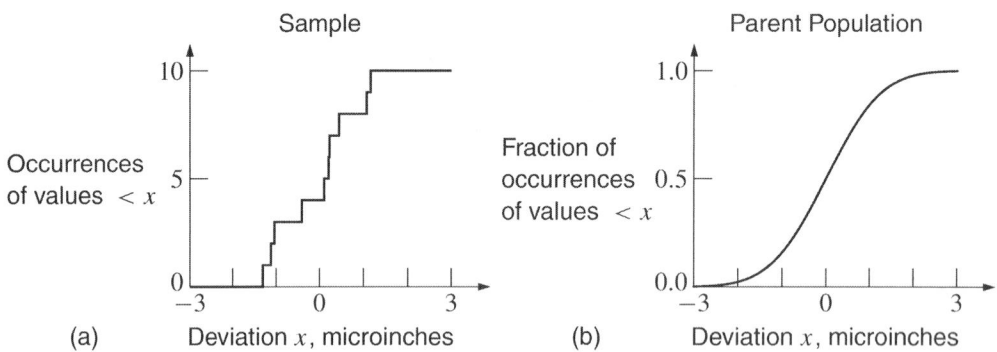

Figure 13.2   Cumulative distribution functions for the sample (a) and parent (b) populations of high-precision ball bearing diameter deviations. In the sample, 4 out of the 10 values are negative, for example, and all fall below 2 microinches. The parent distribution function has a relative (fractional) vertical scale.

Example 13.4
Difference between
measured and
desired values:
quality control

Quality-control tests of electrical resistors are conducted by measuring the resistance of sample resistors taken from the manufacturing line. Variations in the manufacturing process cause random variations in resistance. Monitoring the difference between desired and measured resistance allows a check that the manufactured values are within specified tolerance limits and also allows an estimate of the distribution of resistance values.

Example 13.5
Central value and
spread: radio signals

Noise heard on radio signals is often random owing to imperfections in atmospheric transmission, thermal effects in components, and other interference from other signals. Electronic filtering in the radio receiver extracts an estimated value of the desired signal (the central value) in the presence of noise (spread). A radio capable of scanning frequencies and stopping on good signals uses information about the signal (central value) and noise (spread) to separate clear signals from noisy or weak signals. This information is often expressed as a signal-to-noise ratio.

Example 13.6
Spread in
measurements:
electric power
demand prediction

The spread (variation) of the demand for electric power is required for determining the required overcapacity for an electricity utility. Models of future electric power demand and its estimated spread assist the electrical utility to minimize the cost of generation capacity with a low probability of failing to meet demand.

13.2   **Presentation of measured data: The histogram**

This section deals with one type of graph used to display distributions: the *histogram*, also known as a frequency diagram and, in some computer graphics software, as a bar graph, with zero separation between bars. Figure 13.1 of Example 13.1 on page 191 is an example. The number of occurrences of the data in a range (bin) of defined width is counted and graphed. The left-hand y-axis of Figure 13.1 depicts the count, also called the occurrences or frequency; the right-hand y-axis depicts the count as a fraction of the total count. The relative scale on the right is useful when the shapes of distributions containing unequal numbers of samples are to be compared. The absolute scale on the left is useful when information about the actual number of observations in a particular abscissa (x-axis) bin is needed. Showing both scales is not necessary but is often done when it is not known which scale may be most useful to someone in the future wishing to interpret the distribution of the data.

Choosing bin size   Figure 13.3 illustrates the effect of bin size on a small sample. When the bin size is too narrow, false detail is displayed; when it is too wide, detail is lost. Therefore, choose a bin size to make the corresponding histogram visually meaningful. A rule of thumb is to verify that each of the histogram bars near the central value contains at least five observations. Another rule that applies to bell-shaped curves is to compute bin size h approximately from the formula [6]

$$h = 3.5\, s\, n^{-1/3}, \tag{13.1}$$

where s is the sample standard deviation defined in Section 13.4, and n is the number of measurements.

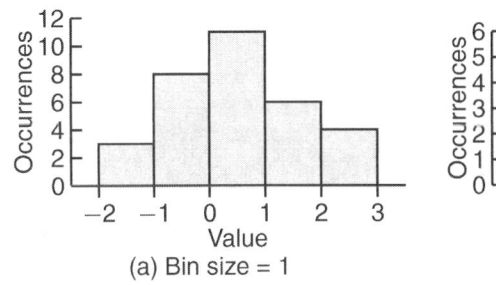
(a) Bin size = 1

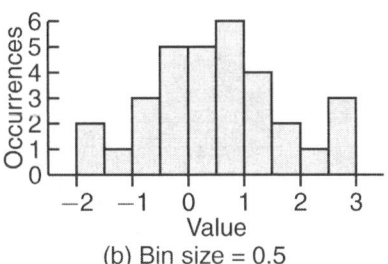
(b) Bin size = 0.5

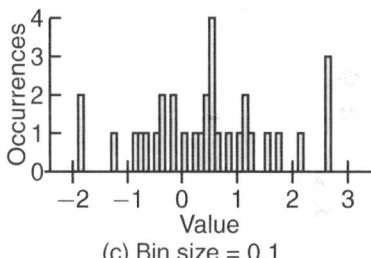
(c) Bin size = 0.1

Figure 13.3   The figures illustrate the effect of bin size and the dangers of small samples. A sample of 32 measurements is shown with a bin size of 1 in (a), 0.5 in (b), and 0.1 in (c). The impression conveyed depends heavily on bin size. Histogram (a) is the most bell-shaped but does not show the peaks in the tails of (b) and (c), which may be statistical anomalies unless confirmed by a larger sample or independent analysis. In (c) there are 50 bins but only 32 measurements so gaps are inevitable and may incorrectly imply that the distribution is not continuous.

13.3   Measures of central value

The mean, median, and mode are the three most common measures of the central value for a set of observations [5]. The word *average* is a synonym for central value and most often means the mean value, but it can also be the median or mode. Be careful in using this word or interpreting it in the statements of others. The mean, median, and mode are discussed and defined below, where we suppose that a sample of n observations

x_i, $i = 1, \ldots n$, has been taken from a parent population of N observations, where $1 \leq n \leq N \leq \infty$.

Mean   Two means will be defined: the *sample mean* and the parent *population mean*.

The sample mean $\bar{x}$ is defined as

$$\bar{x} = \frac{1}{n} \sum_{i=1}^{n} x_i. \tag{13.2}$$

This formula is convenient when all n observations are available, but when a table of occurrences such as Table 13.2 is given, the formula has to be modified. Let the number of occurrences yielding the same value x_i be $c(x_i)$. Then to compute the sum in Equation (13.2), the formula is changed as follows:

$$\bar{x} = \frac{1}{n} \sum_{x_i} x_i \, c(x_i), \tag{13.3}$$

where the subscript x_i under the summation sign indicates that there is a term in the sum for each distinct measured value x_i rather than for each measurement i as in Equation (13.2). If the relative occurrences $p(x_i) = c(x_i)/n$ are known rather than the occurrences $c(x_i)$, then, by taking the denominator n inside the summation in Equation (13.3), the computation becomes

$$\bar{x} = \sum_{x_i} x_i \frac{c(x_i)}{n} = \sum_{x_i} x_i \, p(x_i). \tag{13.4}$$

Example 13.7
A simple mean
calculation

Suppose that in a sample of 10 integers, the number 13 occurs once, 14 three times, 15 four times, and 16 twice. The mean is calculated from Equation (13.3) as

$$\bar{x} = \frac{13(1) + 14(3) + 15(4) + 16(2)}{10} = 14.7,$$

or from Equation (13.4) as

$$\bar{x} = 13(0.1) + 14(0.3) + 15(0.4) + 16(0.2) = 14.7.$$

Example 13.8
Calculating the mean
of student grades

The sum of the grade occurrences in Example 13.1 is 99 868. Using Equation (13.3), the mean can be calculated by multiplying each mark by its occurrences, summing these products, and dividing by total occurrences, as follows:

$$\bar{x} = \frac{0(1) + 1(0) + 2(2) + 3(7) + \cdots 20(3856)}{99\,868} = 14.88$$

Equivalently, each distinct mark can be multiplied by its relative frequency and these products added to obtain the mean mark.

Population mean   If $N < \infty$, then the *population mean* μ is defined to be

$$\mu = \frac{1}{N} \sum_{i=1}^{N} x_i. \tag{13.5}$$

When the population is infinite, then the limit of this sum as $N \to \infty$ must be found. It turns out that a definite integral analogous to the sum in Equation (13.4) is required:

$$\mu = \int_{-\infty}^{\infty} x\, p(x)\, dx, \tag{13.6}$$

where $p(x)\, dx$ is the fraction, that is, probability, of obtaining a measured value between x and $x + dx$. Equation (13.6) is often used as the definition of the mean of an infinite population with distribution described by $p(x)$, but its detailed discussion is beyond the scope of this book. Do not be concerned if you haven't yet studied integral calculus; integrals are used in this book only for illustration and will be part of your engineering vocabulary very soon.

Best estimate of the mean   Often the value of the population mean is required but only a sample is available. The sample mean $\bar{x}$ can be shown [2] to be the best estimate of the population mean μ, where "best" means that it minimizes the mean squared error, as discussed in Section 14.4.1. However, the sample mean is not always the most meaningful value for describing central value; sometimes the median, discussed next, or mode is better suited.

Median   The *median* is the middle value when data is ordered (sorted) by value. That is, excluding the median itself, half the observations will be less than the median and half will be greater.

Suppose that n is odd, and write the data, sorted from most negative to most positive values, as

$$x_1, \ldots x_{m-1}, \ x_m, \ x_{m+1}, \ldots x_n,$$

where $m = 1 + (n-1)/2$. Then the median is defined to be the middle value x_m. If n is even, then the sorted data is

$$x_1, \ldots x_m, \ x_{m+1}, \ldots x_n,$$

where $m = n/2$, and the median is taken to be $(x_m + x_{m+1})/2$, which is midway between the two middle values. The median is always an observed value if n is odd but is not an observed value if n is even and $x_m \neq x_{m+1}$.

The median is typically computed when a central value is sought for a distribution that is significantly asymmetric, as shown in Example 13.10 below. An asymmetric distribution is said to be *skewed*.

Example 13.9
A family with five children

A family has five children of ages 8, 9, 12, 13, and 19. The median age is 12. If the 19-year-old child is not living at home, then the median age of the remaining children is taken from the sample 8, 9, 12, 13, giving the value 10.5, the midpoint of the two central observations 9 and 12.

Example 13.10
Salaries in a small company

The salary distribution in a company is shown in Figure 13.4. The outliers on the right are the salaries of the two company owners. Salaries have been grouped into bins $5000 wide. The mean is not a representative salary, since only two of the 14 people earn more than the mean, and the salaries of these two are more than three times the mean. The median is a more representative central value; the two modes are also representative of the two distinct groups.

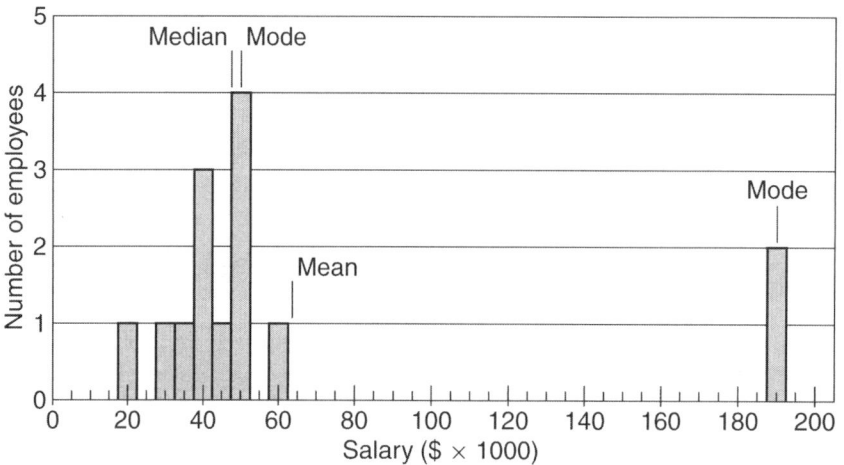

Figure 13.4   The distribution of salaries in a small company. The principal mode applies to employees, the secondary mode to owners.

Mode

The *mode* of a distribution is the value that is observed with the greatest frequency relative to its neighbours or, when there is only one distinct maximum frequency, relative to all members of the population. The mode with greatest occurrence is very simple to compute and identifies the most commonly observed value. For the bins of width $5000 in Figure 13.4, for example, this mode is $50 000.

A distribution is said to be unimodal if it has one peak, bimodal if it has two, and multimodal if it has several well-separated peaks. Figure 13.5 illustrates unimodal and bimodal distributions. Continuous functions such as illustrated will be discussed in more detail in Chapter 14. A single mode usually results from observations of a single value corrupted by random influences in which large errors are less likely than smaller errors. When two or more well-separated peaks occur in a large sample, there are usually systematic influences on the observations.

For quantitative data containing different qualitative types, such as the number of different types of cars rolling off assembly lines daily or enrollments in courses of different subject matter, the mode is effectively the only useful measure of central value.

The effect of skew

A distribution is said to have a negative skew when the distribution has a tail that tends toward lower values than the central value, as illustrated in Figure 13.5(b). The skew of

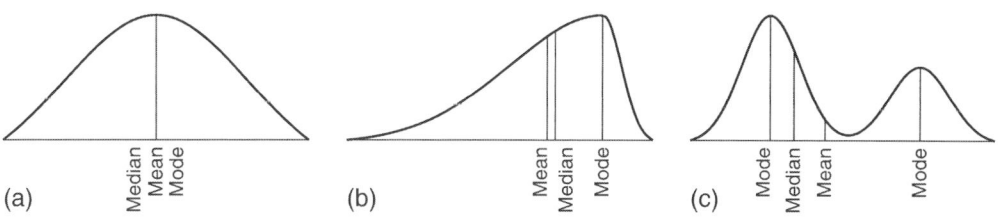

Figure 13.5   Example distribution functions.   Distribution (a) is unimodal and symmetric; therefore, the mean, median, and mode coincide. It is typical of repeated measurements of a single quantity corrupted by random error. Distribution (b) is unimodal and negatively skewed, representative of student marks. Distribution (c) is bimodal and positively skewed.

a distribution can be measured by the ratio of the mean to the median. A ratio greater than 1 corresponds to positive skew; a ratio less than 1 to negative skew.

Qualitative summary   In summary, the median divides the area under the distribution curve into equal parts, and the mode identifies the most frequent or most probable observation [5]. The mean can be informally visualized as the balance point of the area under the distribution curve.

Example 13.11
Student grades: a
negatively skewed
distribution

Student examination grades such as illustrated in Figure 13.1 are often negatively skewed since the majority of students are expected to receive a grade above 50 %; consequently, the tail extending to 0 % is longer than the tail to 100 %. For this example, the mean is 14.88 and the median is 15, and the ratio of mean to median is less than 1 as expected.

13.4   Measures of spread

The central value partially describes a distribution as discussed above but often must be supplemented by describing how the observation values differ from each other. This section discusses range, variance, and standard deviation; the interpercentile ranges defined in Section 13.5 can also be employed.

Range   The *range* is simply the interval bounded by the values of the most positive and most negative observations.

Frequently employed in tests such as quality control on a manufacturing line, the range is sensitive to the extremes of the data but not to the shape of the distribution. The variance and standard deviation, described below, are consequently preferred for shape-sensitive analyses, especially for observations from symmetric or approximately symmetric distributions.

Variance   The *variance* of a distribution is a measure of the deviations of observations from the central value. One might propose a measure of spread that incorporates information about all observations by calculating the mean of the deviations $x_i - \mu$, where the x_i are the observations and μ is the population mean. However, the mean of these deviations is zero. The mean of the absolute values $|x_i - \mu|$ might be tried, but functions involving absolute values are cumbersome to manipulate. A more suitable function is obtained by averaging the square $(x_i - \mu)^2$ of the deviations.

Let N be the number of elements in a population. Then the *population variance* is defined as

$$\sigma^2 = \frac{1}{N} \sum_{i=1}^{N} (x_i - \mu)^2, \tag{13.7}$$

when $N < \infty$; otherwise, it is the limit of the right-hand side as $N \to \infty$. The population mean μ must be known, as well as the total population size N, which is often infinite.

In most cases μ is unknown; hence σ^2 cannot be calculated. Substituting the known sample mean $\bar{x}$ for μ and sample size n for N in Equation (13.7) gives the quantity

$$\text{MSD} = \frac{1}{n} \sum_{i=1}^{n} (x_i - \bar{x})^2 = \frac{1}{n} \sum_{x_i} (x_i - \bar{x})^2 c(x_i), \tag{13.8}$$

where MSD stands for *mean-square deviation;* the rightmost formula is used when occurrences $c(x_i)$ are known, similarly to formula (13.4) for the mean.

The MSD gives the spread (dispersion) of observations about a sample mean but can be shown to introduce a systematic error when used as an estimate of the population variance. A somewhat intuitive explanation of this bias begins by observing that Equation (13.8) appears to calculate the mean of n items. However, these items are not independent, since $\bar{x}$ is calculated from $x_1, \ldots x_n$ and therefore knowledge of $\bar{x}$ and the first $n - 1$ terms allows the nth term to be calculated. Since the n items are not independent, the calculated MSD underestimates the spread.

The difficulty associated with the MSD is removed by calculating the *sample variance*, defined as

$$s^2 = \frac{1}{n-1} \sum_{i=1}^{n} (x_i - \bar{x})^2, \tag{13.9}$$

which approaches the value of σ^2 in Equation (13.7) as $n \to N$ for sufficiently large N. The sample variance is the best estimate of the population variance when the population mean must be estimated using the same sample data, as is generally the case. The MSD remains the best estimate when the population mean is known from theory or from a completely independent sample.

Standard deviation Although variance is a measure of the spread of a distribution, the units of variance are the square of the units of the measurement, preventing direct comparison with the mean, median, or mode. The square root of the sample variance is therefore calculated. The quantity obtained from Equation (13.9) as

$$s = \sqrt{s^2} = \sqrt{\frac{1}{n-1} \sum_{i=1}^{n} (x_i - \bar{x})^2} = \sqrt{\frac{n}{n-1} \text{MSD}} \tag{13.10}$$

is called the *sample standard deviation.* Similarly, the *population standard deviation* σ is given by $\sigma = \sqrt{\sigma^2}$ from Equation (13.7). Standard deviation is the generally preferred measure of the spread of a distribution.

Simplified calculation

Computing the MSD by the formula given in (13.8) or the variance given in (13.9) requires the difference between the sample mean and each observation value to be squared, which is simply done by computer but subject to error when performed by hand. A simplified formula is obtained by expanding the square in Equation (13.8) as follows:

$$
\begin{aligned}
\text{MSD} &= \frac{\sum_{i=1}^{n}(x_i - \bar{x})^2}{n} = \frac{\sum_{i=1}^{n}\left(x_i^2 - 2\bar{x}\,x_i + \bar{x}^2\right)}{n} \\
&= \frac{\sum_{i=1}^{n} x_i^2}{n} - \frac{2\bar{x}(\sum_{i=1}^{n} x_i)}{n} + \frac{n\,\bar{x}^2}{n} = \frac{\sum_{i=1}^{n} x_i^2}{n} - 2\bar{x}^2 + \bar{x}^2 \\
&= \frac{\sum_{i=1}^{n} x_i^2}{n} - \bar{x}^2 = \overline{x^2} - \bar{x}^2,
\end{aligned}
\tag{13.11}
$$

where $\overline{x^2} = (\sum_{i=1}^{n} x_i^2)/n$. It is often simpler to calculate the MSD by subtracting the square of the mean from the mean of the squared observations, as derived above. The calculation of s^2 can be similarly simplified.

Example 13.12
Standard deviation of sample in Example 13.7

By multiplying the square of each data value in Example 13.7 times the number of occurrences and using Equation (13.11), the sample standard deviation is calculated as

$$
s = \sqrt{\frac{10}{9}\left(\frac{13^2(1) + 14^2(3) + 15^2(4) + 16^2(2)}{10} - (14.7)^2\right)} = 0.95.
$$

13.5   Measures of relative standing

When observations can be ordered by value, the distribution can be described succinctly by first partitioning the observations into groups of equal number and then listing the values that separate the groups. The simplest example of this concept is the median, which divides the observations into two equal groups as discussed in Section 13.3. Partitioning the data into more than two groups leads to the following measures.

Quartiles

The *quartile* values divide the observations into four groups of equal number. The value that divides the lowest quarter of the observations from higher values is called the first or lower quartile (LQ). Similarly, the third or upper quartile (UQ) divides the highest quarter of the observations from lower values. The middle quartile is identical to the median.

The spread of the distribution can be described by the interquartile range (IQR), which is defined as

$$
\text{IQR} = \text{UQ} - \text{LQ}.
\tag{13.12}
$$

Half of the data, including the median, is in the IQR. When the median is near the centre of the IQR and a measure of spread as a distance from the median is desired, then the semi-interquartile range (SIQR) is often used and is defined as

$$
\text{SIQR} = (\text{UQ} - \text{LQ})/2 = \text{IQR}/2.
\tag{13.13}
$$

The IQR and SIQR are functions of the quartiles, which, like the median, are independent of the values of the extremes of the distribution.

Deciles *Decile* values divide an ordered set of observations into 10 groups of equal number. Of particular interest are the first or lower decile (LD), which divides the lowest 10 % of the observations from the rest, and the ninth or upper decile (UD), which divides the highest 10 % of the observations from the rest. The interdecile and semi-interdecile ranges are defined from these two values.

Example 13.13
A salary survey

Figure 13.6 is an example of how deciles, quartiles, and the median can be used to show central value and spread. Ontario engineers and engineers in training have been surveyed periodically by PEO and most recently by the OSPE [4]. Yearly salary, year of bachelor's degree graduation, and other information is requested. Such data do not represent measurements of a true value corrupted by systematic and random error, so the mean and standard deviation are not the most appropriate statistics; instead, the median, interquartile range, and interdecile range tabulated in the survey are shown in the figure.

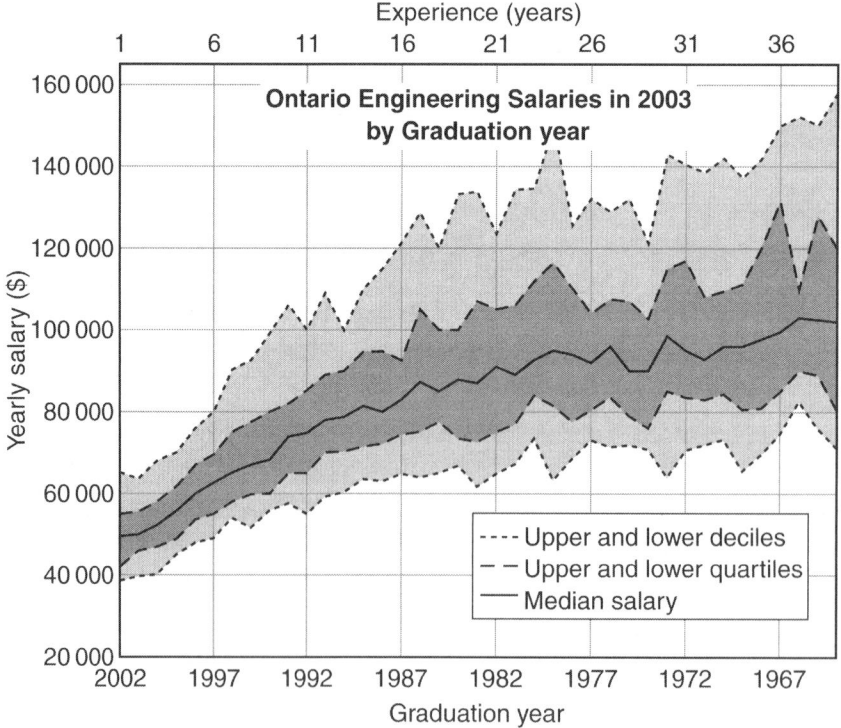

Figure 13.6 The graph shows the salaries of survey respondents in the year 2003. For any graduation year, 50 % of the salaries lie within the vertical range corresponding to the inner grey region, and 80 % of salaries lie within the larger light grey range. The ranges are asymmetric about the median, with greater spread above the median than below it. The graphs indicate a faster rise of salaries in the first years after graduation than later.

Percentile calculations *Percentile* values divide a set of observations into 100 groups of equal number. The median, quartiles, deciles, and percentiles are easy to calculate and useful for describing skewed distributions. However, there must be a sufficient number of observations to make their determination meaningful and, for finite samples, a consistent method such as follows for determining the required values. Consider a set of n observations x_1 to x_n ordered by value. Then the pth percentile is a number y such that $p\,\%$ of the observations are less than or equal to y. Let $u = (p/100) \times n$, let $\lceil u \rceil$ be the least integer greater than or equal to u, and let

$$
y = \begin{cases} (x_u + x_{u+1})/2 & \text{if } u \text{ is an integer} \\ x_{\lceil u \rceil} & \text{otherwise} \end{cases} .
\tag{13.14}
$$

Then, as required, the first decile is the 10th percentile, the first quartile is the 25th percentile, the median is the 50th, and so on.

13.6 Further study

1. Choose the best answer for each of the following questions.

(a) For the set of 10 sample observations $\{2, 4, 3, 5, 5, 9, 6, 7, 7, 7\}$, the ratio $(\text{MSD})/s^2$ of the mean squared deviation (MSD) to the variance is

    i. 10/9   ii. 9/10   iii. 11/10   iv. 11/9   v. 1

(b) For the sample data in part (a), the value of the standard deviation is

    i. 7   ii. 55   iii. 2.4   iv. 9   v. 5.5   vi. 2.1

(c) For the sample data in part (a), the value of the mode is

    i. 7   ii. 9   iii. 2.4   iv. 5.5   v. 55   vi. 2.1

(d) For the sample data in part (a), the value of the median is

    i. 5.5   ii. 2.4   iii. 55   iv. 7   v. 9   vi. 2.1

(e) The bin size of a histogram should show

    i. no gaps in the distribution.

    ii. relevant information while not introducing false detail.

    iii. a bell-shaped curve as nearly as possible.

(f) The mean of a distribution equals

    i. the sum of products of the population values times their probability of occurrence.

    ii. the sum of the population values.

    iii. the sum of occurrences of the population values.

(g) The sum of the relative frequencies of a distribution histogram equals

    i. n, the sum of occurrences.   ii. the mean value.   iii. 1.

    iv. the maximum observation.

(h) The standard deviation of a sample of n observations with mean $\bar{x}$

    i. is an estimate of the true central value of the population provided deviations are independent and distributed symmetrically.

    ii. is given by the formula $\sqrt{\left(\sum_{i=1}^{n}(x_i - \bar{x})^2\right)/(n-1)}$.

    iii. is the middle value of ordered data.

    iv. is given by the formula $\frac{1}{n}\sum_{i=1}^{n}(x_i - \bar{x})^2$.

    v. is the value that has the maximum probability of occurrence.

(i) A sample of n observations is taken from a population. The population value that has the maximum probability of occurrence is estimated by

    i. the sample mean.   ii. the sample mode.   iii. the sample median.

    iv. the sample interquartile range.

(j) A distribution with positive skew has a mean which is

    i. greater than the median.   ii. equal to the mode.   iii. equal to the median.

    iv. less than the median.

---

2.   Find the mean, median, mode, range, semi-interquartile range, variance, and standard deviation for the following set of 17 grades for an examination marked out of 100:

    25, 55, 60, 61, 63, 70, 72, 72, 73, 74, 74, 74, 76, 78, 82, 85, 90

---

3.   Find the mean, median, mode, range, variance, standard deviation, semi-interquartile range, and semi-interdecile range for the language examination marks summarized in Table 13.2. Use a spreadsheet program. Your spreadsheet may include built-in functions for some of the required values. Check your calculations with these functions.

13.7   References

[1] J. S. Bendat and A. G. Piersol, *Random Data Analysis and Measurement Procedures*. New York: John Wiley & Sons, 1986.

[2] P. R. Bevington and D. K. Robinson, *Data Reduction and Error Analysis for the Physical Sciences*. New York: McGraw-Hill, second ed., 1992.

[3] J. D. Petrucelli, B. Nandram, and M. Chen, *Applied Statistics for Engineers and Scientists*. Upper Saddle River, NJ: Prentice Hall, 1999.

[4] Research Dimensions Limited, "2003 membership salary survey detailed report," tech. rep., Ontario Society of Professional Engineers, Toronto, ON, 2003.

[5] J. L. Phillips, *How to Think About Statistics*. New York: W. H. Freeman and Company, sixth ed., 2000.

[6] D. W. Scott, "On optimal and data-based histograms," *Biometrika*, vol. 66, pp. 605–610, 1979.

Chapter

14

Gaussian Law of Errors

Conclusions or predictions based on measurements require a model that describes properties of the inevitable measurement errors. In the early 1800s, Gauss developed the normal (Gaussian) distribution (Figure 14.1) as a model for the errors in astronomical observations, resulting in its early description as the "law of errors." Since then, many other measurement distributions have been observed to follow the Gaussian curve. In addition, a fundamental result called the *central limit theorem* of probability states that if measurement errors are the sum of many independent random effects, then, under reasonable assumptions, the shape of the distribution describing the measurement is Gaussian.

The normal distribution in statistics and the Gaussian distribution in engineering and the pure sciences are the same. In the special case when $\mu = 0$ and $\sigma = 1$ in Figure 14.1, the distribution is called the *standard normal* or *standard Gaussian* distribution, as discussed in Section 14.2. Social scientists often refer to the Gaussian distribution as the "bell curve."

The Gaussian distribution provides a theoretical justification for the method of least squares, which allows the prediction of "best" estimates of a measured value. Such models are fundamental to understanding measurements, which are at the core of the physical sciences and engineering.

In this chapter, you will learn about

- the universality of the Gaussian law of errors in the statistical analysis of measurements, and the conditions that must apply for this model to be appropriately employed as a model of uncertainties,

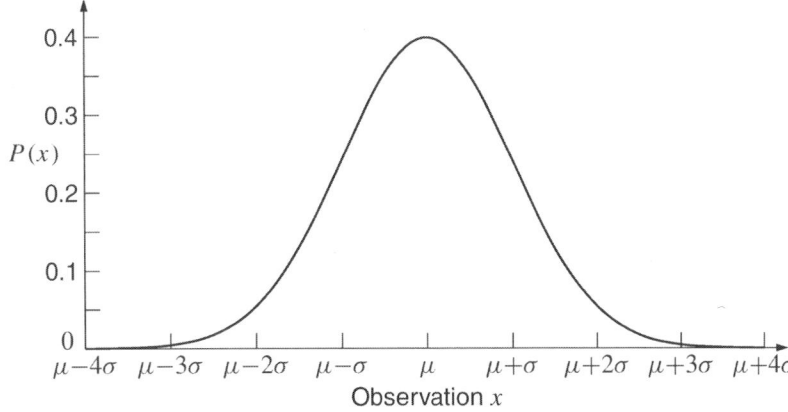

Figure 14.1 The Gaussian (or normal) distribution function with mean μ and standard deviation σ. The curve is a function of x and of two parameters, the constant σ is a horizontal scale factor, and μ is the horizontal location of the peak value of the function.

- the relationship between probabilities, confidence intervals, and the Gaussian distribution,

- a best estimate of population mean, a best fit line to a set of data, and when an observation can be rejected,

- the determination and interpretation of correlation coefficients as a descriptive statistic to measure the relationship between two sets of observations or between observation and theory.

14.1   Conditions for applying the Gaussian model

This chapter concentrates on continuous distribution functions mentioned briefly in Chapter 13 and illustrated in Figure 14.2. As is often done, in this chapter the word *distribution* is taken to be synonymous with the function describing the distribution.

Probability histogram

Figure 14.2(a) is a histogram such as might be drawn to show the values of repeated measurements subject to random error. The measurements have been grouped into bins of suitable width w, and a relative vertical scale is used so that the sum of the bar heights is 1. Then these heights can be taken to be probabilities, and the function that assigns height $f(i)$ to each bin i is sometimes called the *probability mass function*. The area of each bar is $w \times f(i)$ and the sum of the bar areas is w.

Probability density function

Figure 14.2(b) is a continuous function representing the infinite population from which the sample on the left was obtained, scaled so that the area under the curve is 1. The height $P(x)$ does not represent probability but the probability of an observation value x falling in any interval $a \leq x \leq b$ is $\int_a^b P(x)\,dx$, which is the area under the curve in this interval. The function $P(x)$ is called the *probability density*.

The Gaussian function is the most important continuous probability density function and occurs often. However, the following conditions should exist before a distribution is taken to be Gaussian:

Independent measurements

1. Measurements (observations) are independent and equally reliable. The implication is that the measurement errors are random and independent.

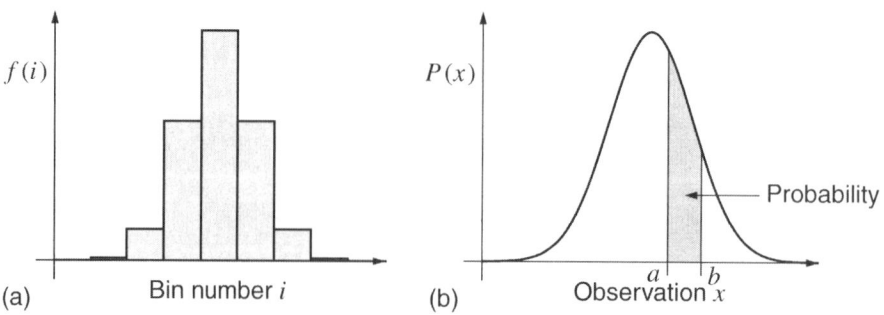

Figure 14.2   (a) The histogram of a probability mass function and (b) a continuous probability density function. The histogram has a relative vertical scale, so that the sum of the bar heights is 1 and the total area is 1 times the bin width. The area under the continuous curve is 1.

Symmetry   2.   Positive and negative errors of the same magnitude are equally likely; that is, errors are distributed symmetrically about the mean value.

Small errors are most probable   3.   Large errors are less likely than small errors; that is, the probability of an error of a given magnitude decreases monotonically with its magnitude.

No cusp   4.   The derivative of the error distribution exists and is continuous everywhere including the mode; hence, the derivative has a value of zero at the mode.

Mode equals mean   5.   The condition that the mode equals the mean follows from Conditions 2 and 3 but is worth mentioning separately because of its usefulness in the absence of a histogram of the errors.

These conditions are sufficient to derive the shape of an error distribution and were used by Gauss to do just that; the result is described in more detail in Section 14.2. Moreover, the Gaussian model is often a useful approximation when the conditions are not precisely met. For example, a Gaussian distribution often provides a good description of examination grades near the mean, even though the distribution is bounded and negatively skewed as discussed in Section 13.3.

Caution   Gaussian-like distributions frequently occur for samples from parent populations of unknown distribution. However, the Gaussian law of errors does not always apply as a model for randomness, and proofs that it applies exist only in special cases. A Gaussian error model is often assumed when detailed information is unavailable and the conditions discussed above are true. As Lippman (quoted in reference [1]) remarks, "Everybody believes in the Gaussian law of errors: the experimenters because they think it can be proved by mathematics; and the mathematicians because they believe it has been established by observation."

Observations that can only attain non-negative values must be treated with care because the symmetry condition cannot be satisfied exactly. Counting is one case of this kind, since the data values are non-negative integers. Examples are counts of persons, vehicles, class sizes, traffic statistics, and the number of earthquakes or Internet messages per unit of time. Other quantities are not integers but still cannot be negative: human height, weight, or lifespan, for example.

To decide whether Gaussian statistics can be applied approximately to non-negative observations, calculate the ratio $\bar{x}/s$ of the sample mean to the sample standard deviation. For small values such as $\bar{x}/s < 5$, the distribution is significantly skewed, and a Gaussian model may only apply in a small neighbourhood of the mean. However, as this ratio becomes large, the conditions for applying the Gaussian model over a significant range may be satisfied.

Example 14.1
Two-outcome measurements: the binomial distribution

The simplest of all measurements are those with only two possible values. Consider the following examples:

- The parts produced on an assembly line either pass a quality-control test or fail it.

- Repeated readings taken with a digital voltmeter are identical except for the least significant digit, which displays one of two values that differ by 1.

• Each sample of air either contains a rare pollution molecule or does not.

In these cases, repeated trials (measurements) are taken and the occurrences of one of the specified outcomes are counted. The outcome probabilities of one trial and the number of trials determine the distribution of the sum.

Tossing a coin is another two-outcome example. Suppose that the measurement of interest is the number of heads in a sample of n unbiased coin tosses. Since heads is just as likely as tails, the probability distribution for a single coin flip is flat, as shown for $n = 1$ in Figure 14.3. The histograms for $n = 2$, 3, and 20 are also shown in the figure; these are examples of the *binomial distribution*. For increasingly large sample size n, the histogram demonstrates the central limit theorem by approaching a Gaussian shape. The theorem states that the sum of a large number of independent, identically distributed observations with finite variance will approach a Gaussian distribution regardless of the underlying distribution for one measurement. A similar result holds for the three other examples above, even though the outcomes of one trial are not equiprobable as for an unbiased coin toss.

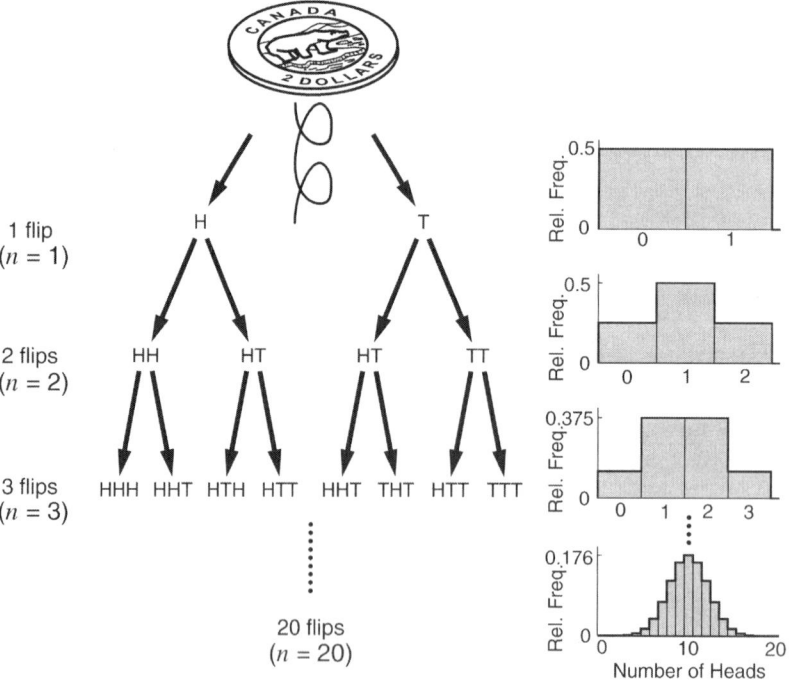

Figure 14.3 The number of occurrences of heads observed in a sample of n coin flips demonstrates the central limit theorem. The relative-frequency histograms on the right approach the Gaussian distribution as n becomes large.

Example 14.2
Size of a
manufactured object

Errors in engineering measurements and fluctuations in other samples of observed data typically result from numerous, often very small, independent, random perturbations. A good example is the measured size of almost any manufactured object, such as the width of a brick. Such quantities are the result of numerous small effects such as raw material characteristics, levels of contamination, oven temperature variations, machine speed variations, vibration during conveyance, measurement errors, and other factors.

14.2   Properties of the Gaussian distribution function

The Gaussian (or normal) curve shown in Figure 14.1 is defined by the formula

$$P(x) = \frac{1}{\sigma\sqrt{2\pi}}e^{-\frac{(x-\mu)^2}{2\sigma^2}},\tag{14.1}$$

where μ is the mean and σ is the standard deviation of the distribution. By making the change of variables

$$z = \frac{x-\mu}{\sigma} \quad \text{or} \quad x = \mu + \sigma z,\tag{14.2}$$

the simpler formula

$$Y(z) = P(\mu + \sigma z) = \frac{1}{\sqrt{2\pi}}e^{-\frac{z^2}{2}}\tag{14.3}$$

results. This function is called the *standard Gaussian* (or *standard normal*) distribution function. The mean, mode, and median of z are all zero, and the standard deviation is 1. This function simplifies the calculation of probabilities as areas under the curve as discussed in Section 14.1; thus

$$\int_a^b P(x)\,dx = \int_a^b \frac{1}{\sigma\sqrt{2\pi}}e^{-\frac{(x-\mu)^2}{2\sigma^2}}\,dx = \int_{(a-\mu)/\sigma}^{(b-\mu)/\sigma} \frac{1}{\sigma\sqrt{2\pi}}e^{\frac{z^2}{2}}\sigma\,dz$$

$$= \int_{(a-\mu)/\sigma}^{(b-\mu)/\sigma} Y(z)\,dz,\tag{14.4}$$

where the change of variables (14.2) has been made. The functions $P(x)$ and $Y(z)$ cannot be integrated in closed form using elementary calculus, so computers or pre-calculated tables of the integrals in Equation (14.4) must be used. The algorithms and printed probability tables need only consider $Y(z)$, which is a function of z alone and not of parameters.

For each of the functions defined above in (14.1) and (14.3), the area under the curve from $-\infty$ to ∞ is 1, as required of a probability density function.

14.2.1   Probability intervals

Although statistics based on Gaussian distributions can be used for many other purposes, here the determination of measurement error bounds such as were required in Chapter 12 will be considered. The probability of an observation of a Gaussian-distributed variable falling within several intervals is given in Figure 14.4. From these intervals the error in a single observation and in the sample mean can be estimated.

One measurement   Suppose, for example, that a single measurement x_1 has been taken and that the population standard deviation σ has been estimated beforehand from theory or other data. Then it is possible to state that, with a confidence level of 68.3 %, the population mean μ is in the range $x_1 \pm \sigma$. Similarly, from Figure 14.4, the true value (the mean) is within the approximate range $x_1 \pm 2\sigma$ with a 95 % confidence level. In both cases, the numerical value of σ would be substituted in the range. The confidence level should always be given, but it need only be specified once for a group of similar readings.

Half-interval d vs
Probability p

d	p (%)
0.67σ	50
σ	68.27
1.28σ	80
1.96σ	95
2σ	95.45
2.58σ	99
3σ	99.73

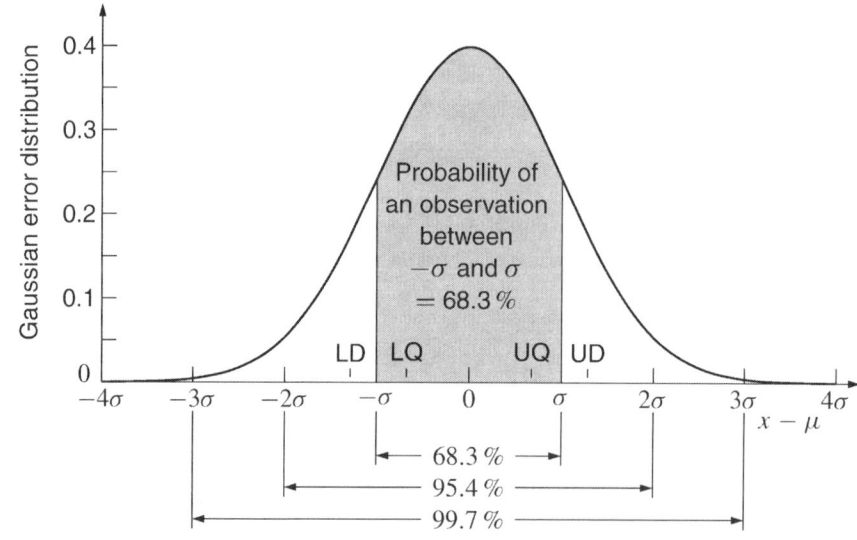

Figure 14.4   Interval probabilities for the Gaussian distribution. The table shows the probability p of any measurement x falling in the interval $\mu - d \leq x \leq \mu + d$. These probabilities correspond to areas under the Gaussian error distribution function as illustrated. The upper and lower quartiles are 0.67σ from the mean and the upper and lower deciles are 1.28σ from the mean.

Example 14.3
A temperature
measurement

Suppose that the standard deviation of readings taken from a mercury thermometer is known to be $0.3\,°C$. Then a single reading of $22.4\,°C$ could be recorded as "$22.4\,°C \pm 0.6\,°C$ with 95 % confidence."

Uncertainty of the
mean of repeated
measurements

Now suppose that n repeated measurements of a quantity have been taken. A typical first step is to verify approximately that the distribution of measurement errors is Gaussian by plotting a histogram such as in Figure 14.2. The mean $\bar{x}$ and standard deviation s are then computed for the sample as estimates of the population values μ and σ, respectively. The uncertainty of the mean of n measurements is less than the uncertainty of one measurement. To show this, let the parent distribution of the measurements x_1, x_2, $\ldots x_n$ have standard deviation σ. Assume that there is no systematic error and that the n measurement errors are independent and symmetrically distributed. Then applying an argument very similar to the propagation of error formula in Section 12.3.1 shows that the quantity

$$\bar{x} = \frac{x_1 + x_2 + \ldots x_n}{n} \tag{14.5}$$

has variance

$$\sigma_{\bar{x}}^2 = \frac{\sigma^2 + \sigma^2 + \ldots \sigma^2}{n^2} = \frac{\sigma^2}{n}. \tag{14.6}$$

Therefore, the standard deviation $\sigma_{\bar{x}}$ of the mean $\bar{x}$ is

$$\sigma_{\bar{x}} = \sqrt{\sigma_{\bar{x}}^2} = \frac{\sigma}{\sqrt{n}} \simeq \frac{s}{\sqrt{n}}, \tag{14.7}$$

where s is taken as an estimate of σ. Then it is reasonable to state that to a given confidence level, the population mean μ is

$$\mu = \bar{x} \pm k \left(\frac{s}{\sqrt{n}} \right), \tag{14.8}$$

where k corresponds to the confidence level chosen. For example, $k = 1$ for a 68.3 % confidence level.

Warning about statistics   Equation (14.7) should not always be interpreted to mean that error in the sample mean can be reduced indefinitely by increasing sample size. Time and resources may limit the number of measurements, and small amounts of bias due to imperfect instruments or operator error are often present.

Example 14.4
Multiple temperature readings   Suppose that multiple readings from the thermometer of Example 14.3 are taken but that nothing is known in advance about the spread of the measurements, as is often true. The readings are 22.4, 22.9, 21.9, 22.6, 22.2, 22.2, 22.4, 22.6. Then the sample mean is 22.4 and the standard deviation is 0.31. For a 95 % confidence level, the probability interval is approximately 2 standard deviations, or $2(s/\sqrt{n}) = 2(0.31/\sqrt{8}) = 0.22$, and the result can be stated as "22.4 °C $\pm$ 0.22 °C with 95 % confidence."

Example 14.5
Election polling   Surveys to determine the percentage of votes that a political candidate will receive are commonly expressed as being "accurate to within 3 percentage points 19 times out of 20." If a particular candidate is expected to receive 35 % of the votes, then this statement is equivalent to saying that it is estimated that there is a 95 % probability that the actual percentage of votes received will be within the range $35(1 \pm 1.96\sigma)$ %. In effect, this 95 % confidence level takes into account the limited size of the sample.

14.3   Fitting a Gaussian curve to sample data

The histogram provides one graphical method to test data to see whether the data follow a Gaussian distribution. This visual test, or validation, involves plotting the sample data and then superimposing the Gaussian distribution to see how closely it fits. Such a basic visual test can be supplemented by statistical tests. The following example illustrates the visual method.

Example 14.6
A metal-fatigue experiment

A group of engineering students conducted an experiment to find a basic model for the fatigue failure of steel. A sample of steel paper clips was obtained. Each paper clip was bent and then restored to shape in a systematic way, repeating until the clip fractured. A complete cycle of a bend from the starting position and fully back without breaking counted as one bend.

The data are summarized in Figure 14.5. The intent was to visualize how closely this histogram follows the shape of a corresponding Gaussian parent distribution $\hat{P}(x)$. We can observe immediately that the histogram cannot be exactly Gaussian because only positive data values are allowed, whereas the Gaussian curve extends indefinitely to the left and right. Nevertheless, the sample mean $\bar{x}$ and standard deviation s can be calculated and substituted as parameters in (14.1). The histogram is a function giving occurrences with respect to x, and its area is the sum of the bin heights multiplied by bin width, or $n\Delta x$. Therefore, the Gaussian curve $\hat{P}$ must have the same independent variable but it must also be scaled to have the same area, that is,

$$\hat{P}(x) = \frac{n\Delta x}{s\sqrt{2\pi}} e^{-\frac{(x-\bar{x})^2}{2s^2}}. \tag{14.9}$$

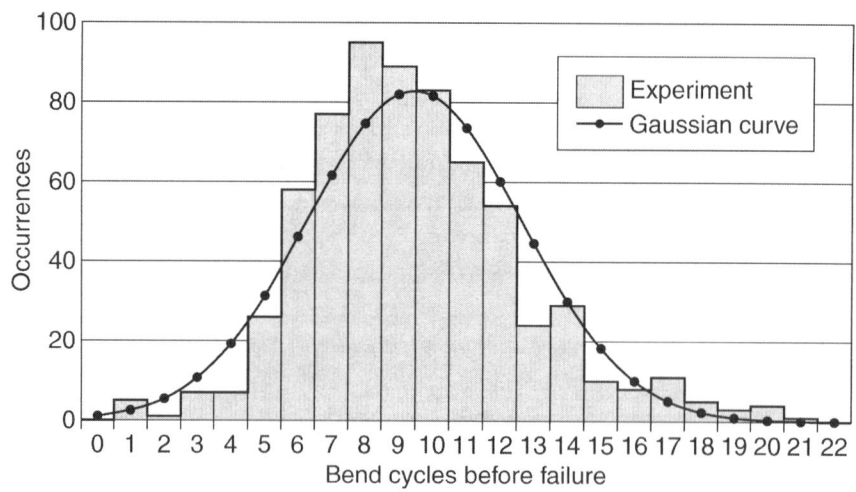

Occurrences n_i of x_i bend cycles before failure

x_i	n_i	x_i	n_i	x_i	n_i
0	0	8	95	16	8
1	5	9	89	17	11
2	1	10	83	18	5
3	7	11	65	19	3
4	7	12	54	20	4
5	26	13	24	21	1
6	58	14	29		
7	77	15	10		

Figure 14.5 Table of observations from 662 paper clips fatigue-tested to failure, with corresponding histogram. The Gaussian distribution $\hat{P}(x)$ with identical mean and standard deviation is superimposed for comparison. The estimated values are defined only at the dots, owing to the discrete nature of the histogram data.

Figure 14.5 shows $\hat{P}(x)$ superimposed on the histogram. For the given data, the calculated values are $n = 662$, $\bar{x} = 9.45$, and $s = 3.19$. Only discrete integer values for the abscissa variable (bend cycles before failure) exist, and the Gaussian distribution gives an approximate model of bin height at these integer values. If the bin size Δx were different from 1, then the scale factor $n\Delta x$ would change, but the smooth curve would still provide one estimation point per bin as in the figure. The mean is only approximately 3 standard deviations from the origin, so the Gaussian curve cannot be expected to predict values near zero precisely.

14.4 Least squares

The basic least-squares techniques in this section are used when a simple model for explaining or predicting measured values is required. Models contain parameters, and the calculated values of these parameters (statistics) characterize the measured sample. There will generally be discrepancies between measured values and the values predicted by the model. When this error has a Gaussian distribution, parameter values that minimize the sum of the squared errors produce the best predicted values.

More advanced discussion of the statistical topic known as *regression analysis* is found in textbooks [2] and statistics courses.

14.4.1 Optimality of the sample mean

The result of a very simple least-squares calculation has already been extensively used. Suppose that a single parameter m is required to predict the values $x_1, x_2, \ldots x_n$ of a sample. Then the error in each predicted value of x_i is $x_i - m$. The sum of the squared errors is

$$e^2 = \sum_{i=1}^{n}(x_i - m)^2, \tag{14.10}$$

which is to be minimized by choosing m optimally. Since the data values x_i are constants, e^2 is a function of m. To find the value of m that minimizes e^2, the derivative is calculated:

$$\frac{de^2}{dm} = \sum_{i=1}^{n} 2(x_i - m)(-1) = 2nm - 2\left(\sum_{i=1}^{n} x_i\right). \tag{14.11}$$

Equating this formula to zero gives

$$m = \frac{1}{n}\sum_{i=1}^{n} x_i, \tag{14.12}$$

which implies that the best value of m is the sample mean $\bar{x}$.

The second derivative of e^2 with respect to m must be tested to ensure that $\bar{x}$ corresponds to a minimum and not a maximum. Differentiating both sides of Equation (14.11) gives $\frac{d^2 e^2}{dm^2} = 2n$, which is positive, confirming that m minimizes e^2.

14.4.2	**The best straight line**

The least-squares method for obtaining the best straight line to fit a set of observed data points provides a mathematical alternative to drawing the line "by eye." The line is "best" in the sense of minimizing the square of the errors between the straight line $y(x)$ and observation pairs x_i, y_i for $i = 1, \ldots n$. Let the equation of a straight line be given by

$$y(x) = ax + b, \tag{14.13}$$

where a is the slope and b is the y-axis intercept. Given n observation pairs, the method seeks to minimize the quantity e^2, which is the sum of the squared deviations, defined as

$$e^2 = \sum_{i=1}^{n}(y_i - y(x_i))^2 = \sum_{i=1}^{n}(y_i - ax_i - b)^2, \tag{14.14}$$

assuming that the errors $y_i - ax_i - b$ are independent with zero-mean Gaussian distributions and have equal standard deviations.

Its derivation will be deferred to later, but the solution is obtained by solving the following two simultaneous linear equations for unknowns a and b:

$$\left(\sum_{i=1}^{n} x_i\right) a + nb = \sum_{i=1}^{n} y_i \tag{14.15a}$$

$$\left(\sum_{i=1}^{n} x_i^2\right) a + \left(\sum_{i=1}^{n} x_i\right) b = \sum_{i=1}^{n} x_i y_i. \tag{14.15b}$$

In these equations, the right-hand sides and the quantities multiplying a and b are known constants, since n and the values of x_i and y_i are known. Solving for a by any method gives the following, which can be written in several equivalent ways:

$$a = \frac{n \sum_{i=1}^{n} x_i y_i - \left(\sum_{i=1}^{n} x_i\right)\left(\sum_{i=1}^{n} y_i\right)}{n \sum_{i=1}^{n} x_i^2 - \left(\sum_{i=1}^{n} x_i\right)^2} = \frac{\overline{xy} - \bar{x}\,\bar{y}}{\overline{x^2} - \bar{x}^2} = \frac{\overline{xy} - \bar{x}\,\bar{y}}{\mathrm{MSD}(x)}, \tag{14.16}$$

where $\overline{xy} = (\sum_{i=1}^{n} x_i y_i)/n$, and $\mathrm{MSD}(x)$ is the MSD of the x_i from Equation (13.11). Using a calculated above, the parameter b is obtained from Equation (14.15a) as

$$b = \frac{\sum_{i=1}^{n} y_i}{n} - a \frac{\sum_{i=1}^{n} x_i}{n} = \bar{y} - a\bar{x}, \tag{14.17}$$

where $\bar{y} = (\sum_{i=1}^{n} y_i)/n$.

Equations (14.15a) and (14.15b), and the solution of these equations can be generated automatically from sets of pairs x_i, y_i by common software tools such as spreadsheets. The method has to be modified when the errors $y_i - ax_i - b$ have unequal standard deviation [3].

Obtaining Equations (14.15a) and (14.15b)   To derive Equations (14.15a) and (14.15b) from (14.14), we can use the fact that at a minimum the derivatives of e^2 with respect to a and b are simultaneously zero. This is easy to see: suppose that for the parameter values $a = a_0$, $b = b_0$, the error $e^2(a_0, b_0)$

is a minimum but the quantity $\frac{\partial e^2}{\partial a}$ evaluated at $a = a_0$, $b = b_0$ is nonzero. Then by changing a slightly to $a_0 + \Delta a$, we can make $e^2(a_0 + \Delta a, b_0)$ less than $e^2(a_0, b_0)$, which contradicts the assumption that the function was at a minimum. A similar argument holds for changes in b. Thus, differentiating with respect to b and equating the derivative to zero gives

$$\frac{\partial(e^2)}{\partial b} = \sum_{i=1}^{n} \frac{\partial}{\partial b}(y_i - b - a\,x_i)^2 = \sum_{i=1}^{n} 2(y_i - b - ax_i)(-1)$$

$$= 2\left(-\left(\sum_{i=1}^{n} y_i\right) + nb + a\left(\sum_{i=1}^{n} x_i\right)\right) = 0. \tag{14.18}$$

A second equation is obtained similarly by differentiating with respect to a:

$$\frac{\partial(e^2)}{\partial a} = \sum_{i=1}^{n} \frac{\partial}{\partial a}(y_i - b - a\,x_i)^2 = \sum_{i=1}^{n} 2(y_i - b - ax_i)(-x_i)$$

$$= 2\left(-\left(\sum_{i=1}^{n} x_i y_i\right) + b\left(\sum_{i=1}^{n} x_i\right) + a\left(\sum_{i=1}^{n} x_i^2\right)\right) = 0. \tag{14.19}$$

Dividing these equations by 2 and rearranging gives Equations (14.15a) and (14.15b).

Example 14.7
Fuel consumption and
engine displacement

Government agencies regularly require tests of vehicle fuel consumption as a guide for consumers. The relationship between fuel consumption and engine displacement is given in Figure 14.6 for a small set of vehicles. The scatter plot in the figure shows data points considerably dispersed about the best-fit straight line. The intermediate values calculated from the data are $n = 11$, $\sum_{i=1}^{n} x_i = 32.6$, $\sum_{i=1}^{n} y_i = 122.0$, $\sum_{i=1}^{n} x_i y_i = 372.5$, and $\sum_{i=1}^{n} x_i^2 = 103.0$, from which the slope is calculated as $a = 1.71$ and the intercept as $b = 6.03$. The graph shows that such a straight line cannot be used as a precise model for single vehicles, but an approximate relationship between displacement and consumption appears to exist.

Consumption y_i (litres per 100 km)
vs Displacement x_i (litres)

x_i	y_i	x_i	y_i
2.0	8.7	3.0	11.8
2.2	9.0	3.5	12.4
2.4	9.0	3.5	13.1
2.4	10.7	3.8	11.8
2.5	11.2	4.6	13.1
2.7	11.2		

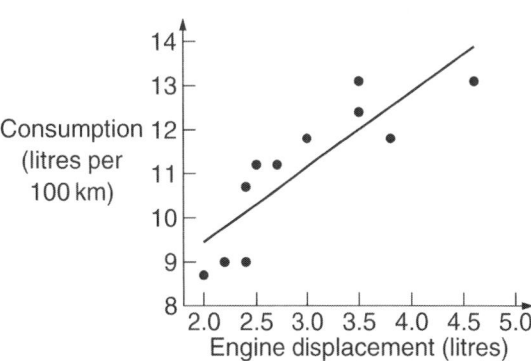

Figure 14.6   Tabular data, scatter plot, and best straight line for fuel consumption and engine displacement of a group of passenger vehicles.

Warning   Fitting a straight line to data is not always justified; the following must be considered.

Unequal errors   If the data points have unequal uncertainties, then the quantity e^2, given by Equation (14.14), has to be modified. Figure 14.7 illustrates how equal data uncertainties become unequal uncertainties when the data are transformed using a nonlinear function such as the logarithm. It is preferable not to transform the data, but to minimize the error between the data and a curve within a family of curves of known shape.

Extrapolation   If the data have been measured over a limited range, it is dangerous to predict the value of $y(x)$ outside this range.

Incorrect model   Computing the coefficients in Equations (14.15a) and (14.15b) and solving the equations to obtain constants a and b is temptingly simple, but it does not guarantee that a straight line is a realistic model of the data.

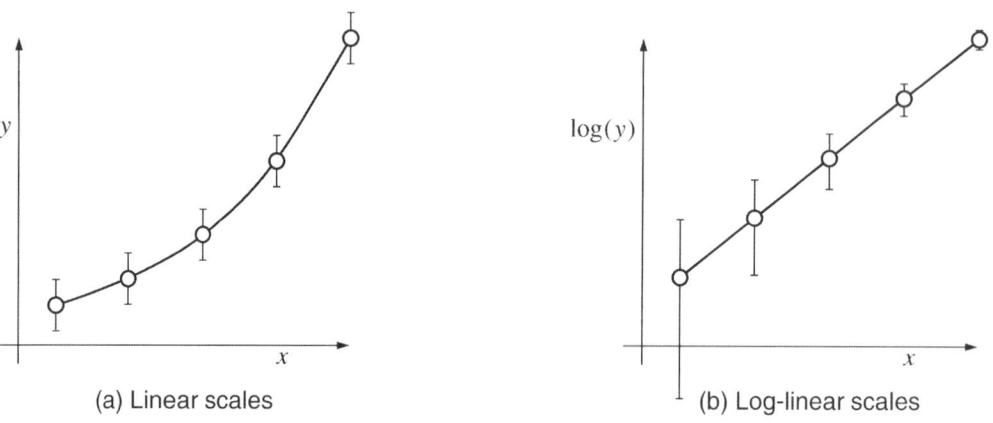

(a) Linear scales   (b) Log-linear scales

Figure 14.7   The equal uncertainties of the data points in (a) become unequal under a nonlinear transformation, such as for the log-linear scales in (b).

14.4.3   Correlation coefficient

The parameters a and b in Equation (14.13) are easy to calculate from formulas (14.16) and (14.17), but these parameters should not be computed blindly; the data should be plotted to check that a straight line will convey the correct message. As illustrated in Figure 14.8, for example, the data may contain invalid points, a straight line may be an invalid physical model, or many different straight lines fit the data almost equally well. In the figure, the calculated slope a and intercept b are identical for the three sets of data illustrated, but the figures are very different and all deliver invalid messages. The invalidity of graphs (a) and (b) is not remedied by applying statistics, but a numerical test can assist in detecting cases such as (c). A dimensionless statistic called the *correlation coefficient* quantifies the closeness of fit of the straight line to the data.

From a set of pairs x_i, y_i, $i = 1, \ldots n$, the correlation coefficient r is calculated using the following formula, which is derived in references such as [2] and [4]:

$$r = \frac{\overline{xy} - \bar{x}\,\bar{y}}{\sqrt{\text{MSD}(x)}\sqrt{\text{MSD}(y)}}. \tag{14.20}$$

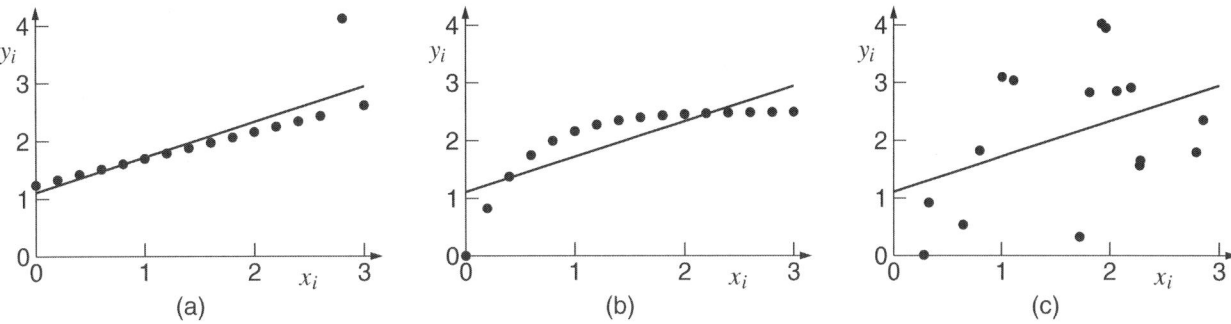

Figure 14.8   What not to do: the data in (a) appear to contain an invalid data point; in (b) a straight line is evidently an incorrect model; the great dispersion and small sample size in (c) make the calculated slope and intercept very sensitive to data error and many straight lines fit the data almost equally well.

As defined previously, $\overline{xy} = (\sum_{i=1}^{n} x_i y_i)/n$ and the MSD quantities are calculated for the x_i and y_i values from Equation (13.11).

By comparing formulas (14.20) and (14.16), the correlation coefficient r can be seen to equal the slope a of the best-fit straight line scaled by a positive value. Close inspection (and an algebraic result called the Cauchy-Schwartz inequality) shows that $-1 \leq r \leq 1$. Furthermore, the values of r can be interpreted as follows:

- If $r < 0$, then the best-fit straight line has negative slope and the data are said to be negatively correlated; $r > 0$ implies positive slope and positively correlated data.

- The value $r = 0$ indicates no dependency between y_i and x_i; the data are described as uncorrelated.

- The magnitude $|r|$ is a quantitative measure of the degree of correlation; data for which $|r|$ is near 1 are said to be highly correlated.

- The value $|r| = 1$ indicates that all data points are on the straight line; the relation satisfied by pairs (y_i, x_i) is exactly described by a and b.

Example 14.8
Coefficient for
Example 14.7 and
Figure 14.8

The correlation coefficient for Figure 14.8(c) is $r = 0.37$, indicating a poor fit, whereas the fit is somewhat better for the graph of Example 14.7, for which the calculated coefficient is $r = 0.86$. Both coefficients are positive and the corresponding graphs have positive slope.

Example 14.9
Children's height

An example of positively correlated data is the increase in children's height with increasing age.

Example 14.10
Predators and prey

In a natural habitat, the decrease in the rabbit population seen to accompany an increase in the wolf population is an example of negative correlation.

| 14.5 | **Rejection of an outlying point** |

Experimental data may contain "outlying" observations, that is, observations that seem highly improbable or unusual. Such outlying points generate concern about their validity because they may be a mistake or an improbable event that is not representative of the rest of the data.

The first step in analyzing outlying points is always to investigate the possible causes of an unusual observation. The unusual observation may signal an experimental process or natural phenomenon that is unexpected. If the observation clearly proves to be a mistake, then the observation should be rejected and the mistake corrected. If no mistake can be found, statistical tests are available to help decide whether one is justified in rejecting the outlying observation. When a point is rejected for statistical reasons, it is not simply deleted from the data set; instead, it is recorded and then marked as "rejected," with the rejection criterion stated briefly.

Statistical rejection of an outlying observation is justified when it is sufficiently unusual that it adversely affects the calculations of the sample mean, sample variance, and sample standard deviation, *even if the observation is valid!* Validity implies no apparent cause for the outlier except an improbable random fluctuation or "bad luck."

Example 14.11
Rejecting the results from a coin flipping experiment

A fair coin should show 50 % heads and 50 % tails if it is flipped a sufficient number of times. However, suppose that we flip a coin six times, with heads resulting each time. This is unusual, because the probability of this happening is only one event in 64, or 1.6 %. Therefore, although this result is valid, we would intuitively reject it. It would not alter our belief that a legal coin should show 50 % heads and 50 % tails if it is flipped. Of course, we might want to check the coin to see whether it has two heads. Similar logic is the basis for rejecting valid, but unusual, outliers.

| 14.5.1 | **Standard deviation test** |

The *standard deviation test* is a common criterion for the rejection of outliers. It is based on the estimated probability of the occurrence of the outlier within the parent population, without regard for the sample size of the observation set. For suspected outlier x_i, let $z = \frac{x_i - \bar{x}}{s}$, which is similar to the definition of the standardized variable z in Equation (14.2). In this formula, z is sometimes called the z-score. The sample mean is $\bar{x}$, and s is the sample standard deviation. The standard deviation test characterizes x_i as an outlier to be rejected if $|z| > 3$. By this criterion, any observation more than 3 standard deviations away from the mean is not considered representative of the population from which the rest of the sample is drawn and may be rejected. The discussion of the Gaussian error distribution showed that 99.73 % of Gaussian observations fall within $\pm 3\sigma$ of the mean μ. Therefore, the probability of obtaining an observation more than 3 standard deviations from the mean is 0.27 % for Gaussian random errors.

Caution

If the sample size is large and relevant or the errors do not follow a Gaussian distribution, then the standard deviation test may be inappropriate. To reject outliers requires that you understand your observations.

| 14.6 | Further study |

1.　Choose the best answer for each of the following questions.

(a) The body diameters of bolts from an automated production line are observed to have a Gaussian distribution with standard deviation of 0.08 mm. The probability that a bolt will have a diameter that is more than 0.24 mm too large is approximately

　　i. 0.15 %　　ii. 0.3 %　　iii. 5 %　　iv. 1.0 %

(b) In order to fit a straight line $y = mx + b$ to a set of measured data points by minimizing the sum of squares of the errors in y, the following should be true:

　　i. The (x, y) points should be plotted using log-log scales.

　　ii. The errors must have a Gaussian distribution.

　　iii. The error of fit should have equal value for all measurements.

　　iv. The slope of the straight line must be known beforehand.

(c) An instrumented vehicle rolls along a railway track, automatically measuring the track gauge every 3.5 m. The sensors occasionally fail because of excessive vibration. Along a section of track the gauge is calculated to have a mean of 1435.1 mm and standard deviation of 1.4 mm. Would a reading of 1437.5 mm be rejected by the 3σ standard deviation test?

　　i. no　　ii. yes

(d) The minimization of least squares is

　　i. unaffected by nonlinear transformations.

　　ii. the only way to fit a straight line to a set of measurements.

　　iii. inappropriate if measurement errors have unequal variance.

(e) The percentage of observations that have a normal (Gaussian) distribution and lie within 3 standard deviations of the mean, that is, between $\mu - 3\sigma$ and $\mu + 3\sigma$, is approximately

　　i. 50.0 %　　ii. 68.3 %　　iii. 95.4 %　　iv. 99.7 %

(f) An automobile storage battery of a certain type fails on the average after approximately 3 years, with a standard deviation of approximately 0.5 years, but a vendor offers a free replacement if the battery lasts more than 4 years. Assuming that failures have a Gaussian distribution, what is the approximate probability that a battery would qualify for replacement?

　　i. 10 %　　ii. 5 %　　iii. 2.5 %

(g) The *correlation coefficient* can be computed to show

　　i. how well a straight-line model fits a set of measured data.

　　ii. how equation errors with unequal spread can be incorporated in the model.

　　iii. the effect of nonlinear transformation on the best-fit line.

(h) The normal (Gaussian) distribution

    i. always has unit standard deviation.

    ii. has positive skew.

    iii. is a model of the cumulative result of many independent random effects.

    iv. always has a zero mean.

(i) For a normal (Gaussian) distribution with mean μ and standard deviation σ, the semi-interquartile range is approximately

    i. $\pm\sigma$ ii. $\pm 68.27\,\%$ iii. $\pm 0.67\sigma$ iv. $\pm 2\sigma$ v. $\pm 3\sigma$

(j) The purpose of fitting a straight line to a set of measured data points is

    i. to allow extrapolation beyond the limits of available data.

    ii. to explain or predict measured values simply.

    iii. to add two additional significant digits.

    iv. to overcome the fact that the data has outliers in it.

2. The four sets of data below represent coordinate pairs (x_i, y_i) for four distinctly different relationships. Using the equations in this chapter, calculate the slope a of the best straight line and the correlation factor r for each case. Make a quick plot in the x–y plane and comment on whether these values appear to be reasonable.

    Case 1: $\{(x_i, y_i)\} = \{(-2, 2),\ (-1, 1),\ (1, -1),\ (2, -2)\}$

    Case 2: $\{(x_i, y_i)\} = \{(1, 0),\ (0, 1),\ (-1, 0),\ (0, -1)\}$

    Case 3: $\{(x_i, y_i)\} = \{(-2, -2),\ (-1, -1),\ (1, 1),\ (2, 2)\}$

    Case 4: $\{(x_i, y_i)\} = \{(-2, 0),\ (-1, 0),\ (1, 0),\ (2, 0)\}$

3. The cylinder pressure observed in an internal combustion engine varies from cycle to cycle due to random fluctuations in the combustion process. If 10 measurements of peak cylinder pressure are made, resulting in a mean peak cylinder pressure of combustion of (2.14 ± 0.02) MPa, how many additional measurements are needed to reduce the uncertainty to 0.01 MPa?

| 14.7 | **References** |

[1] E. T. Whittaker and B. Robinson, "Normal frequency distribution," in *The Calculus of Observations: A Treatise on Numerical Mathematics*, pp. 164–208, New York: Dover Press, fourth ed., 1967.

[2] D. C. Montgomery and G. C. Runger, *Applied Statistics and Probability for Engineers*. New York: John Wiley & Sons, third ed., 2003.

[3] P. R. Bevington and D. K. Robinson, *Data Reduction and Error Analysis for the Physical Sciences*. New York: McGraw-Hill, second ed., 1992.

[4] J. S. Bendat and A. G. Piersol, *Random Data Analysis and Measurement Procedures*. New York: John Wiley & Sons, 1986.

Part IV

Engineering Practice

Figure IV.1   A construction site exemplifies the practice of civil, mechanical, electrical, and other engineering specialties. Project management is common to the practice of all engineering disciplines, and all engineers are responsible for ensuring public and worker safety.

Engineers, regardless of discipline, have something in common: the ability and responsibility to solve problems, particularly problems of engineering design. Part IV introduces

- the creative design process and how to stimulate creativity,
- the interaction of engineers with the business world,
- the protection of intellectual property that results from design,
- the planning and scheduling of engineering projects,
- the assessment of the critically important need for safety in engineering,
- the management of risk.

These key aspects of engineering practice are discussed in the following chapters.

Chapter 15 **Fundamentals of engineering design:** Whether a design is totally different from a previous design or only a minor modification, the problem-solving technique followed by the design engineer usually follows a well-known series of general steps. This chapter describes the steps in a typical engineering design project and illustrates the process with examples.

Chapter 16 **The engineer in business:** Most engineers are employees, but some are consultants and some have their own companies. This chapter discusses the basics of business structures and a method for evaluating a small company, either during its creation or shortly thereafter.

Chapter 17 **Intellectual property:** Engineering work creates results that may qualify for legal protection as intellectual property. The principal types of intellectual property are patents, trademarks, copyright, industrial designs, and integrated circuit topographies. This chapter defines these types, explains how to protect them, and also discusses how to use patent literature to stimulate ideas for future designs.

Chapter 18 **Project planning and scheduling:** This chapter briefly explains two basic, commonly used project-management tools: Gantt charts and the critical path method.

Chapter 19 **Safety in engineering design:** As an engineer, you will be responsible for the safety of any design that you approve. This chapter gives a basic introduction to safety requirements in engineering design, together with guidelines for eliminating workplace hazards. The importance of engineering codes and standards is also described.

Chapter 20 **Safety, risk, and the engineer:** This chapter further examines safety in engineering design and describes techniques for evaluating and managing risk in complex engineering projects.

Chapter 21 **Environmental sustainability:** We live in a time of changing global climate. This chapter explores the consequences of the burning of fossil fuels, and discusses challenges and guidelines for engineers.

Chapter

15

Fundamentals of Engineering Design

"Scientists investigate that which already is; engineers create that which has never been."—Albert Einstein

Design is a fundamental activity that distinguishes engineering from disciplines based on pure science and mathematics. Engineers must have solid, basic knowledge of science and mathematics, but must also be able to apply their talents to improve or create new products, processes, devices, and systems. In Canada, all engineering students are expected to have significant design experience as part of their undergraduate program [1]. Some engineering programs introduce design experiences as early as first year; other programs wait until the student has mastered the fundamentals of applied science before assigning design projects. Whether you experience design early or later in your curriculum, you will find that there are basic principles and steps that are common to all engineering design projects. The commonality of the design process enables engineers from different disciplines to work together productively on complex projects, such as the solar car shown in Figure 15.1.

Figure 15.1 Design of a solar car involves collaboration across engineering disciplines—chemical, civil, computer, electrical, mechanical, and systems design engineering. (Courtesy UW Midnight Sun solar car team.)

By the end of this chapter, you will be familiar with the following:

- working definitions of engineering design and related concepts;
- a basic design process that can be applied in all engineering disciplines;
- techniques for generating and evaluating alternative design solutions;
- characteristics of an effective design team.

15.1	**Defining engineering design**

The word *design* can be confusing, because it can represent both an activity and the result of the activity. In this chapter, *design* refers to the activity of designing, and *solution* refers to the outcome of the design activity.

Engineering design   *Engineering design* may be defined in simple terms as the process of developing workable plans for the construction or manufacture of devices, processes, machinery, or structures to satisfy some identified need. A more formal definition is published by the Canadian Engineering Accreditation Board [1]:

> Engineering design integrates mathematics, basic sciences, engineering sciences and complementary studies in developing elements, systems and processes to meet specific needs. It is a creative, iterative, and often open-ended process subject to constraints which may be governed by standards or legislation to varying degrees depending upon the discipline. These constraints may relate to economic, health, safety, environmental, social, or other pertinent interdisciplinary factors.

Both of these definitions emphasize that engineering design is the creative application of technical knowledge for some useful purpose. Engineering design is as creative as artistic design and differs only in the requirement for scientific, mathematical, and technical knowledge to carry it out properly. The application of engineering design should not only serve a useful purpose, but also protect public safety and welfare.

Looking around the room you are in now, can you identify at least 10 useful items that would not have been in existence in their present format 50 years ago? How have those products affected the quality of your life? The book that you are reading was written and formatted by the authors using computers—technology that was not readily available for public use 50 years ago. Phone calls made from one author to another made use of technology that was available 50 years ago, but additional features and services, such as call display and cell phones, were not.

15.1.1	**Types of engineering design**

Not all design solutions are completely new or original. There are two main types of design processes: *evolutionary* design (redesign) and *innovation* (original design).

Evolutionary design   Many engineered solutions are based on improvements to existing solutions. As technology improves, solutions that were not possible in earlier eras can be realized. For example, the form and features of current passenger cars have come a long way since Henry Ford's Model T, yet the same basic functions are there: propulsion, steering, braking, seating, and others.

Competitive analysis, benchmarking, reverse engineering   Three terms that are often associated with evolutionary design are *competitive analysis, benchmarking,* and *reverse engineering. Competitive analysis* is the process of comparing a design to a similar design or product. This process is helpful for establishing design criteria. *Benchmarking* is the determination of how well a function is performed, usually for later competitive analysis. For example, the time taken by a computer program to perform a standard computation may serve as a benchmark in evaluating other programs or designs. *Reverse engineering* is the process of decomposing an existing solution to understand how it has been constructed and where its design limitations lie. The identification patterns shown in Figure 17.1 on page 257 were discovered by a company that reverse-engineers electronic integrated-circuit chips.

Innovative design   An innovation is a new or original idea and implies a novel way of solving a problem. If the innovation applies to technology, then it may be considered an invention. An *invention* is the realization of a new and useful product, process, or system. Design engineers often seek to patent inventions to control their use. However, not all new solutions are patentable, and some may infringe on others that have already been patented (see Chapter 17).

Combining evolution and innovation   Much complex design relies on a combination of evolutionary and innovative design. In his book *Invention by Design*, Henry Petroski presents some entertaining case studies of how the form and function of common products, such as the zipper, the paper clip, and the pop can, have been developed over time by innovative and evolutionary design [2].

15.2   Characteristics of good design practice

While common sense, or common knowledge, does play a role in design, many design problems require innovative solutions that lie outside the general realm of common knowledge. The following are good engineering design practices that enable difficult design problems to be solved.

List criteria, requirements, and constraints   Establish the context of the design by writing lists of criteria, requirements, and constraints to be considered, preferably in order of importance. *Criteria* are general factors that can be listed, that directly affect the solution or the process of obtaining a solution, and that may aid in choosing the best solution. Requirements and constraints are criteria with more specific meanings (see Figure 15.2). *Requirements* (or *objectives*) are properties that a solution must possess, often expressed in terms of functions or tasks that must be performed (e.g., "The airplane will cruise at 400 km/h"). *Constraints* are limitations

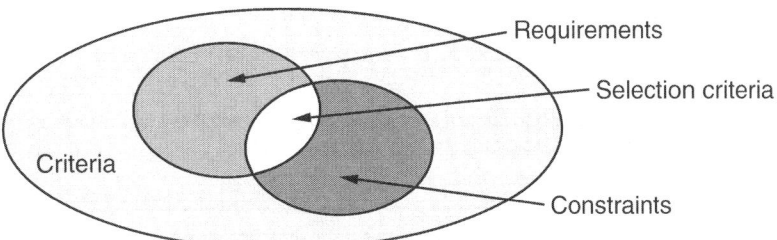

Figure 15.2   The set of solution criteria includes the design requirements and constraints. Requirements must be met to produce acceptable performance. Constraints typically are criteria that must not be exceeded by an acceptable solution and design process. When multiple solutions are possible, selection criteria are used to choose a solution.

on the solution or the design process, such as "The design process must not exceed its budget." Requirements and constraints should be objective and quantitative whenever possible, so that decisions about whether they have been met are defensible, as will be discussed.

Statements of criteria may be very general initially, but as the design progresses, they normally become more specific. For example, an initial requirement for a food processor might be that it must chop and purée food, and a refined version might specify that it must safely chop a cup of raw vegetables in less than 2 s.

Identify users and their tasks

Novice designers often make the mistake of focusing so much on the technical details of a solution that they fail to consider the effect that a design might have on potential users and their ability to perform necessary tasks. Even the most highly automated system, such as NASA's Mars Rovers, requires a human operator to perform tasks such as starting, stopping, monitoring, and maintaining the system. Responsible designers consider all of the potential users in the life cycle of a product or system and the tasks that those users may need to perform. For example, cars are primarily designed for drivers to accomplish the task of getting from one place to another. However, the design should also account for other users, including the manufacturing assembly worker, the maintenance mechanic, the pedestrian who needs to detect the vehicle from a distance, and the used-parts operator who must eventually strip and recycle the vehicle. The overall design of the vehicle can affect the way each of these users performs his or her tasks. The discipline of human factors engineering and ergonomics is devoted to the study of the capabilities and limitations of human users so that designs can meet requirements while preserving safety and public welfare [3].

Identify effects on the environment

All engineering works consume resources and affect their environment, which may be the natural environment, but it could also be an organizational or corporate environment. Predicting the environmental effects of a design can be very challenging—especially if the technology is not well understood. For example, designers of refrigeration units in the 1930s did not predict that chlorofluorocarbons would affect the ozone layer in the atmosphere. Keeping abreast of the scientific research related to an application area is vital to responsible engineering design.

Generate multiple solutions

Generate more than one possible solution to consider. Unlike mathematical problems that have a single "correct" solution, design problems typically allow many possible solutions. Engineering analysis and decision-making techniques are then applied to select the most appropriate solution for the given conditions. The criteria that determine the best solution among those that meet the requirements and constraints are sometimes called *selection criteria.*

Select optimal solutions

Engineering design requires the use of resources to solve a problem. Those resources may be in the form of materials, procedures, or human resources such as creative energy, labour, and related services. Since engineers have a responsibility to their employers, clients, colleagues, and society, good designers look for optimal solutions—those that maximize benefits while minimizing costs. One method for considering a diversity of factors is discussed in Section 15.5.1.

Make defensible decisions Engineering design decisions must be defensible and not based solely on opinions. Ideally, an engineer should be able to defend—in a court of law, if necessary—every design decision from scientific and other perspectives. It is the engineer's responsibility to understand the relevant applied science in order to predict outcomes with a degree of certainty determined by best practice in the discipline, as described below. To ensure public safety, designs must also be tested and validated before products or constructed facilities are released for public use.

Use best practice Learn and apply best practice for arriving at solutions in your area of competence. Engineering design should be based on recognized methods, procedures, codes, and standards, as discussed in Chapters 19 and 20. Best practice evolves as applied science and associated technologies advance.

15.2.1 Design heuristics, guidelines, standards, and specifications

The goal of a design project is usually to produce a list of specifications that allow further development or manufacture of the solution. Several terms that tend to have specialized meanings when applied to design will be defined and used consistently in this chapter.

Heuristics A *heuristic* is a very general rule of thumb that can be applied in a wide variety of situations. Many design principles are really heuristics. An example of a heuristic is "make products easy to use," which is a good guiding principle, but it doesn't tell the designer how to actually find a solution that is easy to use.

Guidelines A *design guideline* provides more information than a heuristic but is still general advice. An example guideline might be "Physical characteristics of products to be operated by the general public should be designed to accommodate the 5th percentile female to the 95th percentile male." Heuristics and guidelines aid in establishing project requirements.

Standards A *standard* provides more direction than a guideline, by stating technical requirements that normally must be met for a particular industry (the standard dimensions of a credit card, for example), but a standard does not provide a complete solution. You must become familiar with the codes and standards associated with your own engineering discipline as discussed in Chapter 19.

Specifications A *specification* is a description of the technical requirements for a specific project, in sufficient detail that someone else can build or otherwise implement what has been envisioned by the designer (see Section 6.1.5).

Example 15.1
Specifications for LED power-on indicator light

You have been asked to help establish the specifications for a small power-on indicator light that is to be part of a 5 V electronic power supply. The design requirement is that the light must be bright enough in daylight to show that the power is on. Your constraints are to specify readily available components of standard value, with a further criterion of minimizing power consumption.

By investigating similar power supplies, you find that typical practice is to specify a red light-emitting diode (LED) in series with a resistor as shown in Figure 15.3. LEDs available to you are sufficiently bright when the current is approximately 20 mA, at

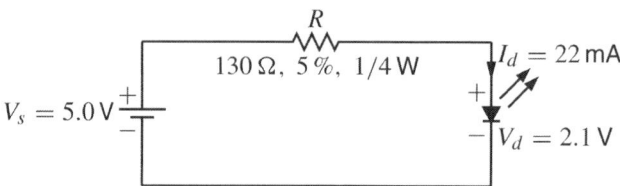

Figure 15.3 A trial design for a power-on indicator light. The resistor has a sufficient power rating and a standard resistance that results approximately in the required diode voltage and current.

which the voltage is approximately 2.1 V. Additional criteria are that the supply voltage may vary by approximately $\pm 5\,\%$ and that resistors have a tolerance rating as well as a power rating (typically $\pm 5\,\%$ and 1/4 W at reasonable cost).

Doing a preliminary analysis, you calculate the nominal resistance as $R = (5 - 2.1)$ V$/(20\,\mathrm{mA}) = 145\,\Omega$, but allowing for 5 % supply voltage reduction and from a catalogue search for standard resistor values, you refine your design to a nominal resistance of 130 Ω. A further calculation of power dissipated in the resistor and of minimum and maximum current values within the range of voltage and resistance variations leads you to the following specifications: (1) circuit configuration as shown in the diagram; (2) 130 Ω, 5 % tolerance, 1/4 W resistor; and (3) red LED as normally available.

15.3 The engineering design process

To achieve optimal solutions, experienced designers use a systematic approach to design. In this case, *systematic* refers to the fact that the general steps of the engineering design process are repeatable from one project to the next. This does not mean that designers give up creativity; rather, the process of design integrates creative activities with information gathering, engineering analysis, testing, and validation.

The basic engineering design process is based on six main activities, each focused on answering a set of questions.

Needs assessment 1. What is the problem? Whom does it affect? What are existing solutions? Are those solutions appropriate in this case, and if not, why not? What will the solution need to do (what are the requirements)? What are some of the limitations imposed on the solution (what are the constraints)? What else is also desirable (what are additional criteria)?

Synthesis 2. What ideas are there for solving the problem? Is there a creative way to address the limitations of the project in order to meet the requirements? Which solution alternatives should be given priority?

Design analysis 3. Is the design idea proposed feasible? Does the solution incorporate best practices? What is the predicted performance?

Implementation 4. How will the solution be built? Do we need to develop a prototype or simulation in order to test the solution?

Testing and validation 5. How will we objectively evaluate the solution to make sure that it is meeting, or will meet, the design requirements? How will the results be measured? What are acceptable outcomes? Are those outcomes met?

Recommendations 6. Are we ready to clearly state the design specifications for construction or manufacturing? If not, what do we need to know and do to improve the solution?

These six steps make up a design loop or cycle. They may be repeated as often as necessary. There are two main approaches to design iteration [4]: the spiral method, shown in Figure 15.4, and the waterfall method, shown in Figure 15.5.

For industries working with short development times and uncertain solutions, a *spiral approach* to design is recommended, especially for products geared for commercial use, like interactive software and home appliances, where the human–machine interface is viewed by the consumer as the product [5].

In his now-classic text, Asimow [6] points out that design gets more complete and more detailed with iteration. His experience is that at least three iterations of the design loop are required: feasibility study, preliminary design, and detailed design. As illustrated in Figure 15.4, once the design problem (DP) is understood, the feasibility study begins (D1 to D2). The goal is to answer the question "Is a solution possible?" Sketches, low-fidelity prototypes, or very basic models (P1) are used to communicate and evaluate the appropriateness of the proposed solution. During the preliminary design cycle (D2 to D3), the purpose is to establish a general plan to guide the detailed design. Critical functions are included in prototypes (P2) for testing. By the third iteration (D3 to D4), prototypes and models are fully functional (P3), and design specifications for

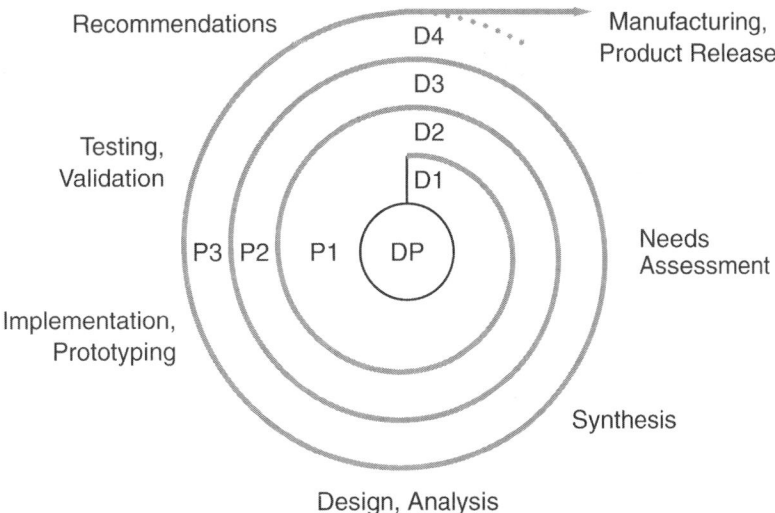

Figure 15.4 The use-centred spiral approach ensures that testing is an integral part of design development. DP represents the design problem and includes the users, the task, and associated environments that must be considered. P1, P2, and P3 represent a progression of prototypes or models used for testing. D1 starts the feasibility cycle; D2 starts the preliminary design cycle; D3 begins the detailed design cycle; and D4 represents the final design.

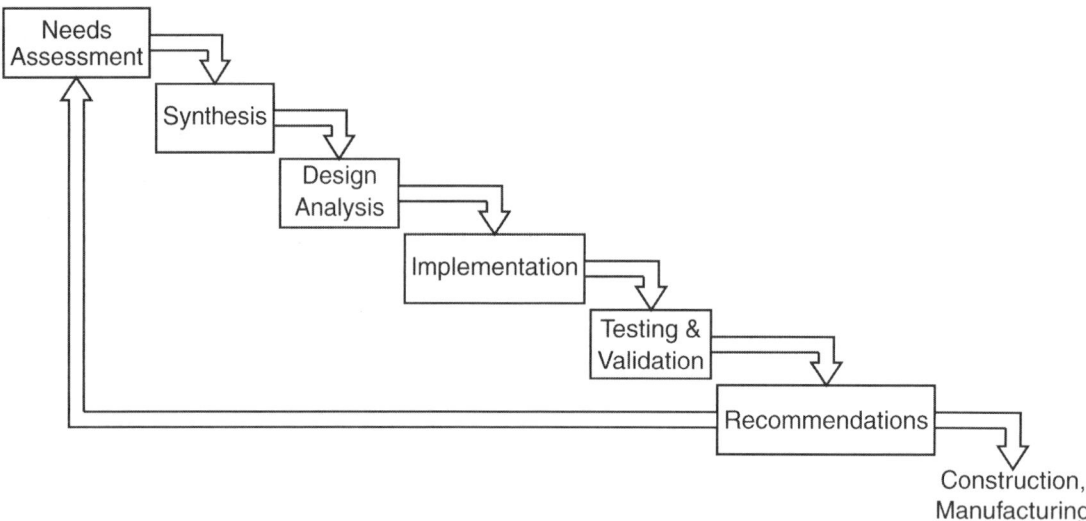

Figure 15.5 Design activities are carried out in discrete stages in a typical waterfall approach, also known as the "drawing board" approach, because designers must go "back to the drawing board" (start over) when design solutions do not work out as anticipated.

manufacturing or product installation are completed.

In a typical *waterfall approach*, the designer carries out a needs assessments and then moves through a sequence of activities, carrying out each activity once. A waterfall approach is often followed when the design solution is well understood or when developing and testing a series of prototypes or models is not practical (e.g., design of a wheelchair ramp for a new building). A disadvantage is that serious flaws in the solution may not be caught until the testing and validation phase, at which point much money and time may already have been invested. This approach is often known as the drawing board approach, because designers must go back to the drawing board (i.e., start over) when design solutions do not work as anticipated. Flaws discovered late in a design process can be many times more costly to fix than those detected in early activities [7].

Complex design projects involving many subsystems that must come together to form a functional unit, like a car, often employ combinations of the two approaches, with some subsystems designed using a waterfall approach (the exhaust system, for example) and others designed using a spiral approach (driver control panel design, for example).

15.4 **Design skills**

Perhaps the most difficult tasks associated with the design process are adequately defining the design problem, establishing requirements and constraints, generating more than one possible solution, and building appropriate models to test the design solution. The following are some tips for carrying out these tasks.

| 15.4.1 | **Clearly defining the problem** |

Whether the design project is the result of an internal company request or a response to an external customer, the start of each design project begins by trying to answer two important questions: Is there a real problem? If so, whose problem is it? [8]

Understanding the context

To answer these questions, you need to understand the context of the overall problem. You can do this through the following tasks:

- *Identify and interview users of existing solutions.* The client (the person paying for your design services) and the actual users of the existing solution may be different people.

- *Distinguish between symptoms and contributing factors to problems.* For example, the need to clean up a polluted lake may be obvious from the symptoms of dying fish and waterfowl and scum-laden banks, but it would be short-sighted to simply begin designing a new sewage treatment plant if the problem is caused by industrial waste or excess lawn fertilizer.

Stating the problem

- *State the problem in a single sentence format.* A single sentence problem statement serves as a quick summary for the reader of a technical document. It gets to the point of the design project, and it forces the designer to be focused. The following prototype statement has an acceptable format:

  Design a (type of solution; product, device, or system) to be used by (potential target users) to (carry out a particular task or set of tasks) that (meets a specific set of benchmarks).

Example 15.2
Water-purifying problem statement

Design a low-cost water-purifying system for remote areas that can be used by adults and children with limited education and will convert 2 L of ground water into drinkable water within 10 min.

| 15.4.2 | **Generating solutions** |

Designing involves study and knowledge of the stated problem, coupled with creativity, hard work, and an educated application of good design principles. Designers do not have time to wait to be inspired when there are deadlines to meet. However, choosing the first solution that comes to mind lessens the chance of arriving at an optimal solution. The word *optimal* suggests that more than one solution is feasible. When deadlines must be met, designers may need techniques for creatively generating solutions. Some typical approaches are given below.

Explaining the problem

- Explaining the problem to someone requires you to state it clearly, and this simplified statement may trigger possible solutions.

Brainstorming

- *Brainstorming* is the process of generating as many ideas as possible without evaluating them in terms of feasibility or practicality. Outrageous ideas are encouraged because they can often spark ideas that may be more meaningful. After a fixed interval of time, usually 20 to 30 min, the solutions can be ranked in order of preference. Brainstorming can be done by individuals or in groups. In a group, a "no criticizing" policy is important until all ideas are presented.

Assumption smashing

- *Assumption smashing* is the innovative challenging of assumptions that a solution has particular characteristics. Designers often have prior assumptions about the requirements and constraints that frame a project.

A major roadblock to creativity is the fear of looking foolish by suggesting an unworkable idea. In fact, most experienced designers have suggested many unworkable ideas [9]. A willingness to suggest unworkable ideas keeps design exciting. Creative activity should be fun. If you and your design teammates are not laughing during an idea-generating session, then you may be restricting your creativity and inhibiting potential innovation.

15.4.3   Building models, simulations, and prototypes

Once you have obtained one or more design solutions, you must find a way to test their feasibility. Large-scale projects usually require the design of a system—an entity with hierarchical components that interact to manipulate inputs and create outputs. The complexity of systems often requires the construction of simulation models to predict how the systems will work before building them. Simulation models contain parameters that can be varied to investigate the corresponding performance.

Types of models

Models can come in many formats. *Concrete* models are physical objects; for example, automobile designers build clay models of concept cars, and industrial designers make physical mockups of workstations out of cardboard, paper, and glue. *Abstract* models are not physically touchable. They may consist of equations, computer simulation models, or computer-assisted drawings (CAD) of machine parts, for example. *Iconic* models are visually equivalent to solutions but are non-functional. *Symbolic* models are higher abstractions of reality, such as sets of equations in which parameters may be manipulated. *Analog* models are functionally equivalent to the real object but may contain different physical quantities or conditions to be tested. *Deterministic* models have cause–effect behaviour that is predictable in principle, whereas *probabilistic* models contain inherent uncertainty. Probabilistic models yield statistics such as averages. For example, the study of rush-hour traffic patterns requires probabilistic models.

Ockham's razor

When building models, keep in mind *Ockham's razor*, which states *"Non sunt multiplicanda entia praeter necessitatem,"* or "Things should not be multiplied without good reason" [10]. In other words, focus on answering the specific questions that have been posed; thereafter, the simpler, the better.

Prototypes

A *prototype* is a physical mockup of the intended solution, of sufficient functionality to allow the design to be tested. The nature of the prototype varies depending on the purpose of the evaluation. Prototype construction has two main parameters:

Fidelity

- *Fidelity* is the accuracy with which the model represents the proposed final design, in terms of materials and construction.

Functionality
- *Functionality* defines the extent to which the model is able to perform the basic functions of the proposed final design.

Low-fidelity prototypes are used in the early design phases and are made out of cheap materials that can be easily changed or discarded. For example, in software design, blank paper and sticky notes of various sizes are often used to mock up proposed screen shots of the user interface that can be discussed with clients, members of the design team, or potential users. Simulations can be used to mock up designs for large-scale projects, such as the design of a railway locomotive or the structure of a highway system. Obviously, paper and sticky notes are not the same as a functional software application, and computer-based simulations are not the same as an actual building or highway. Low-fidelity prototypes allow designers to evaluate design alternatives before investing greatly in the actual detailed design that will lead to the final product.

Often it is with a prototype that we develop an appreciation of how the design might be manufactured, which materials should be used, and how access to components may be included for future maintenance.

15.5   Systematic decision-making

As a member of a design team in a company, you may need to justify and defend your decisions to others, including company executives. Your opinions may not carry much weight, especially if you are a junior engineer. So, how, you may ask, will you be able to convince non-engineers or more senior engineers that your decisions are justifiable?

Basic decision-making
You must first be very familiar with the context and extent of the problem as well as the state of the art of prior solutions. Information literacy and patent searches come into play. Once you have a good understanding of the context of the problem, you can then establish criteria for the decision. Here are the steps for basic decision-making:

- List all possible choices or courses of action.

- List, separately, all of the possible factors or criteria that could affect the decision.

- Compare the two lists (courses of action and criteria). Many choices are obviously impractical, so remove these from the list.

- List the advantages and disadvantages for each choice (or course of action). All of the criteria should appear in the list. The best decision is usually obvious at this point as the choice with the most advantages and the fewest disadvantages.

If the best alternatives are truly equal, a solution can be selected at random or additional criteria can be applied, but what is to be done if the best choice is not obvious? One method is to apply the computational decision-making technique described next. In a report, the resulting table is useful for showing the factors considered in choosing the best alternative, even if the decision was initially reached in some other way.

| 15.5.1 | **Computational decision-making** |

The tabular decision-making method to be described finds the best choice or choices among several known design alternatives by maximizing or minimizing a quantitative function. The method applies in business, for example, where engineers are often asked to develop solutions that maximize net profit. Decision-making of this kind, when done systematically, requires a thorough comparative evaluation of alternative solutions.

A numerical payoff value is calculated for each alternative, and the alternative with the largest payoff is chosen. A cost function to be minimized can be calculated instead, but these two methods are equivalent in any context where a positive cost is a negative payoff. A general method will be described, followed by an example.

Suppose there are m alternative solutions. Then the payoff values denoted by f_j must be calculated for $j = 1, 2, \ldots m$.

Suppose further that there are n selection criteria for judging the alternatives. In the design context, each of the alternative solutions satisfies the design requirements and constraints, so only the selection criteria that further distinguish between solutions need be included. Each of the n criteria is assigned a relative importance, or weight, w_i, for $i = 1, \ldots n$, such that sum of the weights is 1, or 100 %.

A set of mn ratings r_{ij} are defined so that r_{ij} estimates how well alternative j meets criterion i, for $i = 1, \ldots n$ and $j = 1, \ldots m$. The maximum rating is exactly 1. Then the total payoff for alternative j will be the sum

$$f_j = \sum_{i=1}^{n} r_{ij}\, w_i. \tag{15.1}$$

A variation of this method that may be simpler to interpret is to rate the alternatives with respect to a criterion using an arbitrary numerical scale, and to divide by the largest value to obtain a unitless rating. Thus, for each criterion i, assign coefficients c_{ij}, $j = 1, \ldots m$, and define C_i to be the maximum of their absolute values; that is, $C_i = \max\{|c_{i1}|, |c_{i2}|, \ldots |c_{im}|\}$. Then $r_{ij} = c_{ij}/C_i$. Other rating variations are possible.

A table is prepared, showing the alternatives, criteria, ratings, and payoff values, as illustrated in Example 15.3 below.

The difficulties associated with this quantitative method are in determining reasonable weights and precise coefficients. The reasons for the choice of numerical values should be carefully recorded. A decision that stands up to changes of ratings and weights is said to be *robust*. The robustness of a solution can be checked by using a spreadsheet program to construct the table. You can then easily adjust the weights and ratings or add additional criteria to see how they affect the outcome.

Computational decision-making is not a substitute for testing and validation of working solutions. However, even if confidence in the numerical values is lacking, this formal method requires all of the relevant factors to be clearly stated and carefully evaluated.

Example 15.3
Student travel

A student lives approximately 2 km from the university, and must decide which mode of travelling this distance is best. There is no bus service, so the alternatives are walking and riding a bicycle. However, buying a motorcycle or automobile might be considered if these options are not too expensive. The student applies the tabular decision-making process. The criteria are operating cost, time, and safety, which are weighted 30 %, 40 %, 30 %, respectively; thus time is slightly more important than the other two criteria. There are $m = 4$ alternatives and $n = 3$ criteria, so that $mn = 12$ rating coefficients must be defined. Table 15.1 summarizes the computations. The entries in the table are obtained as follows:

- Operating cost is in the first numerical row: walking has zero cost. A car is estimated to cost $1000, a motorcycle $500, and a bicycle $100, all used and obtained at bargain prices. The denominator of the ratios $r_{1j} = c_{1j}/C_1$ is therefore $C_1 = 1000$.

- Time is in the next row. Walking takes 35 min, the bicycle takes 20 min, and the car and motorcycle each take 8 min.

- A somewhat arbitrary safety estimate is used. Walking is neutral, but the motorcycle is judged to be five times as dangerous as the bicycle, with the car somewhere in between.

- The partial scores, which are the products of the ratings r_{ij} times the weights w_i, are computed in the s_{ij} (score) columns. The sums f_j of the partial scores are entered in the bottom row.

Table 15.1 shows that the least-cost alternative for travel to the university is the bicycle. This result may change if any of the numerical values change; for example, the safety of the bicycle in winter weather may not be as favourable.

Table 15.1   Cost computation for the student-travel problem. The four alternatives have been compared using three criteria: operating cost, travel time, and safety. The ratings are $r_{ij} = c_{ij}/C_i$ and the partial scores are $s_{ij} = r_{ij}w_i$. The values in the grey areas of the table are computed automatically after the other values are entered. With the weights and ratings shown, the bicycle costs the least for travel to the university.

| | | Alternatives ($m = 4$) | | | | | | | | | | | |
| | | Walking | | | Bicycle | | | Motorcycle | | | Car | | |
Criteria ($n = 3$)	w_i (%)	c_{i1}	r_{i1}	s_{i1}	c_{i2}	r_{i2}	s_{i2}	c_{i3}	r_{i3}	s_{i3}	c_{i4}	r_{i4}	s_{i4}
Cost ($)	30	0	0.00	0.0	100	0.10	3.0	500	0.50	15.0	1000	1.00	30.0
Time (min)	40	35	1.00	40.0	20	0.57	22.9	8	0.23	9.1	8	0.23	9.1
Safety	30	0	0.00	0.0	1	0.20	6.0	5	1.00	30.0	2	0.40	12.0
Totals f_j				40.0			31.9			54.1			51.1

15.6   A tale of two design solutions

Junior engineers are often assigned very specific design projects with well-defined requirements, constraints, and other criteria. However, both students and engineers sometimes confront design problems that are less well defined, and part of their task is to refine the design constraints. The following case study describes an actual design problem given to engineering students.

15.6.1 The requirement

Students were given the following design challenge: "Design a physical device, or a system with a physical component, to assist students with an everyday task during an electrical power blackout." They were required to work in groups of five or six, and were allowed 10 weeks to follow a spiral design approach and implement a functional prototype. The final prototype did not have to be in a commercially viable form, but it did have to demonstrate all critical functions. Further constraints placed on the project included specified milestone dates for low-fidelity, medium-fidelity, and functional prototypes, and a $40 budget for the entire project.

When given a problem such as this, novice designers often make the mistake of identifying the form of solution first (solar power, for example) before performing a thorough needs assessment. This approach results in the designer trying to find a task to fit the solution rather than following a proper engineering process to find a useful solution to the client's problem.

15.6.2 Feasibility cycle

To begin the design process, the teams obtained a better understanding of the situation and scope of the problem by defining a particular task and associated functions, user group, and environments of operation. Many of the students had recently experienced a blackout, so were able to identify everyday tasks that were affected by a lack of power. To ensure that they were not limiting their needs assessment, teams interviewed other students to find out which tasks should have high priority for students living on- or off-campus. Typical needs included a lack of light during the evening for studying and travelling around campus, a need to heat or cool food, and a need for alarm clocks to work. All teams conducted patent searches and literature reviews to better understand existing solutions and relevant applied science.

Problem statements We will follow the solutions derived by two different teams: team A and team B. The problem statements put forward by each team are given below.

Team A Design a reliable, portable, light-providing device that allows students (ages 6 years and older) to safely study in groups (e.g., read and write) for a minimum of 8 h.

Team B Design a physical device (or system with physical components) that safely assists a university student to wake up effectively in a dormitory room during a blackout.

All teams carried out brainstorming sessions to generate at least three potential solutions to their design problem. They then evaluated each solution by considering the available technology and applied science before selecting one solution to pursue. Team A considered various light-emitting components and sources of power. Team B considered existing alarm systems and timers for incorporation into a pillow.

During this feasibility stage, the teams created detailed sketches and models, similar to those shown in Figure 15.6, to describe the solution they were planning to pursue. All

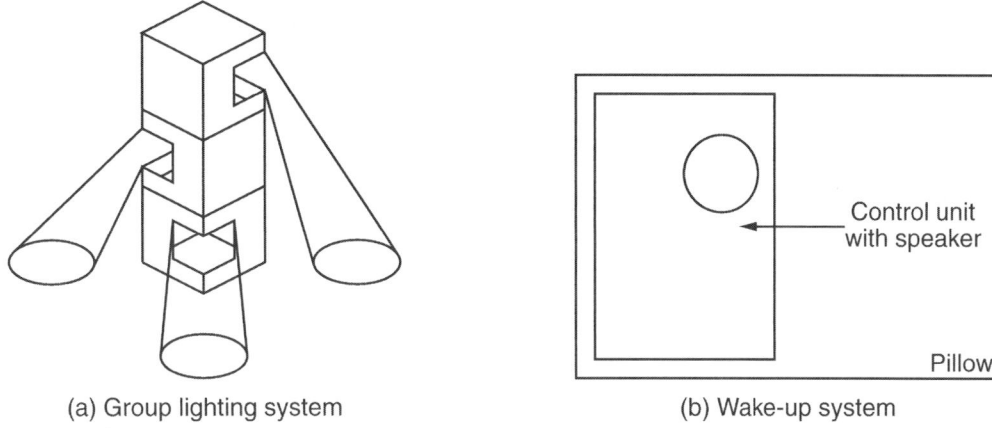

| (a) Group lighting system | (b) Wake-up system |

Figure 15.6   Two very different proposed solutions to the initial design problem.

teams conducted design walk-throughs with their sketches or models. A design walk-through involves asking potential users to carry out or imagine carrying out critical tasks with the proposed solution and to comment.

Feasibility lessons learned, team A   Team A learned that their proposed design solution seemed feasible if they incorporated white LEDs with reflectors, since the power source required by LEDs would be expected to last for several hours. The team kept open two options for powering the LEDs: solar cell technology and alkaline batteries. Their design process recommendations were to continue to explore power options as well as the configuration of the reflector box during the preliminary design phase.

Feasibility lessons learned, team B   Team B learned that users had difficulty understanding how the pillow wake-up system would work. The controls, placed inside a pocket in the pillow, were not obvious or intuitive, and users questioned whether a pillow containing the device would be comfortable for sleeping. Revising their needs assessment, team B concluded that the real problem was the lack of an alarm as the result of a power failure while the student is asleep. If the power fails before the user goes to sleep, then alternative wake-up arrangements can be made (set the alarm on a clock or have a friend call, for example). Potential users also felt that the inconvenience of being awakened by a device that is set once is less than the inconvenience of setting a device every night. This led team B to refine their problem statement as follows:

> Design a system to operate on regular household current that will produce an audible alarm to awaken students within the room in the event of a power failure.

15.6.3   Preliminary and detailed design cycles

The teams followed two more iterations of the design cycle of Figure 15.4. They were responsible for planning and conducting validation experiments to determine the performance of their solution. Design detail and functional performance improved with each iteration, and the teams met the requirements they had established during their feasibility studies.

15.6.4   Final design solutions

Final design, team A   Team A refined their original idea by modifying the shape of the light box so that four units could easily fit together to shine light on a greater work surface, and a glow-in-the-dark feature was added to the power switch. Incorporated into the clear plastic outer shell was a holder for pens or pencils. The use of basic materials and construction techniques kept expenses to a minimum.

Final design, team B   Team B's final solution incorporated an audible alarm with volume of 90 dB, together with a flashlight bulb, to be activated when the house current failed. A switch would turn the unit completely off for removal from the power socket or to change the battery. The flashlight feature was added after a design walk-through revealed that a user, awakened in the dark, would want to be able to find the unit to shut it off.

Lessons learned   Both teams benefited from following a spiral design approach, with team B benefiting the most. Team A's original idea proved to be feasible but needed refinement. If team B had followed a waterfall approach, considerable time and resources would have been spent attempting to implement the original alarm-in-pillow solution. For example, with the proximity of an audible alarm so close to the user's head, determining acceptable decibel levels that would wake sound sleepers without damaging hearing would have been challenging and time-consuming. Having at least two other solutions to return to enabled team B to quickly refocus their design efforts, to develop a solution that safely met the functional requirements of a wake-up system, and to meet project deadlines.

15.7   Design documentation

Deliverables and milestones   Design *deliverables* are those artifacts that the designer must produce to give to someone else so that the design can move from an abstract idea to a useful product or system. Deliverables have due dates associated with them, often referred to as project *milestones*. As a member of a design team, you might be responsible for delivering the design specifications for a component or subsystem of an overall design. You might be responsible for building a concept model that can be used for simulation. You might be responsible for creating the evaluation plan for testing the design against established requirements and benchmarks. You might be the project lead responsible for setting the project schedule and milestone dates. The majority of deliverables are in the form of written reports. The following are the types of reports most often associated with engineering design projects. Each of these reports may be subject to approval by the client or, at least, to internal review by members of the design team.

Design proposals   Design projects often begin with a "request for proposal" (RFP) document sent from a potential client, who may be external to your company or another group within it. The client describes the situation of concern (the general problem area, also called *client* *requirements*) and may place constraints on the project itself or potential solutions. For example, the clients usually specify completion deadlines and an upper limit for costs. The client may also specify the general type of solution sought. The design engineer must create a proposal that is then sent back to the client. RFPs are often sent to more than one engineering group; the client decides who will be awarded the work on the basis of the design proposals received. Design proposals include customer requirements, operational concept descriptions, a project plan with milestones, and a budget.

Customer requirements

Customer requirements may be part of the design proposal or may be separate, but must have the clear approval of the client. The requirement statement summarizes the situation of concern from the viewpoint of the client and outlines the conditions that must be met to address the client's design problem.

Operational concept description

A design proposal often includes a high-level description of the designer's early operational concepts, including appearance of the proposed solution, how it will work, and how it will address the problems identified in the request for proposal. It is often worded in a narrative style.

Example 15.4
Voice-based chat service

The operational concept description included in a design proposal for a voice-based online chat service might read as follows:

> The user will be able to chat with other users of the system in real time while browsing the Internet or performing other computer-based tasks. The system will include functions similar to a hands-free phone and will take control commands from a voice-based interface that accepts input through a standard computer microphone attachment. Interface commands and conversational switching will be handled by dedicated application software. Voiced responses from the system and other parties to a conversation will be through standard speakers either integral to the computer or as accessories.

Project plan and milestones

All design projects need detailed schedules describing the steps that will be taken to satisfy the client. Project milestones, or control points, are deadlines incorporated into the schedule to ensure that the overall delivery date is met. Projects that are not kept on schedule can lead to high costs or lost opportunities for other work. The client has requested the design solution for a reason, and invariably there are others waiting for that solution in order to do their jobs. A rule of thumb is to estimate the time required to do the task without interruptions and delays, and then multiply by a factor of 3.

Project budget

All design projects cost money and other resources. With experience, design engineers can estimate the number of people required for a design project, the cost of their services in terms of salary or labour, and the other resources that must be acquired or purchased. Carefully consider the human resources and skills needed to carry out the project. Adding in people at the last minute is expensive because new team members need time to become productive. In *The Mythical Man-Month* [11], Brooks makes a very good case for why late addition of personnel can be very costly. Wise designers include a contingency of 10 % or more in the total resource budget to deal with the unexpected.

Functional specifications
The functional specification document is a "design-to" document. It must clearly state the functions that the design solution must be able to perform to meet the client's needs. Functional requirements are usually quantitative, in contrast with the often more qualitative customer requirements, and are listed in terms of priority with critical functions stated first. This helps the design team determine which features may be dropped from the design if budgets and deadlines become an issue.

Design specifications
The design specification is a "build-to" document, listing the components chosen to satisfy the functional specifications and their quantitative properties. Design specifications are essentially the detailed blueprints for the design solution. The design team keeps a working document of requirements and specifications outlined in the design proposal. As the design progresses, the design specifications are updated. The design specifications document must be presented in a format that allows someone else who is knowledgeable in the field to fabricate the design as intended. Each design discipline has its own way of presenting its design blueprints or diagrams. For example, mechanical engineering designs use CAD drawings to show 3D views of components, electrical engineering designs use circuit drawings, and software engineering designs may use entity-relationship diagrams.

Test and validation plan
Along with design specifications, a detailed test plan should be kept. The test plan outlines how the design will be objectively evaluated to ensure that it meets or surpasses the quantitative benchmarks in the design requirements. Tests may be conducted on components as well as on the complete design, and testing must be an integral part of the design process.

Design reviews
At the completion of milestones or the complete project, projects are often subject to external review by clients or superiors in the company. A progress report is given to the reviewers, often accompanied by formal presentations, listing what has been done, the time and budget status of the project, and any known or anticipated obstacles. Questions are answered, and a design review summary is written to summarize the proceedings. As the result of a design review, the client may decide to discontinue the project.

Design logbook
Engineering design logbooks should be kept for even the most basic design project. This is your record of your contributions to the project. It should include your ideas, analysis, research, sketches, reflections on ideas, and your own notes on team meetings and action plans. It takes time to develop the skills associated with accurate and detailed record-keeping—so start early. Get in the habit of keeping a logbook with you because you never know when you will get an inspiration for a design project, or when a design idea may prove to be the seed of an innovation. Figure 15.7 illustrates acceptable page formats for engineering design logbooks.

Progress reports
Progress (interim) reports are delivered to the client at agreed milestone dates. They outline the progress of the project to date, any obstacles that were not anticipated at the

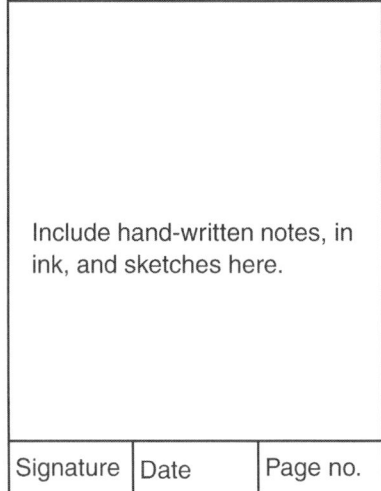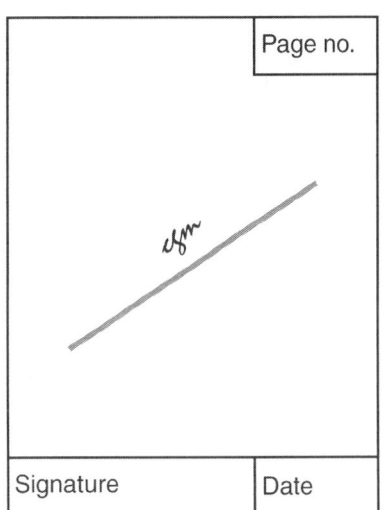

Figure 15.7   Each page of a design logbook must be numbered, signed, and dated. All work should be done in ink. Use a consistent format throughout. Blank space or missed pages should be crossed through and initialled. Either of the above formats is generally acceptable.

time of the design proposal, and the recommendations the design team has for addressing such obstacles (see also page 89).

Final reports   Final reports are the records of the design process and outcome. They should include an overview of the situation of concern (client requirements) as well as an analysis of the relevant research literature and state of the art. This essentially puts the design problem into perspective for the reader and explains the project significance, the alternatives considered, the rationale for the final design decisions, and the final realization of the design. Depending on the size of the project, all of the design specifications may be included in the report. For large-scale projects, the final report may refer directly to separate specification documents. Final reports contain recommendations for future design projects, either of anticipated additions for next versions or of related applications of the design solution. Revisit Chapter 8 for tips on writing technical reports.

15.8   System life cycle

Engineering design should take into account the full life cycle of the product or system, from initial identification of a need through concept development, detailed design, operation and maintenance, and, finally, system retirement [12]. A system life cycle is influenced not only by engineering design decisions but also by competing technologies and the economic marketplace, as well as the political climate and public support. Figure 15.8 shows a typical design life cycle incorporating the three main cycles of the design process discussed in Section 15.3.

Production and deployment   At this point, the design is manufactured or constructed, or a system is implemented. For products or components that are to be mass-produced, the design team may elect to

Feasibility study	Preliminary design	Detailed design	Production & deployment	Operation & maintenance	System retirement

Figure 15.8 A typical design life cycle spans the time a need is identified until the system is retired due to wear or obsolescence. If a need for a new or revised solution exists when the system is retired, then a new system life cycle begins. Grey areas represent the common iterative phases of the engineering design process for product development, discussed in Section 15.3.

include a preproduction phase, or a "beta" version of the product may be tested in the field to correct any design issues prior to full production.

Operation and maintenance During this phase, designers can learn lessons about the design, in terms of what has worked well and what may need to be changed in the next version. This is a common starting point for redesign projects.

System retirement When the system or product is worn out or outdated, it may need to be replaced with a new or redesigned product. This is not a trivial exercise. Engineers responsible for decommissioning a product must be aware of environmental regulations for recycling or product or component disposal.

15.9 Organizing effective design teams

Design is not just a creative process; it is a social process [13]. This may seem like an odd statement, until we consider the role of the engineer in society and the interactions that a designer must have with others, including clients, employers, design team members, and product users. In fact, the ability to communicate and work with others may be as valuable to the professional engineer as the technical skills acquired through an engineering education.

A recent survey by Engineers Canada focused on the engineering practices of its members since 1999. At least 80 % of the respondents indicated that they have been involved in job-related teamwork [14]. This is not surprising given that much of today's technology is too complex to be designed by a single person. Even the Wright brothers did not work alone on the design and construction of the Wright Flyer that flew at Kitty Hawk. Many employers want to hire engineering graduates who can bring to the job strong skills in engineering design, communication, ethics and professionalism, project management, team building, and teamwork.

Concurrent engineering In competitive markets where product design and delivery must be done very efficiently, many companies are adopting concurrent engineering practices, which require the separate design-cycle activities to be performed simultaneously by groups under the direction of a lead group. The lead group is likely to have members who are responsible for conceptual design, marketing, manufacturing and assembly, packaging, and customer or technical support. Efficient communication between groups increases the likelihood of producing a high-quality product on time and at affordable cost. Concurrent engineering requires individuals who are flexible thinkers, who can easily adapt to change, and who are able to work well in multidisciplinary environments.

Design can be an enjoyable process, and membership in a design team that communicates well and works cooperatively can be a very rewarding experience. Experienced designers will tell you that the pressure of budgets and deadlines can quickly take the fun out of a design project if the team does not work at communicating well. Competencies associated with teamwork include [15]

- clearly defining the team's project goals,
- establishing and performing allocated tasks,
- creating and maintaining a supportive team culture,
- planning and managing time effectively,
- implementing operating procedures to ensure effective team interactions, and
- establishing incentives and rewards for team and individual achievement.

From a client's perspective, the success of a design team is measured not just by the ability to meet project requirements, but also by the ability to meet deadlines and clearly communicate the results. From a team perspective, success is also measured by the ability of the team members to cooperate, support one another, learn, and grow as individuals to better undertake the next project.

15.9.1   Becoming a design engineer

"Design can be easy and difficult at the same time, but in the end, it is mostly difficult" [16].

Everyone has a creative instinct; however, it is stronger in some people than in others. What is the key characteristic that makes a person creative? In his excellent inspirational book on creativity, Roger von Oech [9] cited a psychological study carried out to identify the key characteristics of creative people. The study, which was commissioned by a major oil company and conducted by qualified psychologists, came to the simple conclusion that creative people thought that they were creative, and less creative people did not think they were creative. The key to creativity is the desire and conviction that you are creative, but you will need to practise to develop your skills. Your degree program contains opportunity for creativity in design projects. Some further ideas for design practice are found in the questions in Section 15.10 below.

15.10   Further study

1.   Choose the best answer for each of the following questions.

(a) What are the main fundamentals of engineering design?

    i.   applied science and common sense

    ii. mathematics, science, and creativity

    iii. engineering science and constraints

    iv. creativity and common sense

(b) What are the main types of engineering design?

    i. reverse engineering and competitive analysis

    ii. redesign and benchmarking

    iii. evolutionary design and innovation

    iv. innovation and invention

(c) When should designers consider the needs of human users and the environment?

    i. only when the solution involves human users

    ii. as the design problem is being defined

    iii. once a functional solution is available for testing

    iv. throughout the design life cycle, since all solutions involve human users

(d) Why should requirements be quantitative?

    i. They contribute to defensible decisions about whether they have been met.

    ii. It is required by law.

    iii. They are used to set performance benchmarks for the solution.

    iv. They are required in choosing the best solution.

(e) Which are the main activities in a basic engineering design process?

    i. needs assessment, idea generation, implementation, testing documentation

    ii. problem identification, analysis, synthesis, implementation

    iii. needs assessment, synthesis, analysis, implementation, testing recommendations

    iv. problem identification, analysis, implementation, testing, and validation

(f) Which design approach is best suited for complex systems?

    i. the waterfall approach

    ii. the spiral approach

    iii. reverse engineering

    iv. combination of waterfall and spiral depending on the components

(g) How should Ockham's razor apply to prototypes and models?

    i. to keep the design simple for implementation

    ii. to build models to answer specific questions

    iii. to reduce costs

    iv. to make sure that the solution is as simple as possible

(h) Why is it important to optimize design solutions?

    i. to get the best performance possible

    ii. to select the correct solution

    iii. to minimize project time

    iv. to maximize benefits and minimize costs

(i) Who prepares a request for proposal?

    i. the client

    ii. the senior design engineer

    iii. the project manager

    iv. a technical documentation expert

(j) Why is engineering design a social process?

    i. it is often carried out in teams,

    ii. it involves interaction and communication with others,

    iii. it must be carried out in the best interest of public welfare,

    iv. all of the above.

2.   Practise problem definition by choosing a broad area in which you are interested. For example, you might choose robotics, water resource management, or biomedical engineering. Explain a situation of concern related to your chosen area and be sure to reference appropriate sources from the applied science literature. Narrow down the problem so that it can be clearly defined. *Tip:* Good designers develop excellent information literacy skills in order to know what has already been done and to identify leading-edge solutions for an area of engineering. Your university or college should have librarians specializing in engineering and science literature who can focus your literature searches so that you find the material you need for your design projects (and their services are often free of charge).

3.   The intent of the engineering design process is to create a device, material, process, or system that helps others carry out activities. While entertaining, design solutions that follow principles of *chindogu* (the Japanese term for "unuseless" design) [17] would be considered a wasteful use of resources if they were allowed to proceed in a commercial or industrial setting. To practise your creativity skills, find some examples of *chindogu* (the Internet is a good place to look). Identify the problem that the designer was trying to solve, and comment on why the solution is "unuseless." Brainstorm possible alternative solutions. Create a low-fidelity prototype and carry out a design walk-through with at least two other people to test the usefulness of your solution.

4.   A good way to find design tips is to study examples of bad designs. Michael Darnell has developed a web site devoted to examples of bad or hard to use designs. Visit Darnell's site <http://www.baddesigns.com> and make a list of the main factors that tend to contribute to poor solutions.

5.   Visit the web site for the next Canadian Engineering Design Competition categories. Brainstorm at least five ideas for the entrepreneurial or corporate design category. Set up a computational decision-making chart to help you identify a design project for entry in your provincial-level competition.

| 15.11 | **References** |

[1] Canadian Engineering Accreditation Board, *Accreditation Criteria and Procedures*. Ottawa: Engineers Canada, 2007. <http://www.engineerscanada.ca/e/files/report_ceab.pdf> (March 9, 2008). Pre-publication data at <http://www.engineerscanada.ca/e/acc_programs_2.cfm> (September 10, 2008).

[2] H. Petroski, *Invention by Design*. Cambridge, MA: Harvard University Press, 1996.

[3] C. D. Wickens, S. E. Gordon, and Y. Liu, *An Introduction to Human Factors Engineering*. New York: Addison-Wesley, 1998.

[4] K. Otto and K. Wood, *Product Design*. New York: Prentice Hall, 1993.

[5] J. Raskin, *The Humane Interface*. New York: Pearson Education, 2000.

[6] M. Asimow, *Introduction to Design*. New York: Prentice Hall, 1962.

[7] B. W. Boehm, "A spiral model of software development and enhancement," *IEEE Computer*, vol. 21, no. 5, pp. 61–72, 1988.

[8] D. C. Gause and G. M. Weinberg, *Are Your Lights On?* New York: Dorset House Publishing, 1990.

[9] R. von Oech, *A Whack on the Side of the Head; How to Unlock Your Mind for Innovation*. New York: Warner Books, 1983.

[10] Encyclopedia Brittanica, "Ockham's razor," in *Encyclopedia Brittanica Online*, Chicago: Encyclopedia Britannica, Inc., 2004. <http://www.britannica.com> (March 9, 2008).

[11] F. P. Brooks, Jr., *The Mythical Man-Month*. Reading, MA: Addison-Wesley, anniversary ed., 1995.

[12] A. Chapanis, *Human Factors in Systems Engineering*. New York: John Wiley & Sons, Inc., 1996.

[13] K. Smith, *Teamwork and Project Management*. Toronto: McGraw-Hill Ryerson, second ed., 2003.

[14] M. Mastromatteo, "Team player," *Engineering Dimensions*, vol. 25, no. 3, pp. 38–40, 2004.

[15] K. L. Gentili, J. F. McCauley, R. K. Christianson, D. C. Davis, M. S. Trevisan, D. E. Calkins, and M. D. Cook, "Assessing students' design capabilities in an introductory design class," in *Frontiers in Education Conference*, vol. 3, pp. 13b1/8–13b1/13, San Juan, Puerto Rico: Institute of Electrical and Electronics Engineers, 1999.

[16] H. Petroski, *Small Things Considered: Why There Is No Perfect Design*. New York: Vintage Books, 2004. page 128.

[17] K. Kawakami and D. Papai, *101 Unuseless Japanese Inventions: The Art of Chindogu*. New York: W. W. Norton & Company, first American ed., 1995.

Chapter 16

The Engineer in Business

Most practising engineers are employees of businesses. Some engineers are partners with other professionals, and some are self-employed. Figure 16.1, although an exaggeration, shows one difference between a professional, such as an engineer, and a company manager. The important fact is that most engineering graduates and many undergraduates work for businesses permanently or as interns, and should have a basic understanding of businesses and how to get a job in one. Therefore, this chapter begins with an overview of the employment process, continues with a description of company structures, and ends with issues involved in starting a small company.

In this chapter, you will learn

- how to prepare a personal résumé,
- how to prepare for an employment interview,
- the basic forms of business organizations,
- the role of company officers and of individuals,
- an overview of the requirements for a business plan.

Businesses, like the professions, are the subject of academic study and training as well as of much practical advice. Bookstores have entire sections containing self-help books offering advice about how to succeed in the company milieu, how to manage others, and how to organize your personal time. Two of the many possibilities are reference [1], which is written specifically for engineering students, and reference [2], which takes a direct approach to survival in a competitive corporate culture.

The Internet is an increasingly important source of information from governments, which offer advice about starting and running businesses, from legal and consulting

DILBERT

Figure 16.1   The engineer in business. This cartoon oversimplifies the requirements for business leadership but contains a grain of truth: there are no formal qualifications required of company management (Dilbert reprinted by permission of United Feature Syndicate, Inc.).

firms, which offer their services, and from the promotional material of businesses themselves.

<table>
<tr><td>**16.1**</td><td>**Seeking a job: résumés and interviews**</td></tr>
</table>

All universities have placement departments for assisting graduates and undergraduate co-op students or interns to find suitable jobs; your career-service office is a source of literature, advice, and knowledge of job opportunities in your region.

Personal contact is often cited as an important means of obtaining a long-term job. You can meet potential employers by attending activities of the local chapter of the engineering Association or of a technical society (the IEEE, ASME, and other discipline-specific societies mentioned in Chapter 4). Many regions are home to industry-specific associations of mining, petroleum, "high-tech," or other companies, and students are often welcome at their functions.

Although undergraduate job placements and graduate career opportunities are both subject to the vagaries of the Canadian economy and outside forces, the possibilities for engineering students and graduates have been very favourable over recent decades. Furthermore, numerous long-established companies anticipate many retirements in the early part of the next decade, and are actively seeking ways to interest engineering students in their companies.

You will probably seek several jobs over your career, perhaps as a student, but certainly as a graduate. The following sections provide some hints to aid you when you look for a job. Emphasis will be placed on the circumstances of students who are seeking their first jobs, such as internships or co-op placements.

Résumés   A résumé is a document that announces your availability and summarizes your suitability for a job. Sometimes if a company is bidding for a contract, the résumés of engineers within the company are submitted to show that qualified persons are available, but the most common need for a résumé occurs in the context of finding a job or changing jobs.

Potential employers often have many résumés to sift through in order to select candidates to be interviewed. The purpose of your résumé, therefore, is to obtain the interview for you and to provide material for the beginning of the interview conversation. The appearance of your résumé is therefore very important: it should be very clear, professional, and normally not longer than one page or, at most, two pages. Simplicity and elegance count: a document that has been embellished with fancy fonts or colours will not impress.

Your résumé will change as you progress through your undergraduate studies and obtain employment as an engineering graduate. Many people change jobs several times during their professional careers. Therefore, plan to update your résumé regularly, even if you are not seeking a new job. The update can provide the occasion for reflection and satisfaction as you catalogue your accomplishments.

When you begin to prepare your résumé, ask yourself two questions: who will read it, and what should it include? It is best to customize your résumé to each company that you contact, if possible. Suppose you know that a certain company wishes to fill several job positions for which you believe yourself to be qualified. You will wish to emphasize on your résumé how your experience and interests are related to the position you seek. To help with the content, make a list of the following, not all of which will necessarily be included:

- your education, including as appropriate, that you are a second-year (for example) student in the civil engineering (for example) degree program at your university;
- any classes completed that are relevant to the job;
- academic grade average. If it is very low, do not include it but be prepared to discuss it in an interview;
- evaluations received on any previous job experience such as a co-op job;
- projects in which you have participated;
- work experience: dates, company, job title, and responsibilities;
- honours, awards, and scholarships;
- memberships in relevant organizations, such as student membership in your provincial Association or in technical societies such as mentioned in Chapter 4;
- technical skills (programming languages, for example) and certificates;
- leadership and extracurricular activities;
- community service that may demonstrate leadership, responsibility, or hard work;
- interests and hobbies.

A first-year student seeking a first co-op job might not have job experience, and so would include relevant information from high school, but the farther along you are in your undergraduate career, the more selective you should be about including pre-university material. If you were a competitive swimmer at the provincial level, for example, you might keep its mention in your résumé to show your perseverance and hard work; otherwise, recent experience is the most important.

Figure 16.2 shows a possible format and some of the content of the résumé of a student in the early years of a degree program. The résumé should be tailored as much as possible to the company and job that you are seeking. Personal information such as age, height, marital status, health, or personal background such as religion should not be included.

The "Summary of qualifications" item shown in the figure need not be included for a single-page résumé. References, the names of persons who have agreed to recommend you to prospective employers, are not normally given. If asked in an interview, however, it is always desirable to have the names and contact information of referees available. Be sure to ask the referees in advance if they would be willing to be contacted.

In describing your work experience and other activities, use action verbs such as *designed* or *researched*, and be as specific as you can about the details.

Interviews A successful interview may get you a job offer but, in the case of a permanent job, it may only be the first step to further interviews or a probationary appointment. There are two cardinal rules: be yourself, and be prepared.

Terri Sanko

928 King Street East   (519) 888-4567
Peterman ON, N2L 3G1   T.Sanko@svcprovider.ca

Summary of qualifications
    Summarize the points most relevant to the position sought . . .

Work experience
- Parts manager assistant, Qualcan Wholesale, May–August 2008: inventoried and assigned company part numbers for incoming stock
- IT assistant, Pilz Pharmaceuticals, June–August 2007: designed company web pages and recommended software purchases

Education
- Candidate for B.Eng. in Civil Engineering, Lookout University (completing second year May 2009)
- Cumulative average: 3.6/4.5
- Courses related to the position sought . . .
- Projects or laboratories related to the position sought . . .

Awards
- Lookout University entrance scholarship

Volunteer activities
- Counsellor-in-training, Trillium Sailing School, May–August, 2006: taught sailing skills, monitored student safety

Memberships
- Student Member of Professional Engineers Ontario and Canadian Society for Civil Engineering

Language fluency
    English, French

Interests
- Competitive curling (reached provincial junior playdowns)

Figure 16.2   Skeleton of a typical chronological résumé. Some of the items shown are examples; others indicate the type of information that may be included. The farther along in the degree program, the less high-school activities are mentioned. Work experience is normally given in reverse chronological order.

Before you attend the interview, find out as much as you can about the company, the job, and, if possible, the people with whom you hope to work. Look the company up on the Internet and read any publications you can find about the company's goals, products or services, and plans.

Begin the interview by shaking the interviewer's hand firmly but briefly, and make eye contact as you do so. If you have no special skills or experience relevant to the job, remember that an eagerness to learn, an ability to work, and the social skills to work well with others may be what the interviewer is seeking. Make a list of questions you will need the answers to in order to accept an offer. You have the right to know what to expect. It is appropriate to inquire about salary, for example, but it should not be the first question you ask. At the end of the interview, briefly summarize your interest in the job and thank the interviewer for talking with you.

16.2   Business organizations

What is a business? The answers are as diverse as the locations where businesses operate and the people in them. This diversity is recognized in law, where the identity, rights, and obligations of a business are treated similarly to those of an individual.

When a business offers or provides engineering services to the public, the business must possess a certificate of authorization when required by the appropriate professional Association, as discussed in Section 2.3.4.

Business organizations can be classified in several ways, but their legal structure, described in the next section, is perhaps most basic.

16.2.1   Legal business structures

Unincorporated sole proprietorship

A person may conduct a business with a minimum of formality, buying and selling for profit on a street corner, perhaps. Many successful businesses were very small when they began. For example, university classmates William Hewlett and David Packard started the company associated with their names in a garage.

In a sole proprietorship, a person carries on business themselves: all profits and liabilities are personal and unlimited. There may be laws or city by-laws restricting or licensing the business activity or requiring its registration; otherwise, the responsible entity is the person involved. Such businesses are normally financed and sustained personally or by friends and family. No shares can be sold as for a corporation.

A personal business is generally simple to begin and to report for taxes. If the business is to have a name other than that of the person, then it must be registered with a provincial agency.

Partnership

A partnership is essentially two or more sole proprietors conducting a business together. Many professional businesses are partnerships, including legal, medical, and consulting engineering businesses. Starting such a business does not generally require government approval, and registering it is easily done for the payment of a small fee.

The advantage of a partnership is the combination of resources, ideas, and effort of the partners. Should the business fall into difficulty, however, each partner is responsible for all partnership liabilities, no matter which partner caused them.

For taxation, financial income or loss from the partnership is combined with other personal income, except that special regulations may apply when there are more than five partners.

If any partner dies or becomes bankrupt, the partnership is automatically dissolved unless a partnership agreement specifies otherwise. Any serious partnership should be based on a written agreement that specifies partner rights, obligations, share of profits or losses, and ability to dissolve the partnership.

Some partners may wish their obligations to be limited, for example, to the amount of money they contribute to the business. Such limited partnerships must be registered and must contain at least one general partner along with the limited partners. Limited partners must not manage the company, and their names must not be included in the partnership name; otherwise, they may be liable for all company debts like a general partner.

Incorporated business   An incorporated business is a legal entity that has been created by registering it under provincial or federal law. The great majority of businesses that hire engineers are incorporated: some are huge organizations registered in many countries, but a significant fraction of engineering work is performed in small companies.

An incorporated business is responsible for its own income, debts, and taxes. For most purposes, such a business has many of the characteristics of a distinct person and is separate from the persons who established it or own its shares, although in cases of fraud or tax evasion, the directors or managers may be held individually responsible for incurred obligations.

Since the business is a separate person-like entity, no one owns it directly; on incorporation, however, it issues shares, which may be owned by one person or many. A business that offers its shares to the public is called a *public corporation* and must satisfy more elaborate rules concerning company structure and the reporting of its affairs than a private company. Shares may be bought and sold in stock markets, subject to securities legislation and market regulations, including the regular disclosure of company financial performance.

The advantages of incorporating a business include the limited liability of shareholders, transferability of share ownership, potential to raise funds, and continuity of activity. Since corporations are taxed at a lower rate than individuals, incorporating a small company may provide advantages over absorbing all income and expenses personally. The disadvantages are added cost, complexity, and expense of record-keeping and reporting.

Shareholders   In principle, shareholders are distinct from the company itself and are not liable for company debts if there is financial difficulty. However, issuing shares is not a means for persons starting a company to avoid paying its debts, since banks normally will not issue loans to a small company without the personal guarantee of its principal investors.

There must be an annual meeting of shareholders, with specific business to transact. Shareholders vote by number of shares held. They elect the board of directors, who are theoretically responsible for representing shareholders as well as fulfilling specific duties. The minimum size of the board, in some cases, is one person.

Board of directors   Board members need not be company employees. Their main business is to supervise the management of the company and, specifically, to elect a chair and appoint company officers. The appointed officers (i.e., president, treasurer, and secretary) are responsible for day-to-day business and are legally able to bind the corporation to contractual obligations.

If a company becomes insolvent without having fully paid its employees, then in some circumstances, board members may be personally responsible for unpaid taxes and up to six months of employee salaries.

Company hierarchy   Under the board and principal officers of a business are the departments, groups, or other functional divisions, depending on the activities, size, and management style of the company. In a small business, titles may be very informal, and persons may take on multiple roles. In larger organizations, the responsibilities of individuals must be more precisely defined.

16.3   The individual in corporate culture

When you accept a job, you create relationships with the corporate entity and your co-workers. Your co-workers expect friendly professional behaviour, as thoroughly described in references [1] and [2]. However, the relationship is more than social; establishing your place in a company involves an investment by the company and an investment by you.

You will invest time and energy in learning and adapting to new circumstances, with the expectation of satisfying work, reasonable pay, and opportunity for advancement. The company investment can almost always be measured in terms of money. When you were hired, someone in the company justified your position financially. The costs include salary, benefits for a permanent employee, and the overhead cost of providing space and company resources for you to do your job. The return on the investment in a single person is sometimes difficult to measure, but your presence is expected to increase company performance in the long term, if not more quickly.

Increasing the return on the investment in you   Simply doing what your boss tells you may minimally satisfy your job requirements, but you can increase the company's return on its investment in you with a minimum of thought. One duty of senior company management is to define the goals of the company; similarly, your department manager should define the goals of the group or department in which you work. Determine how your presence can or should contribute to the goals of your group and company. When you have finished a task, suggest or ask for another, perhaps with increased responsibility, that clearly furthers the stated goals. Keeping goals in mind and working toward them will visibly increase your value as a company member and colleague.

Finding a job that suits the individual   It is a safe assumption that people will do their best for their employer and themselves if their talents and temperament suit the requirements and environment of the job. Psychologists and guidance counsellors apply this assumption when administering psychological tests that attempt to determine the type of work that would suit individuals. The following material concentrates on job satisfaction rather than broader psychological factors.

Like everyone else, engineers derive satisfaction from their work when it is seen to have significant value, and this perception is generally more important than salary level. However, people differ in the kind of work that satisfies them. Therefore, in choosing a job or assembling a team to work on a project, these differences should be taken into account.

One classification of temperament and talent in technical work is given below. Ask yourself which of the following best characterizes you.

- The *inventor* likes to take unrelated elements to solve new problems, always wanting "better" results.

- The *innovator* implements new concepts or improves existing designs and likes tinkering to make systems "work." Many engineers fall within this category.

- The *entrepreneur* organizes resources and people to fill a void and likes to create organizations.

- The *manager* directs, creates policy and strategy, and obtains satisfaction by developing an organization and motivating people.

- The *administrator* implements strategies and likes to create order and efficiency.

Individuals typically fit more than one of these descriptions at different times; nevertheless, it could be poor strategy to take on a manager's responsibility if you are strongly of the inventor's or innovator's temperament, for example. Forming a team with others can lead to success if the talents of team members are complementary as, for example, when an entrepreneur teams with a financial administrator and an inventor or innovator.

16.4 Starting or joining a small company

Small companies deserve special consideration. You might be offered a job or internship in a business containing only one or two primarily technical positions. You might be asked by an acquaintance or classmate to join them in a small enterprise. You might anticipate starting a company yourself, either immediately or later in your career. This section discusses some of the major considerations involved in these choices.

This section can provide only a glimpse of a vast subject; to probe further, reference [3] is a good place to start. In addition to the aspects of business discussed here, a basic knowledge of business law as discussed in reference [4] and in engineering law courses is useful.

Small businesses often contain opportunity for higher rewards than larger companies, but the personal risks are also typically higher. The ultimate in this trade-off is in starting a business for yourself. In theory, you then have the freedom to profit to the fullest from your own ideas and to be your own boss, with the risk of losing considerable time and money. In practice, however, owners of small businesses often find that they spend much of their time satisfying the demands of others: clients, staff, suppliers, bankers, lawyers, accountants, and governments. Often the risks and rewards are shared, as well, with banks, venture-capital companies, partners, or shareholders. Small businesses tend to have a more informal work atmosphere and a greater variety of opportunity compared to large corporations. Consequently, surveys typically show greater job satisfaction in well-managed small businesses than in large ones.

| **16.4.1** | **The business plan** |

A well-written business plan is the result of many hours of work and revision. It serves as a plan before company start-up and as a performance-measurement reference during operations. On start-up and well beyond, small companies require a written plan in order to keep the many aspects of running a business under control and to convince bankers, investors, clients, and potential employees of the worth of the business. If you are invited to join a small business, ask to see the business plan; if you intend to start a business of your own, write a document containing at least the following items.

Executive summary　Business plans are often skimmed by persons such as bankers or potential investors who need an overview of the business. A brief description of the business should be given, then of its ownership and management, followed by the key factors that may lead to success and the key risks involved. The potential market for the goods or services produced and any competitive advantage the company will have are clearly stated. The strategy for beginning and continuing operations is summarized. Finally and very importantly, the levels of income, expenses, sales, and assets should be projected for the next few months and years. Particular attention should be paid to the three essentials of business: the production of the goods or services, the means of selling them, and the financial management of cash and other assets.

The business venture　As an introduction, there should be a clear and succinct statement of the product or service that the business provides, an indication of the market, and the prospects for producing, financing, and selling the product or service.

The company　A brief history of the company should be given, with its intended size, structure, and current state: its premises, number of employees, and current stage in its development plan.

Management　Small companies are highly dependent on the individuals who manage them. The names of the principal officers and investors, with their experience, training, responsibilities, and financial commitment to the enterprise, should be included. All of the principal skills—production, sales, and finance—should be present, and members of a good management team should possess complementary skills.

The product　A description of the product or service sold should be included, in enough detail for readers to judge potential cost, sales volume, product quality, and risk.

The market　The group of actual and potential clients or customers should be described. An estimate of the total market within reach of the company should be given, along with the target fraction of the market that is expected to be captured.

Sales plan and forecast　The sales to date and the means of achieving sales in the future, typically for three or more years, are included.

Production　The means by which the company is going to produce the product or service sold must be given. The material assets of production (machines), human skills, and required financial resources are described.

Gross profit The income per unit sold, less the costs of production, yield a gross profit margin, which must be adjusted by considering fixed and other costs. Combined with sales achieved and projected sales volume, these figures allow estimates of the scale of company operations required as well as the potential for net profitability.

Risks Since a business plan is a prediction of the future, an honest assessment of the principal threats and uncertainties should be given. A description of competing companies should also be included. Alternative strategies in case of problems should be described.

Financial requirements and cash flow Although personal success or failure is not always related to money, the success of a business can almost always be related to money. Particularly during start-up and expansion, a spreadsheet projection of income and expenses month by month for the next three to five years is necessary, in order to predict the requirements for loans or investments. This section is always read skeptically, and in detail, by bankers or other potential investors, clients, and employees.

Appendices A good business plan is based on specific information as well as on prediction. Back-up information such as summaries of operations to date, research material, or patents is included to support the main body of the plan.

16.4.2 Getting advice

There are many sources of information about starting and running small companies. To encourage employment, the local, provincial, and federal governments offer both management and taxation information and advice. Banks offer pamphlets and, sometimes, personal advice about new ventures. Some lawyers and accountants offer personal and web information related to their role in assisting a company. There are numerous business-development centres sponsored by universities and governments. Technical societies, such as the IEEE [5], offer advice and forums for discussing business ventures. Entrepreneurship clubs offer advice and networking. Books of advice on starting and running a business [6] are perennial sellers. All of these are useful sources of information, but, as always, someone who has started or joined a small company may be your best source of local information.

16.5 Further study

1. Choose the best answer for each of the following questions.

(a) Surveys show

    i. great variability but no consistent difference between employment conditions in companies as a function of size.

    ii. that job satisfaction is greatest in small companies.

    iii. that job satisfaction is typically greater in small, well-managed companies than in large companies.

(b) A limited partnership

    i. consists of one or more limited partners and one or more general partners.

    ii. can be formed in the name of one or more limited partners.

    iii. consists of persons whose obligations to pay debts is limited.

(c) The purpose of a business plan is mainly to attract investors.

    i. false   ii. true

(d) If your business might encounter significant competition from existing companies, then it is better not to mention this risk in your business plan.

    i. false   ii. true

(e) Starting your own business means that you can spend all of your time on your own pursuits.

    i. true   ii. false

(f) If a partner in a business dies or becomes bankrupt, the partnership

    i. must be dissolved by law.

    ii. is automatically dissolved unless a partnership agreement provides otherwise.

    iii. continues as usual, with the remaining partners assuming all assets and liabilities.

(g) All businesses must be registered with a government agency to operate legally.

    i. true   ii. false

(h) An unincorporated sole proprietorship

    i. has one share, owned by the proprietor.

    ii. has no shares.

    iii. has many shares, which are sold publicly on a stock exchange.

    iv. may have many shares, possibly held by different people, but not sold publicly.

(i) Incorporating a small company may provide advantages over a sole proprietorship because

    i. tax rates are lower for companies than for individuals.

    ii. all potential liabilities can be transferred to the company.

    iii. personal guarantees are typically not required when seeking loans to the business.

(j) The chair of the board and the company officers are appointed by

    i. the principal investors in the company.

    ii. the shareholders at the annual meeting.

    iii. the board of directors.

2. Do a web search to find at least one checklist for starting a small business in your province.

3. By web search, find at least one sample business plan for a company that you might like to join, such as a consulting company. How long do you think it would take to write a convincing document, starting with the sample, for a business that you would wish to start?

16.6 References

[1] C. Selinger, *Stuff You Don't Learn in Engineering School: Skills for Success in the Real World*. Hoboken, NJ: Wiley-IEEE Press, 2004.

[2] R. Templar, *The Rules of Work, A definitive code for personal success*. London: Pearson Education, 2003.

[3] CBSC, *CBSC—Home Page*. Ottawa: Canada Business Service Centres, 2004. <http://www.cbsc.org/english/> (March 9, 2008).

[4] D. L. Marston, *Law for Professional Engineers*. Toronto: McGraw-Hill Ryerson, third ed., 1996.

[5] IEEE, *IEEE Career and Employment Resources*. New York: Institute of Electrical and Electronics Engineers, 2003. <http://www.ieeeusa.org/careers/> (March 9, 2008).

[6] J. D. James, *Starting A Successful Business In Canada*. Vancouver, BC: Self-Counsel Press, 2000.

Chapter 17

Intellectual Property

Engineers create designs, processes, ideas, and products, and when this creativity is expressed in tangible form, it becomes intellectual property.

In a broad context, the term *intellectual property* refers to the ideas and creations produced by others and ourselves. How we use and acknowledge the ideas of others is an ethical matter, as discussed in Chapter 3, and how to acknowledge published written work by others is described in Section 8.1.2. However, specific laws, as well as ethics, apply to intellectual property that has been put into tangible form. We are concerned with the legal definitions of intellectual property, our rights to protect what we own (see Figure 17.1), and how to obtain and use the intellectual property of others.

In this chapter, you will learn

- the importance and use of the formally defined types of intellectual property,

- the types of intellectual property for which special laws have been written,

- how to obtain protection for your intellectual property.

Figure 17.1 Engineers sometimes display a sense of humour in their designs. These microscopic patterns and many others [1] have been discovered on commercial microcircuits. Initially, such patterns provided protection against illegal chip copying (see Section 17.7), but now they are an expression of the chip designer's individuality. Some companies currently allow only the designer's initials to be used. (Courtesy of Chipworks, Inc., Ottawa, Canada.)

17.1   Introduction

The intellectual property considered in this chapter is, generally, any creative material that has been put into tangible form. Pure ideas do not qualify, but when they are recorded or written, for example, the recordings or written copies are intellectual property. For example, you might develop a good idea by discussing it with a colleague. You would then be ethically bound to acknowledge that discussion, but neither you nor your colleague has legal ownership until the idea is expressed in written form, for which the author can claim copyright, or in another tangible form as will be described.

The following sections first contrast proprietary property, which is owned by an individual or company, with public-domain property, which is accessible by all. Then the principal types of intellectual property will be described, with a brief description of how the rights of the owner are established. The remainder of the chapter gives more detail about each of the types of intellectual property.

17.1.1   Proprietary intellectual property

Information for which ownership has been established is said to be *proprietary*. The owner may be an individual or a legal entity, such as a corporation or partnership. Intellectual property can be owned, bought, sold, or shared like any other tangible asset.

Government legislation sets the legal rules regarding intellectual property. As a result, legal protections such as patent law do not extend outside the country that passed the law. However, governments sign agreements and conventions to extend proprietary rights between countries.

The following six types of intellectual property are of interest to engineers; Table 17.1 summarizes their properties and the laws that govern them:

- patents,
- copyrights,
- industrial designs,
- trademarks,
- integrated circuit topographies,
- trade secrets.

The formal process of establishing ownership is similar to the process for land: an application is made to a government office, which registers ownership of the property if the applicable legal requirements are met. Copyright protection, however, does not require formal registration. Once registered, a description of the property and the registration documents are public. The Canadian Intellectual Property Office (CIPO), which is part of Industry Canada, registers the first five types of property in the above list. Trade secrets are not registered and do not require disclosure or have any special protection.

All definitions and interpretations related to intellectual property are subject to the re-interpretation of lawyers and courts and are changing rapidly in some areas, as pertinent laws and practices prove to be out of date. Two examples of rapid change will be mentioned. First, electronic media make it very easy to freely distribute many kinds of artistic works, and the applicability of copyright to electronic documents is under active debate. Second, existing laws are difficult to apply to recent developments in bioengineering. These and other cases require both new legal definitions and a re-examination of ethical conduct.

17.1.2   The public domain

Information that has no ownership is said to be in the *public domain,* which means that anyone can use it. Property in this designation includes material

- that is common knowledge,
- for which legal protection has expired,
- that has been placed in the public domain by the owner.

Table 17.1   Types of intellectual property and associated protection.

Intellectual Property	1. Type of legal protection 2. Criteria for protection 3. Term of protection
Patent	1. The right to exclude others from making, using, or selling the invention. 2. The invention must be new, useful, and ingenious. 3. 20 years from date of filing an application.
Copyright	1. The right to copy, produce, reproduce, perform, publish, adapt, communicate, and otherwise use literary, dramatic, musical, or artistic work, including computer programs. 2. The material must be original. 3. The life of the creator plus 50 years to end of the year; except for photographs, films, and recordings (50 years total).
Industrial design	1. The right to prevent competitors from imitating the shape, pattern, or ornamentation applied to a useful, mass-produced article. 2. The shape, pattern, or ornamentation must be original, and the article must serve a useful purpose. 3. Up to 10 years from registration.
Trademark	1. Ownership of a word, symbol, or design used to identify the wares or services of a person or company in the marketplace. 2. The trademark must be used in business, in Canada, before it can be registered and must not include prohibited words: profanity, geographic names, . . . 3. 15-year periods, renewable indefinitely.
Integrated circuit topography	1. Ownership of the three-dimensional configuration of layers of semiconductors, metals, insulators, and other materials on a substrate. 2. The configuration must be original. 3. 10 years from application.
Trade secret	1. Possession of a secret process or product. 2. Employees must sign nondisclosure contracts, agreeing to maintain the secret. 3. Potentially unlimited. However, disclosure or theft must be prosecuted under tort or criminal law.

Most of our knowledge is in the public domain, including mathematical equations, natural laws, and information published in engineering, professional, and scientific journals. Newton's second law, for example,

$$f = ma, \tag{17.1}$$

where f, m, and a are respectively force, mass, and acceleration, is in the public domain.

Standard time is an example of published engineering information entering the public domain. Sandford Fleming was a railway engineer who helped to build the trans-Canada railway at the time of Confederation, when every station was on its own local Sun time. When he missed his train, he decided to eliminate the confusion. Within three months he had published his first paper on the concept of standard time zones and the 24-hour clock. The eventual result was the time zone system we use worldwide today, in which, by the calculation $360°/(24\,\text{h}) = 15°/\text{h}$, $15°$ of longitude correspond to a one-hour increment in standard time. The system was adopted worldwide in 1884, with the primary meridian ($0°$ longitude) passing through Greenwich, England.

A publication—a history book, for example—may contain common knowledge, but a particular written version of it may be the subject of copyright.

When legal protection of intellectual property expires, the property is in the public domain. Shakespeare's plays, for example, are in the public domain.

17.2   The importance of intellectual property

There are two important consequences of the principle of ownership and of the requirements for registering intellectual property:

- The owner can control and possibly profit from the use of the work by others.

- The records of registered intellectual property are vast public storehouses of knowledge.

The first purpose of protecting intellectual property is to ensure that its producer is rewarded for creating it. This protection encourages creativity, by the argument that few people would invest in research, development, or creation if they could not have exclusive use of the results. Under Canadian law, the owner of an intellectual property is permitted to benefit from it and to prevent competitors from using it. The owner may sell the intellectual property or may license it to others to manufacture or use.

A company's intellectual property may be its most valuable asset. For example, the rights for an integrated circuit chip might be worth more than the factory in which it is being manufactured.

It is up to the owner of the intellectual property to initiate proceedings to protect registered property; the Canadian Intellectual Property Office (CIPO) does not protect owners except by registering ownership.

As a practising engineer, you must know how to protect intellectual property; you must also understand and respect the property rights of others. Practising engineers also derive great value from the CIPO as a source of information about current technological developments. Advice about intellectual property can be obtained from the Publications Centre of the CIPO [2].

Most of the CIPO files and databases are open to the public; this information permits you to monitor the progress of competitors and to inspect their intellectual property, as an aid to inspire your inventiveness. Even if you should find that your proposed patent, copyright, industrial design, trademark, or circuit topography is already owned by others, you may be able to license it from the owner, or, when the protection expires, you may copy and use it without permission.

17.2.1   Rights of employers and employees

The relationship of an engineer to an employer must be considered. Who owns the rights to intellectual property developed as part of employment activities? What about ideas developed using company facilities in the employee's spare time, unrelated to regular tasks? When an employee changes employers, how much knowledge can be taken to the new employer? The answers to these questions depend on individual circumstances, particularly the employment contract.

Employment contracts sometimes state that the company owns *any* intellectual property developed by the employee, who cannot publish or exploit intellectual property without the consent of the employer. Other employment contracts limit the company's claim to property related to the job. In many cases, companies are prepared to assume the effort and expense of registering the property for the mutual benefit of company and author. Hiring people away from a competitor is sometimes regarded as a form of industrial espionage (see Questions 5 and 6). An employment contract may prohibit an employee from working for a competitor of the employer for a period, two years, for example, after leaving the company.

The relationship of publisher to author is similar to that of employer to employee. Technical societies, which publish large amounts of engineering work, have created policies for ownership; see reference [3], for example.

17.3   Copyright

The laws concerning copyright are of direct importance to every student: the author of written material produced during a degree program automatically owns the copyright of the work. In addition, all members of a university community make use, usually under licence, of material the copyright of which is owned by others.

According to the Copyright Act as administered by the CIPO [4], a copyright is the right to produce, reproduce, perform, publish, adapt, sell, or lease an original literary, dramatic, musical, or artistic work. Reproduction includes recording, photographing, filming, or using any "mechanical contrivance," including communications technology and telecommunication signals, for delivering the subject matter.

The author owns the rights to an original work, unless the work was created as part of a job or employment or is a commissioned work. If the creation occurs during an employment, the employer usually owns the rights, unless there is an agreement to the contrary.

Protection of a copyright lasts for the author's lifetime plus 50 years in Canada. Photographs, sound recordings, and most films and videos are protected for 50 years from the date of their creation. However, the copyright period is not uniform in all countries. In 1998, the United States extended copyright protection to life plus 70 years or, for works created by corporations, to 95 years. These terms also apply in Europe.

Material subject to copyright may be printed or written but also may be recorded on electronic media. Computer programs, technical reports, and engineering drawings and data are considered to be literary or artistic works and are protected by copyright.

17.3.1   Copyright registration

In Canada, a copyright is automatically possessed by an author or creator, whether or not, as required in some other countries, the material is marked "©" or is registered.

Registration is not necessary to obtain copyright protection. However, you may still want to register the copyright with the CIPO for a small, one-time fee. Registration gives notice that a copyright exists, and establishes the date of creation, the original author's name, and the ownership of the work.

Canadian copyright also gives foreign protection as a result of two international treaties. Automatic ownership of copyright is valid in all countries that adhere to the Berne Copyright Convention. Furthermore, copyright can be extended to all countries that adhere to the Universal Copyright Convention by marking the creation with the copyright symbol ©, the name of the author, and the year. Canada and the United States belong to both conventions.

Violation of a copyright is called infringement, and if it includes misrepresentation of authorship, then it is plagiarism, as discussed in Section 8.1.2 and Section 3.5. Infringement occurs with a substantial quotation or borrowing of another's work without the author's permission, even if the source is attributed properly [3].

Enforcing a copyright is the owner's responsibility. If someone infringes your copyright, you may sue under civil law for the recovery of lost income. However, the Copyright Act also has clauses for criminal infringement for serious, organized infringement or "piracy," such as reproducing a copyrighted work for sale or hire. The penalty for criminal infringement is severe.

17.3.2   Fair dealing

The Copyright Act permits a small amount of copying or reproduction under "fair dealing" provisions. These include short extracts for the purpose of review, criticism, or research. The author and the source of the quoted material must be completely identified. However, the amount that can be copied is not well defined. If copying replaces sales, it is certainly not "fair." Multiple copying always requires the consent of the copyright owner.

Photocopying written work for research and study is allowed but controversial, since there are no exact guidelines for the number of paragraphs or pages that can be copied under fair dealing. Obviously, photocopying an entire work, such as a textbook, is illegal infringement. Universities, therefore, have strict rules about copying. Licensing collectives, such as Access Copyright, have been established to negotiate licences for schools and universities and to pay royalties to authors.

17.3.3   Copyright and computer programs

Computer programs may be protected by copyright. A program can be copyrighted as a literary work, regardless of the manner of storage or recording. Under the Copyright Act, it is an infringement to make, sell, distribute, import, or rent copies of computer programs, without written consent of the copyright owner. As noted in Section 17.4, a process with a computer program embedded in it may be patentable, but the written program itself is a literary work and may be protected by copyright.

Fair dealing, discussed in Section 17.3.2, also applies to computer programs. The owner of an authorized copy may make one backup copy for his or her own use. This copy must be destroyed when the authorization ends for the original. The authorized copy may also be adapted or translated into another computer language, if this is necessary to make it compatible with a particular computer system.

The "piracy," or unauthorized copying of software, is so easily done that people sometimes do not realize that they are breaking the Copyright Act. As a professional engineer, you must avoid unethical activities that might cloud your reputation. Avoid pirated software, and ensure that all your computer software is authorized.

The interpretation of copyright law for computer programs is still developing. For example, a form of reverse engineering, called *program decompilation,* may be much cheaper than obtaining a licence or developing a completely new program, but the legal issue is whether decompilation represents fair dealing. A single decompiled copy may be considered fair dealing if decompilation is used merely to inspect competing software.

17.4   Patents

A patent is the legal right, lasting 20 years, to exclude others from making, using, or selling an invention. For that time period, a patent is a piece of legal property. The first patent in Canada was granted in 1824, for a washing machine. The one-millionth patent was granted in 1976 for biodegradable plastic, useful for garbage bags. The Canadian Intellectual Property Office handles approximately 35 000 patent applications a year.

After the protection of private property, the second major function of a patent is *public notice.* An application for a patent requires a full and clear description, so that a skilled person could reproduce the invention [5]. In return for full disclosure, the owner receives the exclusive right to benefit from the use or sale of the invention. The patent records are a vital resource to industrial and academic institutions. However, the Patent Office does not enforce the patent law. This means that the protection offered by the patent can only be invoked by the patent holder through legal processes, which can be very expensive and time-consuming.

The principal patent legislation is the Patent Act of 1989. Earlier patents are regulated by slightly different rules. The term, or length of protection, is 20 years for a patent application filed after October 1, 1989. The Patent Act allows the rights to an invention to be sold or licensed to others by registering a written document with the Patent Office.

A patent application may be submitted for any of the following, provided it is new, useful, and ingenious:

- a product (of manufacture),
- an apparatus (a machine for making other objects),
- a manufacturing process,
- a composition of matter,
- an improvement for any of the above.

In fact, the last category contains the largest number of applications.

The three key criteria—novelty, usefulness, and ingenuity—are examined for each patent application, a process that requires time and judgment. The invention must be original, must serve a useful purpose, and must not be an obvious improvement over prior art. Patents sometimes appear naïve or bizarre; see Figure 17.2, for example.

The first known patent to be awarded was to Filippo Brunelleschi by the Republic of Florence in 1421, while the first patent law protection for inventors was in the Republic of Venice, 1474 [6]. These events during the Italian Renaissance are evidence of the emerging concept of intellectual property and of the application of the new sciences to applied problems.

Perpetual motion machines have long been the goal of amateur inventors. As recently as 1979, a patent application was made for an "energy machine" that produced more energy than it consumed. Such inventions clearly violate the law of conservation of energy. More subtle attempts conserve energy but violate the second law of thermodynamics by proposing, for example, to rectify a noise voltage or drive a racket from molecular motion (see [7] for a discussion of "Maxwell's demon").

An invention may not be patented if it is intended for an illegal purpose or if it is already in the public domain. Pure ideas, scientific principles, and abstract theorems are not patentable. A process with a computer program embedded in it may be patentable, but the question of whether the algorithmic processes of computer programs

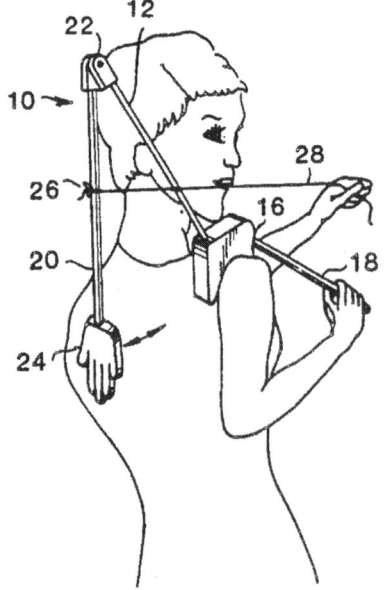

United States Patent [19]

Piro

[11]	**Patent Number:**	**4,608,967**
[45]	**Date of Patent:**	**Sep. 2, 1986**

[57]   ABSTRACT

A self-congratulatory apparatus having a simulated human hand carried on a pivoting arm suspended from shoulder supported member. The hand is manually swingable into and out of contact with the user's back to give an amusing or an important pat-on-the-back.

Figure 17.2 Described as a self-congratulatory apparatus, this back-patting device received U.S. patent number 4,608,967 in 1986.

are themselves patentable is the subject of debate. Computer program files can be protected by copyright as discussed in Section 17.3.

Genetically engineered life forms are patentable, thus protecting the investment made in the development of new biological organisms. For example (see reference [6]), a microbe for oil cleanup was patented in 1980, and a patent for a genetically engineered mouse appeared in 1988. Biotechnology is a new, complex, and disputable patent area; for example, a four-year litigation involving a drug for multiple sclerosis included a claim of over one billion dollars.

Biotechnology and software are two evolving areas where patent criteria are anything but clear. Moreover, the long patent process and the clarifications produced by legal challenges are sometimes outpaced by these rapidly changing technologies.

In recent years, the patent process has come under criticism, particularly in the United States, for delays caused by a major increase in the number of applications. In addition, corporations have been accused of applying for patents principally to impede their competitors through lawsuits rather than to exploit the invention itself. These and other developments are causing patent laws in several countries to continue to evolve.

17.4.1   The patent application process

In principle, a patent application can be prepared and submitted by the inventor, and guidelines are available from the Patent Office [5]; however, preparing the application is complex, and patent agents are usually retained for this purpose.

The four main steps in the patent process are as follows.

- Retain an agent, and search existing patents.
- File an application.
- Request an examination.
- Protect the patent.

The first step in the patent process is the search of existing records to see what is already registered or known. Because of the complexity of the search, a patent agent or attorney is usually retained; their fees can amount to several thousand dollars. The agent is typically not an expert in the subject matter of the patent and must be given careful descriptions and statements as a basis for preparing the application.

The patent application must be submitted within one year from the date of the first public disclosure of an invention. This delay could be risky because priority goes to the first applicant to file. This "first to file" rule, the international norm, is much easier to adjudicate than the "first to invent" rule, which is used in the United States and which can involve litigation to decide who was the first inventor. Applications may take years to process, but they are made public 18 months after filing.

In the period after publication but before issuance of the patent, an invention may be legally made or used by others, or sold. However, the inventor can seek retroactive compensation when the patent is awarded. Many articles are stamped "patent pending" to warn that retroactive compensation may be sought.

In order to establish a filing date, at minimum a petition, a defining description, and a fee must be submitted. The full patent application must be completed within 15 months and must contain the following five parts:

- petition: a formal request that a patent be granted,
- abstract: a summary for publication in the Patent Office Record,
- specification: a description of the invention, including relevant prior art, and its novelty, usefulness, and ingenuity,
- claims: a part of the specification that defines precisely what aspects of the invention the inventor wants to protect,
- drawings: illustrations (and possibly models) of the invention.

The most important part of the application is the scope of the claims. These define the protection provided by the patent and are usually as broad as possible to protect against competitors who produce similar inventions with minor differences.

Once the patent application has been filed, a request must be made to have it examined. Such a request must be made within five years of the filing date; otherwise the application expires. During this delay, the applicant may be engaged in market research and business planning to see whether it is worthwhile to proceed.

There are fees for filing, for examination, for granting the patent, and for annual maintenance. Protecting the patent against infringement entails legal fees, so a patent can be an expensive undertaking in total.

Canadian patents are valid only in Canada, and applications must be filed in each country where protection is desired. However, the international Paris Convention for Patents (1883) makes patent applications in member countries valid as of the date granted in the home country, provided application is made abroad within a year. Most industrialized countries participate in the Patent Cooperation Treaty of the World Intellectual Property Organization in Geneva, which permits a foreign patent to be obtained from within Canada. The standardized filing procedure simplifies the application, so you can apply to over 100 countries with one application in Canada.

The Patent Office cannot assist inventors to lease or sell patents; however, a patent owner may place an advertisement in the Patent Office Record, which is published weekly and is available in any large Canadian public library. Many entrepreneurs, investors, and researchers consult this publication on a regular basis.

17.5 Industrial designs

The aesthetic look of an object can be one of its main selling features and can be registered as an industrial design. Any original shape, pattern, or ornamentation that is not induced purely by the function of the article can be registered. The owner of a registered industrial design has exclusive rights for five years, renewable once for a further five years. Some examples are the decorative pattern on the handles of a cutlery set, a wallpaper pattern, or an unusual but appealing chair shape. Industrial design protection is like copyright but with some differences, as follows.

- Industrial design protection applies to the shape, pattern, or ornamentation of mass-produced, useful articles. The visual aspects of the design are protected.

- Industrial design protection can be obtained only by registration, and the article must be registered during the first year of public or published use.

- The protection has a maximum term of 10 years.

- If the design or pattern is on an article or piece of work that is subject to copyright but is used to produce 50 or more such articles, then it can only be protected as an industrial design.

Industrial design protection applies in Canada only. Separate registration must be made for foreign countries, but under the Paris Convention, a person who applies for protection in Canada has six months to apply for registration in other treaty countries.

17.6   Trademarks

A trademark is a word, symbol, or design intended to identify the wares or services of a person or company in the marketplace [8]. The following types are included:

- *Ordinary marks* distinguish articles or services. Most corporate logos fall into this category.

- *Certification marks* indicate a quality standard for a product or service set by some recognized organization (see Figure 17.3).

- *Distinguishing guise* is a wrapping, shape, packaging, or appearance that serves to identify the product as distinct from other similar products or services. An example is a candy shaped like a butterfly.

The trademark owner benefits from public recognition of the quality and reputation of the product or service. A company name does not necessarily qualify as a trademark

Figure 17.3   Three examples of registered certification marks from the CIPO database are shown. The first is a Canadian Automobile Association mark familiar to consumers. The second is registered by the Air-conditioning and Refrigeration Institute, and the third is used by members of the Association of Professional Engineers and Geoscientists of New Brunswick. (With permission of the trademark holders.)

unless it is used to identify products or services. A trade name or business name used to identify products or services can be registered as a trademark.

The right to register a mark in Canada belongs to the first person to use it in Canada or to a person with a registered trademark in a member country of the Union for the Protection of Industrial Property. The term of registration is 15 years from the date of original registration or from the most recent renewal. To be continued in effect, renewal must take place before expiry, but the number of terms of renewal is unlimited. While an unregistered trademark may be commonly recognized, only registration in the trademark register is proof of exclusive ownership and protection against infringement. The process of registration also ensures that the proposed mark does not infringe existing trademarks.

A proposed trademark is deemed to be registerable unless it falls into one of the following categories:

- It is similar to an existing registered trademark.
- It is the applicant's name or surname unless the name has become associated with a product and it identifies that product.
- It is a clearly descriptive term. For example, "Red" cherries cannot be registered, since all cherries are red, and it would be unfair to other cherry sellers.
- It is a deceptively incorrect term. For example, "Air Courier" is not allowed for a company that delivers parcels by truck.
- It is the name of the product in English or any other language. For example, "Fleurs" could not be registered for a flower shop.
- It suggests government approval; for example, a mark including the Canadian flag.
- It suggests a connection with the Red Cross, the Armed Forces, a university, a public authority, or the RCMP.
- It could be confused with international signs, such as traffic or civil defence.
- It could be confused with a plant variety designation.
- It is the name of a living person or one who has died in the preceding 30 years, unless the person has given written consent.

The Canadian Intellectual Property Office ensures that no two trademarks are the same or confusingly similar.

It is recommended, although not legally necessary, that registered trademarks be identified by the symbols ® or ™ or *Trademark Registered, or MC or Marque de Commerce. Trademark owners must be vigilant and point out any improper use, such as being copied or used as a generic or common name. To reduce this risk, the trademark should always be used as an adjective, never as a noun. As examples of trademarked terms becoming common names, kerosene and cellophane are generic nouns derived from former trademarks for Kerosene fuel and Cellophane sheet.

Registering a trademark protects it only in Canada. For countries in the Union for the Protection of Industrial Property, registration is obtained by filing a copy of the Canadian registration. For a foreign trademark to be registered in Canada, it must be distinctive, non-confusing, non-deceptive, and not otherwise prohibited.

| 17.7 | **Integrated circuit topographies** |

Integrated circuit topographies are defined under the Integrated Circuit Topography Act (1993) [9]. The intellectual property is the three-dimensional configuration of layers of semiconductors, metals, insulators, and other materials that form the circuits in the semiconductor microchip. To be registered, the topography must be original. Registration of the topography can supplement patent protection for a novel circuit itself or its function.

This type of protection is very new, and the laws will undoubtedly change as time passes. Over 20 countries have similar protection for semiconductor chips. Fair dealing provisions allow copying the design for research and teaching and limited reverse engineering.

Registration is required for protection, must be applied for within two years of commercial use of the design, and is valid only in Canada.

The term is to the end of the tenth year from the first commercial use or from the filing of the application for registration, whichever is earlier.

| 17.8 | **Trade secrets** |

Trade secrets are intellectual property; however, the government does not regulate them. A trade secret is a commercially important secret formula, design, process, or compilation of information. The purpose of maintaining secrecy is evident: to retain exclusive use and benefit from a process or product. The benefits may be great, the term of use is potentially indefinite, and there are no fees. However, this form of protection works only as long as the secret can be maintained.

Unlike other forms of intellectual property, there is no specific government act that protects trade secrets. The other forms of protection inherently require disclosure as a condition of protection, which provides a source of information and a stimulus to technological progress.

The action required to protect a trade secret is simply to keep all the relevant information secret. Employees who have access must agree to maintain secrecy, usually as part of their employment contract. If a trade secret is deliberately or accidentally misappropriated, then legal action may be undertaken for damages, but it requires recourse to the laws for theft or tort (a willfully or negligently wrongful act).

Trade secrets are subject to leaks, espionage, independent discovery, reverse engineering, or obsolescence, and the cost of security to maintain secrecy may be very high. Often, the main problems are to control carelessness and to ensure the confidentiality of information held by former employees, perhaps now employed by a competitor!

| 17.9 | **Further study** |

1. Choose the best answer for each of the following questions.

(a) In Canada, the term for protection of trademarks is

   i. 15 years, renewable once.

ii. 20 years from filing.

iii. the life of the creator plus 50 years.

iv. 15 years, renewable indefinitely.

(b) A patent conveys the legal right to

i. exclude others from determining how the invention operates.

ii. exclude others from improving the invention.

iii. exclude others from finding out about the invention.

iv. exclude others from making, using, or selling the invention.

(c) The most suitable protection of the attractive design of a pair of gloves is

i. trademark.   ii. patent.   iii. industrial design.   iv. trade secret legislation.

v. copyright.

(d) The term of copyright protection is the same in all countries that have signed the Berne agreement.

i. true   ii. false

(e) The term for copyright ownership in Canada is

i. the life of the author plus 50 years.

ii. the life of the author plus 70 years.

iii. the life of the author plus 20 years.

iv. 20 years from filing.

v. 50 years from registration.

(f) An employment contract can stipulate that an employee may not work for a competing company for a stated time.

i. true   ii. false

(g) In Canada, the length of time that a patent gives protection to the owner is

i. 20 years from invention.

ii. the life of the inventor plus 20 years.

iii. the life of the inventor plus 50 years.

iv. 20 years from filing.

(h) In Canada, a written work must be registered with the National Library before the author can claim copyright.

i. true   ii. false

(i) There are no specific laws defining industrial designs as intellectual property; they are protected by keeping them secret.

i. false   ii. true

(j) Computer software can be protected by patent law but not by copyright law.

i. false   ii. true

2.   What type, or types, of intellectual property can be used to protect the following?

  (a) a new pneumatic pump that makes athletic shoes more comfortable

  (b) the attractive design of a new pair of athletic shoes

  (c) a logo to identify the new line of athletic shoes

  (d) a slogan to advertise the new line of athletic shoes on television

  (e) a new process for curing rubber used in athletic shoes

  (f) a new microcomputer chip to control the production line for athletic shoes

3.   Engineering reports were discussed in Chapter 8. Compare the headings in a patent application with the headings in a typical engineering report. What are the similarities and differences?

4.   Omar Khayyám, an eminent Persian astronomer who died in 1123, wrote many verses in the Persian language. These were collected and translated into poetic English in 1858 as *The Rubáiyát of Omar Khayyám* by Edward FitzGerald (1809–1883). Imagine that current Canadian copyright law had applied for five centuries in Persia and England.

  (a) When would Khayyám's copyright on his work have expired?

  (b) Could FitzGerald copyright the translation in 1858?

  (c) If the answer to part (b) is affirmative, in which year would FitzGerald's copyright expire?

  (d) If you publish a newly illustrated version of FitzGerald's *Rubáiyát* this year, can you copyright it this year?

5.   While working for the Grand Whiz Company, you imagine a new product. It is not directly related to the projects you have worked on there, but there is no doubt that the working environment, and especially the discussions with your colleagues, has stimulated your imagination.

  Do you: (a) take the idea to the Grand Whiz management for a shared patent, if the company bears all the expense of the patent process; (b) quit, and then set up your own business to exploit your idea; (c) keep your job, and take the idea to your entrepreneurial sister-in-law and share the profits with her?

6.   Competitors often seek proprietary business information. If the methods are legal, they are called market research; if the methods are questionable, they are called industrial espionage. One of the easiest ways to acquire intellectual property from a competitor is to hire someone who has worked there.

  (a) What are some ways of preventing the loss of intellectual property in this way?

  (b) What is the career effect for an employee who easily transfers jobs for this reason?

7. How would you protect the following items of intellectual property?

(a) a company name and logo

(b) a story, poem, or laboratory report

(c) a natural law, if you discovered it

(d) a distinctive design for a mass-produced item

(e) a novel and ingenious mousetrap

17.10 References

[1] H. Goldstein, "The secret art of chip graffiti," *IEEE Spectrum*, vol. 39, no. 3, pp. 50–55, 2002.

[2] Canadian Intellectual Property Office, *Canadian Intellectual Property Office— Welcome*. Industry Canada, Canadian Intellectual Property Office, 2001. <http://strategis.ic.gc.ca/sc_mrksv/cipo/welcome/welcom-e.html> (March 9, 2008).

[3] W. Hagen, *IEEE Intellectual Property Rights*. New York: Institute of Electrical and Electronics Engineers, 2001. <http://www.ieee.org/copyright/> (November 9, 2004).

[4] Canadian Intellectual Property Office, *A Guide to Copyrights*. Ottawa: Industry Canada, Canadian Intellectual Property Office, 2000. <http://strategis.gc.ca/sc_mrksv/cipo/cp/copy_gd_main-e.html> (March 9, 2008).

[5] Canadian Intellectual Property Office, *A Guide to Patents*. Ottawa: Industry Canada, Canadian Intellectual Property Office, 2001. <http://strategis.gc.ca/sc_mrksv/cipo/patents/pat_gd_main-e.html> (March 9, 2008).

[6] B. Bunch and A. Hellemans, *The Timetables of Technology*. New York: Simon & Schuster Inc., 1993.

[7] L. Brillouin, *Science and Information Theory*. New York: Academic Press, 1962.

[8] Canadian Intellectual Property Office, *A Guide to Trade-marks*. Ottawa: Industry Canada, Canadian Intellectual Property Office, 2001. <http://strategis.gc.ca/sc_mrksv/cipo/tm/tm_gd_main-e.html> (March 9, 2008).

[9] Canadian Intellectual Property Office, *Integrated Circuit Topographies*. Ottawa: Industry Canada, Canadian Intellectual Property Office, 2001. <http://strategis.ic.gc.ca/sc_mrksv/cipo/ict/ict_gd_main-e.html> (March 9, 2008).

Chapter 18 Project Planning and Scheduling

Every project, whether it involves design, development, or construction, will benefit from systematic planning and scheduling. In fact, almost every aspect of everyday life could be improved by better planning or scheduling. These two basic terms will be employed as follows:

- *Planning* is the determination of all the activities required to complete a project or similar enterprise and their arrangement in a logical order.

- *Scheduling* is the assignment of beginning and end times to the activities.

Planning, then, is determining *how* to do something, and scheduling is deciding *when* to do it. In any project, the purpose of planning and scheduling is to achieve the minimum time, minimum waste of time and material, and minimum total cost. These objectives cannot always be achieved. For example, achieving the minimum time is typically difficult at minimum cost. However, with proper planning and scheduling, the tradeoff between these objectives can be seen in advance, and optimal choices can be made. The material in this chapter is contained in publications such as references [1–4].

18.1 Gantt charts and the critical path method

Charts can be used to aid the visualization of planning and scheduling activities. There are several suitable formats, but the most common is probably the bar chart, also called the *Gantt chart*, after its inventor. Many project-management software tools assist with the creation and modification of such charts. A basic example is shown in Figure 18.1. The chart shows activities, in order of starting times, on the vertical scale; the horizontal

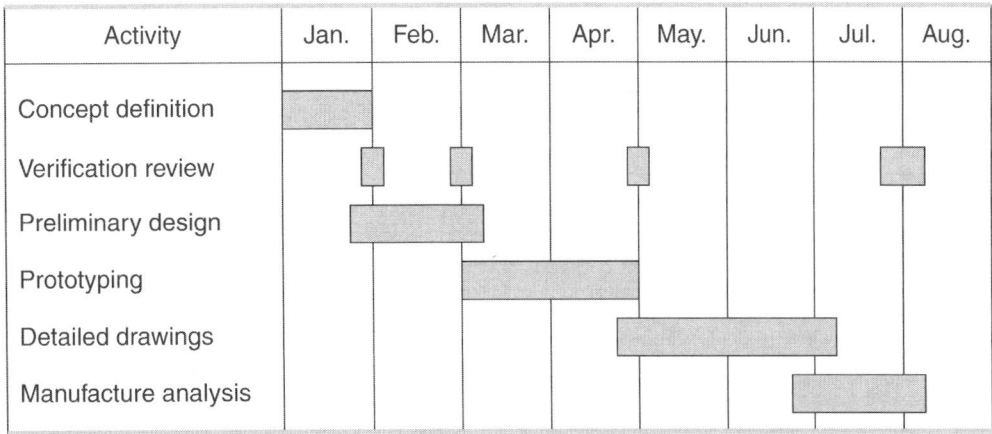

Figure 18.1   A basic Gantt chart for a design project.

scale indicates time. The time intervals required by the activities are shown by horizontal bars, and as the project proceeds, notes may be added or the style of the bars modified to show activities, for example, that are critical or exceeding their planned duration. At any time, the activities in progress and those scheduled to start or end soon can be determined. However, if a problem occurs with an activity (if delivery of material is late, for example), the basic Gantt chart does not show whether other activities will be affected or whether the delay will affect the completion date of the whole project.

Critical path method   A technique called the *critical path method* (CPM) gives detailed scheduling information. CPM is a modified form of the "program evaluation and review technique" (PERT), developed by the U.S. Navy in 1958 to speed the design of the Polaris missile. It proved to be an instant success by cutting 18 months from the completion date of the project, and it has been the standard method of planning used in the aerospace and electronics industries and in many construction projects ever since. CPM is particularly suitable for projects composed of easily defined activities of which the durations can be accurately estimated (most construction projects and some software engineering projects, for example, fall into this category).

Many planners require CPM to be employed to monitor and control the timetable of a project, but they still use simple Gantt charts when discussing or reporting progress.

18.2   Planning with CPM

To plan a project using CPM, follow these two steps.

1. List each activity involved in the completion of the project.

2. Construct an arrow diagram that shows the logical order of the activities, one activity per arrow.

When the arrow diagram is complete, it will show all the activities in the project, in the proper order. The planning phase is therefore complete. The two steps above may actually be repeated several times, since the action of constructing the arrow diagram will remind the planner that an activity has been omitted or perhaps that one activity should be divided into two more specific activities. The list of activities and the arrow diagram are then modified until they are complete and correct. However, before discussing details of the arrow diagram, we must define two terms more accurately: *activities* and *events*.

- An *activity* is any defined job or process that is an essential part of the project. Each activity in the project is represented by an arrow in the arrow diagram such as shown in Figure 18.2. Each activity has an associated time duration Δt.

- An *event* is a defined time. All activities begin and end at events, represented by circles or nodes, as shown in Figure 18.2. In some contexts, events are called "milestones," which are defined steps on the way to completion of the project.

An arrow diagram, therefore, is a graph, in which the nodes are events and the branches are arrows representing activities. To construct an arrow diagram, we start

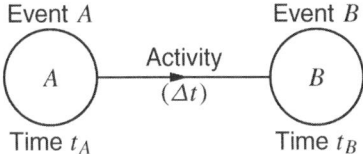

Event A Event B

Activity
(Δt)

Time t_A Time t_B

Figure 18.2   Activities and events in an arrow diagram, with time interval $\Delta t = t_B - t_A$.

by drawing a circle for the *start event*, which is the beginning of the project. The first activity arrow is then drawn from the start event, and the other activity arrows are joined to it in the sequence in which the activities must be performed. The following simple example illustrates the planning procedure.

Example 18.1
Tire changing

For three or four weeks each autumn, when the weather gets cold, tire stores (in some climatic regions) are crowded with people who want to purchase winter tires. If these stores could improve the efficiency of selling and mounting tires, then there would be fewer delays, customers would be more content, and tire sales would increase.

The activities for a typical two-tire purchase are listed in Table 18.1 along with estimated activity times. Consider such a purchase as a project. How would you plan and schedule it for minimum delay? Suppose that there are two cases: a small, one-person tire store, and a large, fully staffed tire store.

The arrow diagram can be created easily, using the list of activities in the table. If the tire store has only one person acting as sales clerk, stockroom clerk, and mechanic, then the simple diagram of Figure 18.3 results. The process cannot be improved if only one person is involved, since the activities must be performed sequentially.

However, if the tire store employs a sales clerk, a stockroom clerk, and at least

Table 18.1   Activities for the sale and installation of winter tires.

Activity	Time (min)
1. Customer drives car into tire-store garage.	2
2. Customer inspects catalogue and selects tires.	30
3. Sales clerk confirms stock on hand.	5
4. Stockroom clerk takes tires to garage.	10
5. Mechanic raises car and removes wheel 1.	5
6. Mechanic mounts new tire on wheel 1 and balances it.	10
7. Mechanic replaces wheel 1.	5
8. Mechanic removes wheel 2.	3
9. Mechanic mounts new tire on wheel 2 and balances it.	10
10. Mechanic replaces wheel 2.	5
11. Clerk writes bill; client pays for tires and installation.	10
12. Mechanic lowers car and drives it out of garage.	2

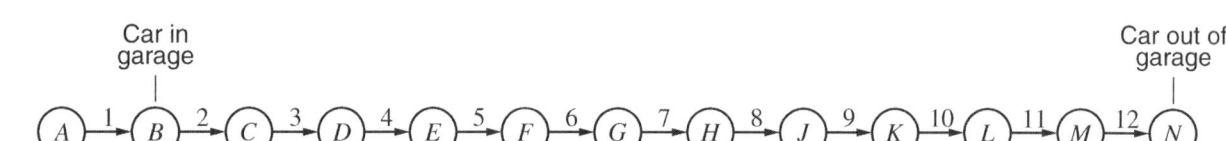

Figure 18.3   A one-operator tire purchase. The label I is often omitted to avoid confusion with the number 1.

two mechanics (and has two tire-mounting machines), then the arrow diagram can be redrawn as shown in Figure 18.4. Since both tires are installed at the same time, the time for the transaction is reduced, as will be seen.

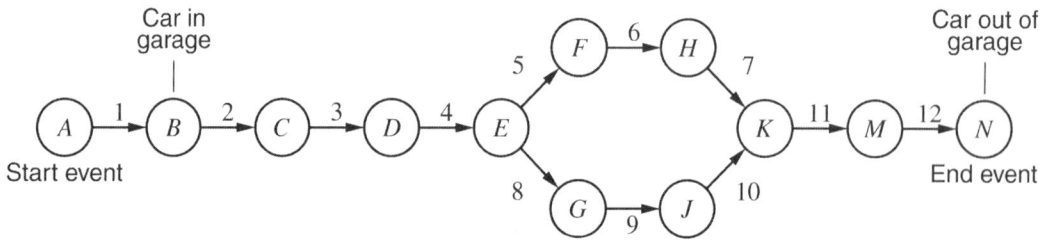

Figure 18.4   A large-store tire purchase, with paths showing independent activities.

We now look at the process critically. The list of activities is in a logical order, beginning with the arrival of the customer and ending with the departure. However, it is not necessarily the most efficient order. For example, why must the car be in the garage while the customer inspects the catalogue or pays the bill? The bill could be paid either before or after the tire change, thus freeing the shop and equipment for part of the time. With these changes in mind, the process can be redrawn as in Figure 18.5, showing that the customer inspects the catalogue and selects the tires (activity 2) and pays the bill (activity 11) before the car enters the garage (activity 1). The stock clerk delivers the tires to the garage (activity 4) while the customer pays and then drives the car into the garage. The dashed lines in this figure are called *dummy activities,* which take zero time but are included to show the proper sequence of events. In this case, they are required in

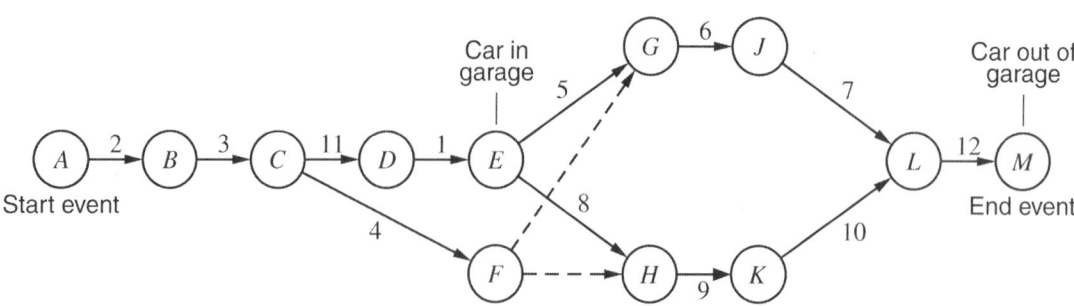

Figure 18.5   The improved procedure, with tires taken from stock simultaneously with payment.

order to show that the tires must be in the garage (activity 4) for mounting to take place (activities 6 and 9). In the next section, the concept of scheduling is introduced, and we can calculate the amount of time saved.

<table>
<tr><td>**18.3**</td><td>**Scheduling with CPM**</td></tr>
</table>

When the initial planning is complete, scheduling can begin. We say "initial" planning because we may, as a result of scheduling, decide to change the plan. Two terms associated with scheduling must be defined.

- The *earliest event time* (EE) for an event is the earliest time at which the activities that precede the event can be completed.

- The *latest event time* (LE) for an event is the latest time at which the activities that follow the event can commence without delaying the project.

The EE and LE times are easily calculated, as shown on the following pages. Since the EE and LE times are associated with events, they are usually included in the event circles (the nodes of the graph), as shown in Figure 18.6, to make the arrow diagram more understandable.

Calculating earliest event times

The EE value at the start event is zero. At any other event, the EE value is the sum of the activity times for the arrows from the start event to that event, but if there is more than one path from the start event, the EE value is the largest of these path times. The calculation is usually done by working from left to right until the end event is reached. The EE value at the end event is the total time T required for the project.

Calculating latest event times

The calculation of LE times is similar to that of EE times, except that the path traversal is carried out in reverse, beginning with the end event. The end event LE value is T. To calculate the LE value for any other event, the activity times for the arrows between that event and the end event are subtracted from the total time T, but if there are two or more paths to the end event, the largest of the path times is subtracted from T. The LE value for the start event must be zero, which serves as a check for arithmetic errors.

The critical path

A path with the longest path time from the start event to the end event is said to be "critical," since additional delay in such a path delays the whole project. Conversely, if the critical path time is reduced, then a reduction of the project time may be possible. All non-critical paths have spare time available for some activities. Events with equal EE and LE values are on a critical path.

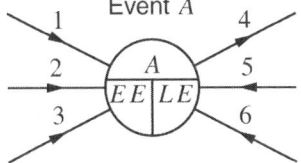

Figure 18.6   The format for writing earliest event time (EE) and latest event time (LE) in an event circle.

Example 18.2
Tire changing
(Example 18.1
continued)

Consider again the tire-changing example discussed previously; the activities and their durations are shown in Table 18.1. We shall find the critical path for both the one-person and fully staffed stores.

The planning graph for the one-person operation, shown in Figure 18.3, has been redrawn in Figure 18.7, showing EE and LE values in the event circles. The activity time Δt is shown in parentheses below each arrow. As discussed previously for the one-person operation, the CPM arrow diagram is a simple sequence of activities, so the relationships between Δt, EE, LE, and T values are simple. The total project time is the sum of the activity times; that is, $T = 97$ min. In this simple case, $EE = LE$ at every event, so every activity and every event is on the critical path. If there is a change in any activity time, there is a corresponding change in the project completion time.

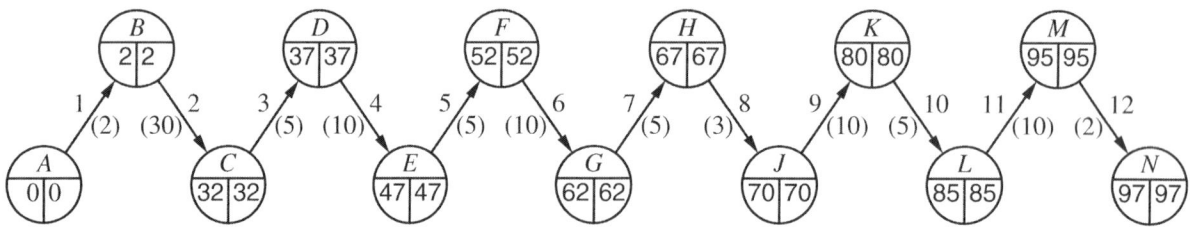

Figure 18.7   The schedule for one operator, from Figure 18.3, showing event times and a total time $T = 97$ min.

Consider the fully staffed tire store of Figure 18.8, which is Figure 18.4 with the scheduling information included. The figure shows that simultaneously mounting two tires (activities 6 and 9) reduces the total project time to $T = 79$ min. The critical path is indicated in Figure 18.8 by thick arrows. To reduce the total project time, it is necessary to examine and improve activities on this path.

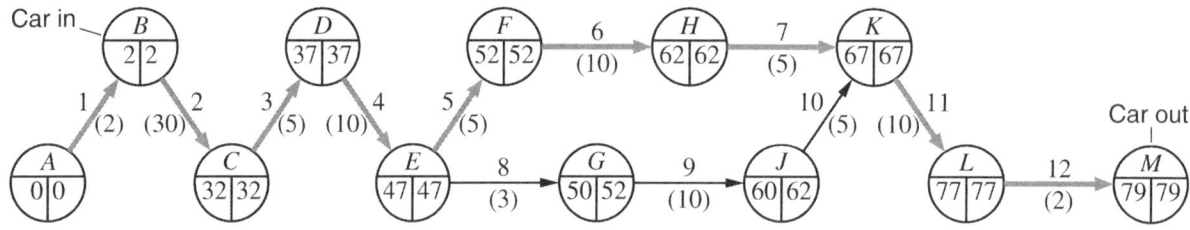

Figure 18.8   The schedule for the large store, from Figure 18.4, showing event times and a total time $T = 79$ min. The critical path is indicated by the thick arrows.

The improved procedure shown in Figure 18.5 is redrawn in Figure 18.9 with the scheduling information included. The total project time has been further reduced to $T = 69$ min. In addition, grouping the sales and billing activities and separating them from the shop activities has required the car to be in the garage from $EE = 47$ to $EE = 69$, a reduced time of only $(69 - 47)$ min $= 22$ min. This is clearly an improvement, freeing the garage for other work. The project could probably be further improved, but this single cycle of planning and scheduling has illustrated how a methodical approach leads the planner, almost automatically, to think about methods of improvement.

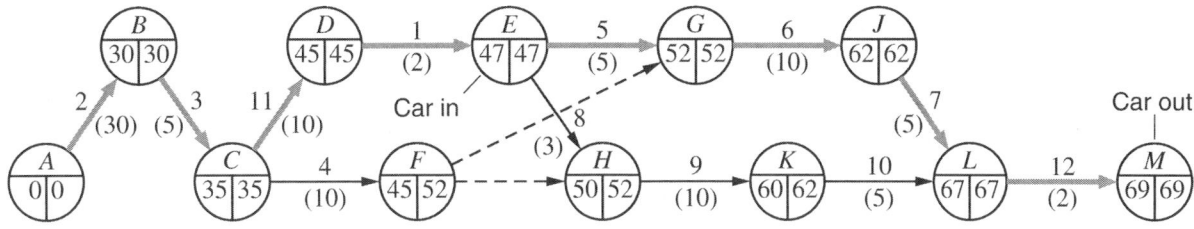

Figure 18.9   The improved procedure, from Figure 18.5, showing event times and a total time $T = 69\,\text{min}$.

Float time   A final important definition will be introduced. Activities that are not on the critical path have some spare time, usually called *float time* or *slack time*. We can calculate this time F as follows: if an activity starts at event A and ends at event B, then F equals LE at event B minus EE at event A, minus the activity time Δt. In equation form:

$$F = LE_B - EE_A - \Delta t. \tag{18.1}$$

As an example, consider activity 9 in Figure 18.9, which can start as early as $EE = 50\,\text{min}$ (event H) and must end before $LE = 62\,\text{min}$ (event K). The difference is $12\,\text{min}$, although activity 9 requires only $10\,\text{min}$. Therefore, we have $(62-50-10)\,\text{min} = 2\,\text{min}$ float time.

The float time is significant for project managers. If a delay occurs during an activity, the completion of the project will not be affected, as long as the delay is less than the float time for that activity. Activities on the critical path have zero float time.

18.4   Refinement of CPM

The tire-changing example of Section 18.3 is rather simple, but it includes all of the basic CPM concepts. It shows how the CPM method forces the planner to ask questions that may lead to improvements. The improvement of an adequate but inefficient plan is called "optimization." In simple cases, the list of activities shows the planner the optimal course of action and the arrow diagram may not be needed.

It may be thought that the techniques for calculating EE and LE times and the total time T are rather formal and mechanical. However, the calculation of the total project time T can be done in no other way, regardless of the scheduling technique. The reverse process, the calculation of LE values and the identification of the critical path, is unique to CPM and distinguishes it from other methods. The critical path and the LE times provide valuable information when delays occur in the middle of a project. The use of CPM can provide insight that may suggest changes to counteract or eliminate the delays.

The arrow diagrams sometimes appear quite complicated, even for simple problems such as the tire-store example discussed above. However, the diagram is a permanent record of the logical decisions made in the planning process. The diagram becomes more important as the number of activities increases. When the activity count is large (approximately 100), it is necessary to use one of the many CPM computer programs developed to calculate EE, LE, T, and float times. These programs typically list the activities in order by start times; some will produce a Gantt chart, which is usually easy to understand. The Gantt chart for the tire-store example is shown in Figure 18.10, where the solid bars are activity times and the grey bars are float times. The chart corresponds to the arrow diagram in Figure 18.9.

Figure 18.10   Gantt chart for the tire-changing example, illustrating the improved procedure of Figure 18.9 with slack times shown in grey.

18.5   Summary of steps in CPM

The following eight steps outline the use of CPM in planning and scheduling a project.

1. List all the activities in the project, and estimate the time required for each activity.

2. Construct the arrow diagram, in which each arrow represents an activity and each event (node) represents a point in time.

3. Calculate the earliest event time (EE) for each event by calculating the maximum of the path times from the start event. The end event EE value is the total project time T.

4. Calculate the latest event times (LE) for each event by subtracting the maximum of the path times to the end event from T. The start event LE should be zero.

5. Calculate the float time for each activity using Equation (18.1).

6. Identify the critical path. Events that have identical EE and LE values have no spare time and are on the critical path. The activities that join these events and have zero float time are the critical path activities.

7. Optimize the project. When the diagram is finished and values of EE, LE, T, and F have been calculated, it is possible to survey the project and look for ways to reduce time and cost, eliminate problems, and prevent bottlenecks.

8. Use the information obtained to keep the project under control. Do not use it just for initial planning and file it away. The main strength of CPM is its ability to identify courses of action when crises occur during the project.

18.6   Further study

1. Choose the best answer for each of the following questions.

(a) The planning component of project planning and scheduling

    i. establishes and organizes the steps to be followed to complete a project.

    ii. optimizes project objectives.

    iii. attempts to minimize time, waste, and cost.

    iv. i and ii.

    v. i, ii, and iii.

(b) A Gantt chart is

    i. a project-planning tool but not a project-scheduling tool.

    ii. a project-visualization tool.

    iii. a project-scheduling tool but not a project-planning tool.

    iv. useful for revealing how delays in one project activity will affect other project activities.

(c) The critical path method

    i. is superior to a Gantt chart.

    ii. simultaneously incorporates project planning and scheduling.

    iii. uses slack time to accommodate uncertainties in activity durations.

(d) The critical path method

    i. yields a project plan that minimizes project time.

    ii. can provide guidance on how to optimize an existing project plan.

    iii. uses either time or cost to identify the critical path of the project.

(e) The critical path method starts by

    i. identifying project objectives.

    ii. identifying the critical path of the project.

    iii. listing project activities.

    iv. building a Gantt chart.

(f) Critical path method events

    i. describe the activities essential to the project.

    ii. are also referred to as milestones.

    iii. measure the time required to complete a project activity.

    iv. are also referred to as critical path activities.

(g) Dashed lines are used in critical path method diagrams to

    i. connect the earliest event time (EE) with the latest event time (LE).

    ii. represent locations of float time in a project.

    iii. represent activities that take zero time.

    iv. identify an activity that is optional or where a choice between activities to follow exists.

(h) In the tire-changing example (Example 18.1) of the critical path method, the time to complete the project was reduced from 97 min to 69 min by

    i. hiring more people.

    ii. recognizing that certain activities could be run in parallel.

    iii. inserting float times into strategic activities.

    iv. i and ii.

    v. ii and iii.

(i) In the critical path method, the slack time of an activity

    i. can always be calculated as the difference between the latest event time (LE) of an event and its earliest event time (EE).

    ii. is the time when workers rest.

    iii. can only be calculated once the critical path method diagram is complete.

    iv. provides opportunities for accommodating unexpected project delays.

(j) Identify the statement or statements that are true for the critical path method.

    i. Achieving a desired minimum time to completion is guaranteed.

    ii. It is impossible for a project activity on the critical path to have a non-zero float time.

    iii. A unique critical path in the project is always readily identified by the CPM.

    iv. The critical path of a project must be identified before float times can be calculated.

18.7 References

[1] K. G. Lockyer and J. Gordon, *Project Management and Project Network Techniques.* London: Pitman Publishing, sixth ed., 1996.

[2] J. J. O'Brien and F. L. Plotnick, *CPM in Construction Management.* New York: McGraw-Hill, fifth ed., 1999.

[3] A. Harrison, *A Survival Guide to Critical Path Analysis.* Oxford: Butterworth-Heinemann, 1997.

[4] S. A. Devaux, *Total Project Control: A Manager's Guide to Integrated Project Planning, Measuring, and Tracking.* New York: John Wiley & Sons, Inc., 1999.

Chapter

19

Safety in Engineering Design

No engineering design or workplace environment can ever be absolutely free of hazards; however, usually with little effort it is possible to reduce the risk of injury or illness to an acceptable level. Although all engineers must be concerned with safety, these matters can also be considered as a technical specialty, and there is a technical society [1] for loss prevention specialists.

To deal with hazardous situations effectively, the engineer needs a methodical procedure for recognizing hazards and introducing remedies. This chapter discusses some basic methods of recognizing hazards and reducing them by introducing

- safety responsibilities of the engineer,

- guidelines and principles for recognizing and controlling hazards,

- applicable safety codes and standards.

19.1 Responsibility of the design engineer

Health and safety issues usually receive public attention only after an accident such as a bridge collapse, a railway derailment, a spill of toxic chemicals, or a gas explosion. These disasters are reported in the news from time to time, although the numbers of such incidents have decreased over the years as education, codes, and standards have improved. For example, fatal boiler explosions were common in the late 1800s, when steam was first being used for engines and heating. However, since the development of the ASME boiler and pressure-vessel design code, boiler explosions have disappeared, even though steam at much higher temperatures and pressures is widely used for electrical power generation and for heating large buildings.

Engineers may be responsible for managing situations that contain potential danger for those present or those who will occupy or use the result of the work. Examples of potential danger are the presence of high voltages, high temperatures, high velocities, toxic chemicals, and large amounts of energy of many kinds, such as mechanical, hydraulic, thermal, or electrical energy. In addition, health or safety hazards may be encountered in construction, manufacturing, or maintenance processes in the factory, worksite, or field installation.

An engineer carries a serious responsibility when supervising a design or construction project, because failure to act to correct a potentially dangerous situation or failure to follow codes and standards is, by definition, professional misconduct. The engineer who neglects health and safety aspects of a design runs the double risk of disciplinary action by the engineer's regulatory Association (as discussed in Chapter 3), as well as potential legal liability if damage or injury should occur as a result of design deficiencies or inadequate safety measures.

There are two groups of people whose health and safety must be considered: the eventual users of the device, product, or structure being built and the workers manufacturing it or its components. The safety of any design or construction project must be examined explicitly with respect to both of these groups.

| 19.2 | **Principles of hazard recognition and control** |

There is a general procedure, described briefly below and explained in more detail later in this chapter, for dealing with a hazard.

1. Identify and assess the hazard.

2. Try to prevent or eliminate the need for creating the hazard.

3. If the hazard cannot be eliminated, then it should be treated as a form of signal that emanates from a "source" and follows some "path" to a "receiver," the human worker or user of the design, where it may inflict harm. The application of this analogy is called *source–path–receiver analysis,* and identifies three locations where action can be taken to prevent the damage. Some examples are shown in Table 19.1.

4. If the above steps prove unsuccessful and the resulting design proves to be unsafe, then remedial action is essential: recall the unsafe devices, notify people of danger, assist the injured, and so on, as appropriate.

The following paragraphs outline a basic set of principles for assessing hazards.

Assess the capabilities and limitations of users
Designers must realize that the equipment or products they are creating will ultimately be used by persons whose physical and mental abilities may vary considerably. Designing an item for a general user is more demanding than creating products for highly select user groups such as airline pilots, for example. General users have wide variations in abilities, and their minimum skill level should not be overrated.

Anticipate common errors and modes of failure
Experience has shown that when a design goes into widespread use, there is validity to Murphy's law: "Anything that can go wrong will go wrong." Thus, designers must try to anticipate what can go wrong with their solutions and modify the design to prevent these unwanted events, or if prevention is not possible, at least to minimize the consequences and warn of the hazard.

How can this be done? In general, it means asking a series of questions such as, "Can a given event or sequence take place? If it can, what will be the result? Can this result lead to illness, injury, or death? If so, what can be done to prevent this?" This process is known as "idiot proofing."

Evaluate designs for safety and health
Evaluation of a design for safety and health risks is usually undertaken as part of a design review. While the process is by no means foolproof, it can help considerably in the elimination of most obvious hazards. It may also prove helpful in establishing a defence against a future product liability lawsuit by demonstrating that the hazardous nature of the product was explicitly considered and prudent control steps were taken.

Table 19.1   Examples of source–path–receiver analysis and hazard control methods.

Hazard	Control Measures		
	Source	Path	Receiver
Mechanical	1. Enclosure guards 2. Interlocking guards (mechanical, electrical) 3. Reduction in speed 4. Limitation of movement	1. Guard by location (rope off area, etc.) 2. Remote control	1. Education 2. Rules and regulations for clothing, etc. 3. Pull-away devices 4. Aids for placing, feeding, ejecting workpieces 5. Two-hand trip switch buttons
Noise	1. Enclosure 2. Surface treatment 3. Reduction of impact forces	1. Building layout 2. Increase distance 3. Acoustic fibres 4. Mufflers	1. Protective equipment: earmuffs, earplugs 2. Limit exposure time
Electrical	1. Low voltage instruments 2. Fuses, circuit breakers 3. Insulation 4. Lock-outs 5. Labelling, test points	1. Grounding 2. Use of ground fault detectors	1. Protective equipment 2. Education
Thermal (heat or cold)	1. Shielding 2. Insulation 3. Painting 4. Ventilation 5. Limiting physical demands of the job	1. General ventilation	1. Select acclimatized personnel 2. Acclimatization program 3. Adequate supply of water 4. Special clothing; ventilated suits 5. Proper work–rest schedules 6. Limiting exposure time
Chemical	1. Isolate or substitute 2. Change of process	1. Ventilation	1. Protective equipment, respirators, etc.

Provide adequate instructions and warnings

Many safety-related incidents can be traced to poorly designed and prepared operating instructions or maintenance manuals, as well as inadequate warning labels. When these items are prepared, designers should carefully consider the people who will be using their products or machines. Warnings should be permanently affixed in bright colours near all danger points in locations where they will readily be seen. If possible, these warnings should be permanently attached to the product in an indestructible form. Pictorial warnings have been found to be more effective than written messages.

Design safe tooling and workstations

One of the most common industrial injuries involves a worker inserting a hand into a machine closure point that is closing. These accidents occur in spite of machine guards when the effect of a guard can be circumvented by a careless operator. It is better to have integral or built-in systems that prevent irregular actuation. If correctly designed, such systems can be almost impossible to disable.

Consider maintenance needs

There are nine basic rules for making certain that a product can be safely maintained and repaired. They are as follows.

1. *Accessibility.* Parts should be arranged so that the most frequently replaced items can be reached with a minimum of effort.

2. *Standardization.* If possible, interchangeable parts, set in common locations, should be used on different models of the product.

3. *Modularization.* It may be possible to replace an entire unit without the need to repair malfunctioning individual components. This technique is particularly useful for field repairs.

4. *Identification.* Marking, coding, or tagging products makes them easier to distinguish. Colour coding of electrical wiring is a good example.

5. *Safety considerations.* When the product is being serviced or repaired, the worker must not be placed in an unsafe situation. For example, sufficient clearance should be provided, particularly in dangerous locations, such as near high-voltage sources.

6. *Safety controls.* Prevent the machine from being turned on remotely (from another switch location) while maintenance is performed.

7. *Storage areas.* Provide separate storage areas for special tools, such as those needed for adjustment.

8. *Grounding.* Ground the machine to minimize electrical hazards.

9. *Surfaces.* Provide proper walking surfaces to prevent accidents due to slipping or tripping.

Provide clear indications of danger

Anticipate incorrect assembly, installation, connection, or operation, and prevent them through design. Prevent a malfunction in any single component or subassembly from spreading and causing other failures. These techniques are part of fail-safe design.

When hazards cannot be eliminated, signs and labels recommended by standards institutes such as the American National Standards Institute (ANSI) or the International Organization for Standardization (ISO) should be displayed. The warnings indicate the nature and severity of the hazard and the means of avoiding it. Figure 19.1 shows elements of standard signs and product safety labels. The principal signal words are as follows.

DANGER

The danger sign warns of extreme or imminent risk of death or injury. The lettering is white on a red background.

WARNING

The warning sign denotes a specific potential hazard that could result in injury or death. This panel has black lettering on an orange background.

CAUTION

The caution sign is used to warn of risks or unsafe practices that could result in minor or moderate injury. The panel has black lettering on a yellow background.

A safety sign or warning label includes a panel with one of the above signal words, and may include explanatory text and one or more pictograms such as illustrated in Figure 19.1. The pictograms have three principal formats as illustrated in the figure:

Figure 19.1 Elements of standard hazard signs and product labels. The background is red for the danger panel, orange for the warning panel, and yellow for the caution panel. Standard pictograms include the yellow warning triangle, the blue circle showing a mandatory action (wearing a hard hat), and the barred red circle on a white background prohibiting an action (no thoroughfare).

the triangle for warning signs, a blue circle showing mandatory action such as wearing protective clothing, and a barred red circle prohibiting an action. Hazard signs and labels intended for international distribution may require symbol-only formats because of the many languages involved, but North American standards do not conform totally with this practice.

Control energy density High energy densities increase the possibility of accidents and should be avoided if possible. For example, include appropriate pressure-relief valves or sensors with automatic shutoff facilities that operate when out-of-range conditions are detected.

Initiate a recall The final remedial action is the product recall, as a result of recognizing a real or potential hazard after a device, structure, or system has been put into operation. Action is taken to prevent the potential hazard from becoming a reality. Recalls may be instituted for one or more of five principal reasons.

1. Analysis reveals the presence of a potential hazard that can result in a pattern of serious incidents.

2. Reports are received from users or others of unsafe conditions, unsafe incidents, or unsafe product characteristics.

3. An incident reveals a previously unforeseen product deficiency.

4. A government standard or similar regulation has been violated.

5. The product does not live up to its advertised claims with regard to safety.

19.3 Eliminating workplace hazards

People are often required to work in environments where conditions are less than ideal. Such work may expose them to a variety of job-related stresses that can affect their health over time or increase their chances of becoming involved in an accident. The definitions of *stress* and *strain* in this context are different from conventional engineering terminology. *Stress* refers to any undesirable condition, circumstance, task, or other factor that impinges on the worker. *Strain* refers to the adverse effects of these stress

Table 19.2 Hazard checklist for machine design.

1. Is the machine designed so that it is impossible to gain access to hazard points while the power is on?

2. Are the controls located so that the operator will not be off-balance or too close to the point of operation whenever actuation is required?

3. Are the power transmission and fluid drive mechanisms built as integral parts of the machine so that the operator is not exposed to rotating shafts?

4. Is the machine designed for single-point lubrication?

5. Are mechanical rather than manual devices used for holding, feeding, and ejecting parts?

6. Are there automatic overload devices built into the machine? Are fail-safe interlocks provided so that the machine cannot be started while it is being loaded, unloaded, or worked on?

7. Is there a grounding system for all electrical equipment?

8. Are standard access platforms or ladders provided for the inspection and maintenance of equipment? Are walking surfaces made of non-slip materials?

9. Are equipment components designed for easy and safe removal and replacement during maintenance and repair?

10. Are all corners and edges rounded and bevelled?

11. Are all sources of objectionable noise minimized?

12. Are all control knobs and buttons clearly distinguishable and guarded so that they cannot be accidentally activated?

sources (or stressors) on performance, safety, or health. For example, extremely high (or low) values of temperature, relative humidity, and workplace lighting as well as excessive amounts of noise and vibration are typical stresses that workers may encounter.

A common approach practised extensively in areas of product safety involves the use of checklists. Frequently such lists take the form of specific questions addressed to the designer. These are intended to prompt the designer to investigate the most common hazards associated with design, choice of materials, manufacturing processes, and functional and maintenance requirements. Examples of checklists are given in Table 19.2 for machine design and Table 19.3 for checking hazards in the workplace.

Another important area of concern for both workers and product users relates to chemical contaminants that may be present in production environments or released by products. As a minimum, the designer should be fully aware of any hazards that may be associated with the product. The designer should know how to measure them, recognize the symptoms they produce, and understand the basic steps needed to reduce or eliminate

Table 19.3   Hazard checklist for workplace layout.

1. Design equipment so that it is physically impossible for the worker to do something that would hurt himself, herself, or others.

   Examples of such protective design include the following:

   (a) A rotary blade that will not start unless a guard is in place.

   (b) The inclusion of interlocks to prevent operation of a machine unless the operator's limbs and body are in a safe position.

   (c) One-way installation: connecting pins that are asymmetrical so that they will fit into a connector in only one way.

2. Cover or guard any moving parts of machinery that could cut a worker or fly off.

3. Make sure that the plant is thoroughly surveyed to detect the presence of noxious gases or other toxic air contaminants.

4. Provide a sufficient number of conveniently located fire extinguishers and an automatic sprinkling system.

5. Use non-slip surfaces on floors and stairways. Eliminate steep ramps used with rolling devices, such as forklifts.

6. Use reliable equipment that will not fail at unscheduled times.

7. Eliminate design features associated with accidents; redesign the workspace to eliminate awkward postures, to reduce fatigue, and to keep workers alert while performing repetitive tasks.

8. Label hazards clearly and conspicuously.

9. Provide warning devices.

any associated risks to health or safety. Such measures can be applied to either the source, the path, or the user, as explained below.

Source control   *Source control* refers to techniques such as capturing, guarding, enclosing, insulating, or isolating a suspected hazard.

Path control   *Path control* means increasing the distance between the source of the hazard and the receiver. Techniques that are useful here are muffling for noise, grounding for electricity, and improved ventilation to remove toxic byproducts from the air.

User control   *User control* refers to providing personal protective equipment to workers, thus reducing their exposure to hazards. Also, schedules can be modified to reduce stress, and improved training can be obtained. A general scheme showing examples of control techniques is shown in Table 19.1 on page 285.

| 19.4 | **Cost–benefit justification of safety issues** |

Generally there is a cost justification of safety programs, although the savings from such programs are often difficult to predict precisely. The savings may come from a number of different areas:

- reduced workers' compensation insurance costs,
- improved productivity from reduced error rates,
- a more stable and content workforce,
- more consistent output, resulting in better quality control,
- less government involvement to enforce safety standards,
- confidence in the operating equipment, yielding higher output,
- improved communication and employee morale.

| 19.5 | **Codes and standards** |

The word *standards* has a general meaning in everyday use, but in the context of engineering design, the term refers to documents describing rules or methods that serve as models of professional practice. Written standards are published by technical societies, as discussed in Chapter 4, but government agencies and commercial organizations also publish standards. Following a standard in design may be optional but is evidence that the work has been conducted to a professional level of competence.

Codes and *regulations*, on the other hand, are parts of or given authority by statutes, by-laws, or collections of them, by which a national or local government requires specific practices to be followed, including adherence to particular standards.

Internationally, there is a trend toward performance-based codes and away from prescriptive codes. In Canada, performance-based codes start by specifying code objectives and hence are also referred to as objective-based codes. Prescriptive codes specify exact physical characteristics to be satisfied, such as maximum physical dimensions, minimum hardness, or minimum strength, for example. Objective-based codes associate a specified characteristic (such as reduction of risk, endurance, aesthetics, or accessibility) with a performance requirement (such as time tolerance, robustness, capacity, or power). A prescriptive requirement is specified when a performance requirement cannot be established or is inappropriate. Objective-based codes may specify example alternatives for acceptable solutions to the performance requirement. The 2005 edition of the National Building Code of Canada uses an objective-based format.

Objective-based codes admit more flexibility and innovation in design, but they also require more technical sophistication of design engineers.

Example 19.1
Fire escapes

An objective-based code might set as an objective the need to protect people from fire by enabling them to reach a place of safety. The associated performance requirement may specify a minimum time to reach a place of safety once a fire is detected, and perhaps also specify a minimum distance to an exit as a prescriptive requirement. As an alternative acceptable solution the code may then offer the use of sprinklers.

Examples of national standards organizations are the following:

- Standards Council of Canada (SCC),
- American National Standards Institute (ANSI),
- Deutsches Institut für Normung (DIN),
- Association Française de Normalisation (AFNOR),
- British Standards Institute (BSI).

National model codes and many national standards have no force in law unless they are adopted (or adapted) by the province, territory, and, in some cases, municipality, with the authority to establish legal codes or regulations.

The Standards Council of Canada (SCC) is a Crown corporation that promotes and coordinates the development and application of national and international standards. Its publications assist in finding international standards [2] and standards referenced in federal legislation. Students and professors at Canadian universities can obtain access to standards without charge by contacting the SCC.

Two international organizations that create and approve standards are the International Organization for Standardization (ISO) and the International Electrotechnical Commission (IEC). The national standards organizations are members of these bodies.

19.5.1   Finding and using safety codes and standards

Many codes and standards have been developed over the years to set minimum acceptable health and safety levels. Applying applicable codes and standards for safety and other factors is not only good professional and business practice, but failure to do so may be judged to be professional misconduct, as discussed in Chapter 3. Therefore, you may be faced with determining which codes and standards apply to your work. The following paragraphs may be of assistance.

Prior practice   You may be designing modifications to products for which adherence to standards has been documented or to products in the same area. Checking similar prior work or consulting experienced engineers in the field is a primary means of determining which standards apply. You should check whether standards used previously have been changed.

No better source of advice can be found than consulting with an engineer qualified in the specialty and place involved in the project.

On-line databases   The most convenient search method is to search the vast standards databases provided by the SCC and ANSI, accredited standards developers such as the Canadian Standards Association (CSA), and more area-specific organizations such as the IEEE and the American Society for Testing and Materials (ASTM).

Catalogues   Not surprisingly, given the large number of published standards and applicable codes, there are catalogues listing them, published mainly by the standards organizations such as listed above and by industry-specific bodies such as the Society of Automotive Engineers (SAE) or by commercial testing organizations such as Underwriters Laboratories (UL). On-line catalogues [2, 3] are becoming the standard source for finding standards. Finally, the cataloguing bodies also provide search services for finding standards that may apply to specific situations.

Search assistance   Some of the same agencies that list standards also have listings or links to government regulations and statutes. References [2] and [4] are good starting places.

The statutes and regulations that originate from national, regional, and local governments contain the codes covering engineering and other professional work, particularly as it affects public safety.

The following list is a sampling of codes by government level.

Federal regulations and model codes

Canada Occupational Safety and Health regulations	Canadian Electrical Code
The Labour Code	The National Building Code
Hazardous Product Act	Canadian Environmental Protection Act

Provincial regulations

provincial building codes	provincial electrical codes
Construction Safety Act	Drainage Act
Employment Standards Act	Labour Relations Act
Fire regulations	Municipal Act
Planning Act	Surveyors Act
Industrial Safety Act	Operating Engineers Act
Elevators and Lifts Act	Boilers and Pressure Vessels Act
Occupational Health and Safety Act	Environmental Protection Act

Municipal regulations   Each town or city may impose additional regulations or laws or modify codes to suit local conditions.

An engineer beginning work on a new project should ask whether the work may be governed by some code or law. In the case of electrical networks, roads, bridges, buildings, elevators, vehicles, boilers, pressure vessels, and other works affecting the public, the answer will be "yes." Consult with more experienced engineers, and determine the codes and laws that apply.

19.6   **Further study**

1. Choose the best answer for each of the following questions.

(a) High temperature, high humidity, or high noise levels in a workplace

    i. represent possible strains for workers

    ii. are not allowed by safety codes or standards

    iii. represent possible stresses for workers

    iv. can be eliminated by use of the source–path–receiver hazard-control method

(b) The source–path–receiver hazard-control method

    i. does not require caution, warning, or danger signs if the receiver is protected from the hazard.

    ii. uses checklists to ensure common hazards are not missed.

    iii. requires the engineer to implement three safety measures, one at each of the source, the path, and the receiver.

    iv. may not eliminate a hazard.

(c) Implementing a health and safety program

    i. often saves money.

    ii. is not required if a company has sufficient workers' compensation insurance.

    iii. is concerned only with worker safety and not product quality control.

    iv. may enable a company to substitute its own safety standards for government safety standards.

(d) A design engineer

    i. must meet all codes and standards.

    ii. may legally be expected to meet a higher safety standard than required by codes.

    iii. must meet only those codes and standards required by law.

    iv. can meet a code requirement by implementing a safety design that can be proven to be safer than the existing code requirement.

(e) The safety codes applicable in Canada are

    i. objective based.

    ii. continually changing.

    iii. the legal responsibility of standards organizations such as the CSA.

    iv. the legal responsibility of the engineering societies such as the IEEE.

(f) A design that meets all safety codes and standards

    i. may still represent a potential safety liability for the signing engineer.

    ii. no longer represents a potential safety liability for the design engineer.

    iii. does not require further safety improvements.

    iv. will never be a risk to the public.

(g) Engineers must follow all codes and regulations that apply to their work

    i. because they are legally bound to do so.

    ii. because they may be charged with professional misconduct.

    iii. because they may be charged in civil or criminal court if they fail to do so, and an accident or failure occurs.

    iv. because their work cannot be patented otherwise.

    v. for the first three of the above reasons.

    vi. for all of the above reasons.

(h) The roof of the Hartford Civic Center, designed with extensive use of computer programs, collapsed under the weight of ice and snow in 1978. Who or what was legally responsible for the failure of the roof?

    i. the snow and ice load

    ii. the author of the computer programs used to design the roof

    iii. the structural engineer

    iv. the computer programs used to design the roof

(i) It is the legal responsibility of the following to report unsafe construction or manufacturing situations.

    i. Canadian Standards Association personnel

    ii. a construction or manufacturing engineer

    iii. an engineering student

    iv. a construction or manufacturing engineer and Canadian Standards Association personnel

    v. none of these

(j) The safety of any design must be examined from the aspect of safety of the user, but the safety of the workers who produce the design for sale to the public is the sole responsibility of the manufacturing company.

    i. false   ii. true

2. Measurements of the human body generally follow the familiar bell-shaped curve of the Gaussian distribution. There are three basic ways the designer can take into account these differences between people, as follows.

(a) Design for an extreme size, for example, by locating the controls on a machine so that the shortest person can reach them, implying that everyone else can reach them too.

(b) Design for an average, for example, setting the height of a supermarket checkout counter to suit a person of "average" height. Both tall and short cashiers will have to accommodate themselves to the single height available.

(c) Design for an adjustable range. This is the idea behind the adjustable front seat of an automobile. From its closest to its farthest position, it can accommodate over 99 % of the population.

Is any one of the three design options always the safest? For each, can you imagine a design situation where it would be safest? least safe? How would you apply the ideas from Chapters 13 and 14 to achieve method (c) so that a design can accommodate over 99 % of the population?

3. Listed below are various methods of reducing hazards. Can you classify each method as a source, path, or receiver control method as in Section 19.3? Do we need additional categories of control?

(a) Prevent the creation of the hazard in the first place. For example, prevent production of dangerous materials, such as nuclear waste.

(b) Reduce the amount of hazard created. For example, reduce the lead content of paint.

(c) Prevent the release of a hazard that already exists. For example, pasteurize milk to prevent the spread of dangerous bacteria.

(d) Modify the rate of spatial distribution of the hazard released at its source, for example, by installing quick-acting shutoff valves to prevent the rapid spread of flammable fluid.

(e) Separate, in space or time, the hazard from that which is to be protected. For example, store flammable materials in an isolated location.

(f) Separate the hazard from that which is to be protected by imposing a material barrier. For example, build containment structures for nuclear reactors.

(g) Modify certain relevant basic qualities of the hazard, for example, using breakaway roadside poles.

(h) Make what is to be protected more resistant to damage from the hazard. For example, make structures more fire- and earthquake-resistant.

(i) Counteract damage done by environmental hazards, for example, by rescuing the shipwrecked.

(j) Stabilize, repair, and rehabilitate the object of the damage, for example, by rebuilding after fires and earthquakes.

4. Find at least one code, regulation, or statute that governs the design of the following products in your community: (i) propane storage tanks, (ii) video recorders, (iii) cell phones, (iv) snowmobiles, (v) highway bridges, (vi) children's rattles, (vii) home plumbing, (viii) automobile gasolines, (ix) ponds, and (x) natural gas pipelines.

5. A software company provides a computer database program that is used to store and transmit medical records. Identify a code, regulation, or statute that applies to this product. Develop one example of an objective-based code requirement that might be applied to such a computer program. Include the following in your model code:

> Overall objective
>> Specific objective
>>> Performance requirements
>>> Prescriptive requirements (at least one)
>> Alternative acceptable solutions (at least one)

Should this medical database program be designed by a professional engineer? Why or why not?

6. Develop a hazard checklist for the design of a telephone. Compare your checklist with another student's list. Did you miss anything important? Revise your checklist as appropriate. How might you ensure that your checklist is as complete as possible without being unwieldy? From your revised telephone hazard checklist generate two checklists, one for the design engineer and one for the manufacturing engineer. Which of these two engineers has a higher level of responsibility for ensuring that a safe product is delivered to the customer?

19.7   References

[1] Canadian Centre for Occupational Health and Safety, *Canadian Society of Safety Engineering.* Canadian Society of Safety Engineering, 2001. <http://www.csse.org/> (March 9, 2008).

[2] Standards Council of Canada, "Standards in focus," 2004. <http://www.scc.ca/en/publications/standards/index.shtml> (March 9, 2008).

[3] American National Standards Institute, *NSSN: A National Resource for Global Standards.* Washington, DC: American National Standards Institute, 1998. <http://www.nssn.org/> (March 9, 2008).

[4] Department of Justice Canada, *Justice Laws Web Site: Consolidated Statutes and Regulations.* Ottawa: Legislative Services Branch, 2007. <http://laws.justice.gc.ca/en/index.html> (March 9, 2008).

Chapter

20

Safety, Risk, and the Engineer

Managing and reducing *risk* and increasing its opposite, *safety,* are paramount engineering responsibilities. Several advanced, computer-oriented techniques are available for analyzing hazards and helping to create safe designs. This chapter discusses the following techniques for risk analysis and management:

- checklists,
- operability studies,
- failure mode analysis and its variations,
- fault-tree analysis.

20.1   Evaluating risk in design

As designs progress in any industry, the specifications tend to require higher performance, lower cost, or both, compared to previous generations of similar designs. Hazards tend to increase because of factors such as greater complexity, greater use of toxic and dangerous substances, higher equipment speeds, temperatures, or pressures, and more complex control systems. In addition, increased consumer awareness of risk results in demands for higher safety standards. The design engineer is caught between the increased hazards and the higher demands for safety.

Risk factors must be considered when evaluating alternatives at every design stage, but particularly during design reviews, which are usually held near the end of the project. The design review is a formal evaluation meeting in which engineers and others examine the proposed design and compare it with the design criteria. The review must include a final check that the design has no serious hazards. Unsafe designs do not occur because questions could not be answered; they usually occur because questions about safety were not asked.

The necessity of answering risk-related questions leads to the study of risk management, which is a structured approach for analyzing, evaluating, and reducing risk. After identifying and ranking the risks, risk-reduction strategies are applied, starting with the most beneficial and continuing until things are "safe enough," the risk has been reduced to an "acceptable level," or available resources are exhausted. This process is examined in the following sections.

20.2   Risk management

A hazard is anything that has the potential to cause injury, death, property damage, financial waste, or any other undesirable consequence. The purpose of risk management is to reduce or eliminate the danger caused by hazards.

The three-step process   We must be able to identify the hazards that are present, estimate the probability of their occurrence, generate alternative courses of action to reduce the probability of occurrence, and, finally, act to manage the risk. Therefore, risk management can be viewed as a three-step process involving analysis, evaluation or assessment, and decision-making. The analytical methods in this chapter deal mainly with analysis, but all three steps are described briefly below.

Risk analysis   The first step, risk analysis, has two parts: identifying hazards or their undesirable consequences, and estimating the probabilities of their occurrence. Hazard identification and probability estimation require logic, deduction, and mathematical concepts, so risk analysis is objective and mathematical. Once the hazards and their occurrence probabilities have been identified, the consequent damage or injury must be evaluated.

Risk evaluation (assessment)   In the second step, alternative courses of action to reduce hazards are generated, their costs and benefits are calculated, the risk perceptions of the persons affected are assessed, and value judgments are made. Therefore, risk evaluation is less objective than risk analysis. Determining possible options and their acceptability is a tricky and often debatable step in the process. History and experience play a major role. Public perception of a product can change: a previously acceptable level of emissions, for example, may become unacceptable. When all factors are brought together, the evaluation forms the basis for management decisions to control or reduce total risk.

Management decisions   The final stage of risk management involves selecting the risks that will be managed, implementing these decisions, allocating the required resources, and controlling, monitoring, reviewing, and revising these activities.

Good practice   In design, the term *good practice* means following established methods for safe design or construction. Good practice is often embodied in industry standards and government regulations that may be mandatory. However, whether mandatory or not, adherence to good practice does not guarantee that a product will be safe in the hands of the ultimate user or that a production line can be designed, built, operated, and decommissioned with an acceptable level of risk. From the perspective of liability, adherence to good practice is the minimum acceptable level of care expected, and more advanced risk-management methods are usually essential.

20.3   Analytical methods

There are many methods, ranging from informal to highly structured, that can be used to identify hazards and to evaluate risk. Some methods simply identify hazards; others help to determine the consequences or the probabilities of the consequences occurring, or both. Still other techniques extend into risk evaluation by ranking risks according to specified criteria. Many companies modify these standard techniques to suit their own circumstances. The techniques are not mutually exclusive; a combination of methods may be best. Four frequently used methods for risk analysis, which is also called *safety engineering,* will be explained briefly.

20.3.1   Checklists

Checklists are particularly useful when a design has evolved from a previous design for which all hazard sources were carefully listed, so that the consequences of the evolution are easy to identify. Preparing a comprehensive, useful checklist can be onerous, and if the product or process changes, the checklist must be revised. Hazards that are not specifically mentioned in the checklist will be overlooked, and hazard definitions can change with new information or evolving risk perception.

Checklists should be developed as early in the project as possible. In fact, ideally they should be part of the design criteria. The checklists in Table 19.2 and Table 19.3 are examples of a typical format. These examples, as well as the generic headings suggested below, can be modified for developing specific design-review checklists.

Typical headings for a design-review safety checklist

1. General standards and standardization
2. Human factors: displays and controls
3. Maintainability in service
4. Instruction for users and maintenance personnel
5. Packaging, identification, and marking
6. Structural materials
7. Fittings and fasteners
8. Corrosion prevention
9. Hazard detection and warning signals
10. Electrical, hydraulic, pneumatic, and pressure subsystems
11. Fuel and power sources.

Checklists are built on past experience. For new or radically altered products and processes, checklists should be supplemented by *predictive techniques*, such as in the following subsections.

20.3.2   Hazard and operability studies

The methods in this subsection are used mainly in the process industries to uncover hazards and problems that might arise during plant operation. The hazard and operability (HAZOP) technique could be called *structured brainstorming*. The technique is applied to precisely specified equipment, processes, or systems; therefore, the design must be beyond the concept stage and in a more concrete form before the technique can be used. Variations of the technique can be used at all stages of the life of a plant, such as at preliminary design, final design, startup, operation, and decommissioning.

The study is carried out by a team of experts who, together, have a full understanding of the process to be examined, such as the basic chemistry, process equipment, control systems, and operational and maintenance procedures. To carry out the study, complete information is needed, including flowcharts, process diagrams, equipment drawings, plant layouts, control system descriptions, maintenance and operating manuals, and other information as necessary.

Some definitions and concepts used in HAZOP studies include the following:

- *nodes:* the points in the process that are to be studied
- *parameters:* the characteristics of the process at a node
- *design intent:* how an element in the process is supposed to perform, or the intended values of the parameters at each node
- *deviations:* the manner in which the element or the parameters deviate from the design intent
- *causes:* the reasons why the deviations occur
- *consequences:* what happens as a result of the deviations
- *hazards:* the serious consequences (note the different meaning of *hazard* in this context)
- *guide words:* simple words applied to the elements or parameters to stimulate creative thinking to reveal all of the possible deviations

The prime purpose of HAZOP studies is to identify hazards and operational problems. The hazards and problems must be well documented for subsequent attention, and there must be an effective follow-up to ensure that the HAZOP decisions are implemented. The first step is to identify the nodes in a plant process; each node is then studied, as in the following example.

Example 20.1
A heat exchanger

Consider a process that includes a heat exchanger in a duct carrying hot gas. Water flows through the coil in order to cool the gas. The water flow into the heat exchanger would be one of the study nodes. The parameters at this node would include flow, temperature, and pressure. The design intent at this node might be to provide water at a specified flow rate, temperature, and pressure. A set of guide words would be applied successively to the parameters. To illustrate, consider the guide words "NO FLOW." The cause of "NO FLOW" could be a closed valve, a broken pipe, a clogged filter, or the failure of the municipal water supply. Each of these causes is realistic, so the deviation could occur and the consequences of the deviation must be examined. The possible consequences could include vaporizing the water in the heat exchanger coil, thus producing dangerously high pressure, or allowing the temperature of the gases to rise, yielding dangerously high rates of reaction elsewhere in the process. Both of these consequences are serious and warrant attention.

The other parameters at the node—temperature and pressure—would also be studied. All other nodes in the process would be studied in a similar manner.

20.3.3 Failure modes and effects analysis

Failure modes and effects analysis (FMEA) is a "bottom-up" process that estimates the reliability of a complex system from the reliability of its components. With the addition of the assessment of the criticality (importance) of failures, a technique called "failure modes, effects, and criticality analysis" (FMECA) results. The technique is inductive, and the analysis is carried out by completing a table, following logical rules [1]. Full details of the component, equipment, process, or complex facility to be analyzed must be available, including drawings, design sketches, standards, product specifications, test results, or other information.

The basic components to investigate first depend on the desired resolution of the analysis. For a chemical plant with several processes, for example, a component might be an individual process in a reactor. If a single process were considered, the components might be the individual pieces of equipment supporting that process. If a specific piece of equipment is be analyzed, say, a pump, the component parts would be the housing, seals, impeller, coupling, motor, and motor control, for example.

FMEA process   The FMEA steps are as follows.

1. List each of the components or subsystems in the unit.
2. Identify each component by part name and number.
3. Describe each component and its function.
4. List all of the ways (modes) in which each component can fail.
5. For each mode of failure, determine the failure effect on other components and on the unit as a whole. Enter the information in the table.
6. Describe how each failure mode can be detected.

FMECA process   In addition to the FMEA analysis, criticality analysis continues with the following:

7. Indicate the action to be taken to eliminate the hazards and identify who is responsible for taking that action.
8. Assess the criticality of each failure mode and its probability of occurrence.

Many forms of tables have been developed for use with FMEA and FMECA techniques. The measure of criticality is the seriousness of the consequence, which may range from a simple requirement for maintenance, through property loss or personal injury, to catastrophe. In the FMECA method, a measure of the consequence is usually combined with the estimated probability of occurrence to weigh the risk associated with each failure.

The potential difficulty in using bottom-up methods for complex products or systems is the large effort required by a large number of components. Another technique, such as the qualitative fault-tree analysis described in Section 20.3.4, may be required to narrow the scope of the analysis. The FMECA method is effective in determining the consequences of single component failures; it is not suitable for determining the effect on the system of the concurrent failure of several system elements.

20.3.4   Fault-tree analysis

In contrast to FMEA, fault-tree analysis is a "top-down" process. First, suppose that a single major event, called the top event, has occurred. Then determine all events that could cause the top event. These contributory events are then analyzed, in the same manner, to determine the events upon which they depend. The process is continued until events are reached that do not depend on other events or need not be further broken down. Independent events are said to be *primary* or *basic*. Events that are not further broken down are called *undeveloped* events. Events can be human errors or failures of equipment.

Fault-tree analysis can be used at any stage in the evolution of a product or process. The events and their relationships are usually represented diagrammatically using standard logic symbols. This results in a tree-like diagram [2], which gives the method its name.

Example 20.2
A hair dryer

A hand-held hair dryer can be used to illustrate the fault-tree technique. Assume that the dryer is made from non-conducting plastic, has no metal parts that extend from the inside of the dryer to the outside, and has a two-conductor power cord. Assume also that the dryer is not connected to a power supply protected by a ground-fault-interrupt circuit breaker, as required in bathrooms.

The top event will be assumed to be the electrocution of the user. The purpose of the analysis is to determine how this top event could occur and to find steps to prevent it or, if total prevention is not possible, to reduce its likelihood.

The fault tree is shown in Figure 20.1. The top event will occur only if all of the inputs to the AND gate are present; that is, only if all of the events in the row immediately below the AND gate occur. The shape of the leftmost box identifies an event that is expected to occur. The dryer uses the normal 120 V house supply, a dangerously high voltage. Two events in the first row are enclosed in diamonds and not developed further.

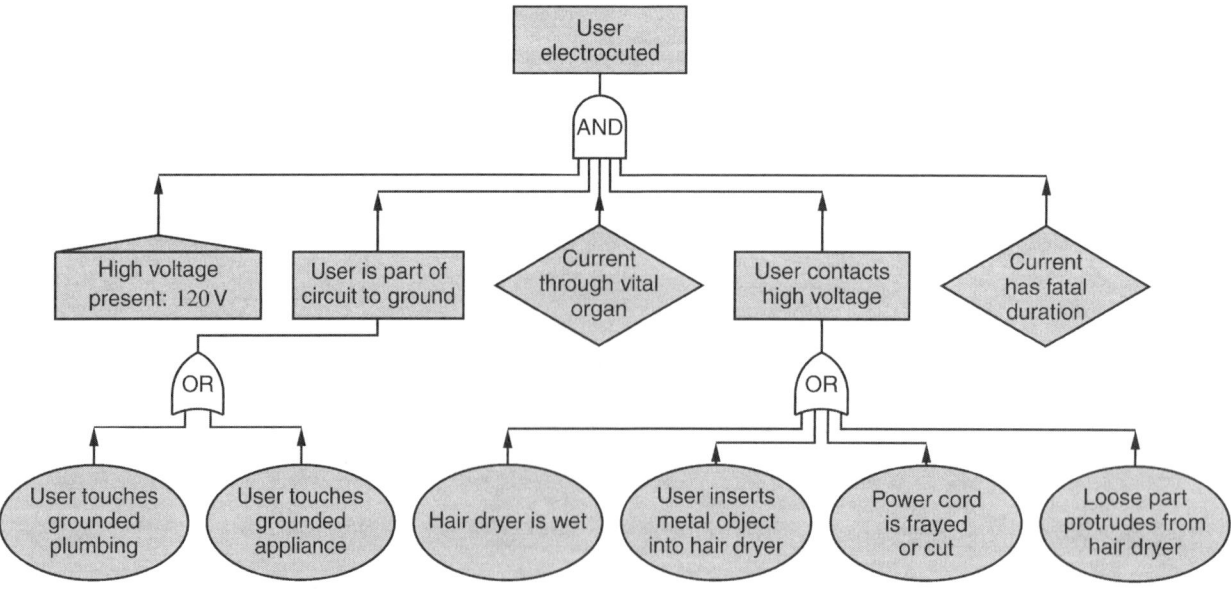

Figure 20.1   Partial fault tree for the hair dryer in Example 20.2. The two diamonds represent undeveloped subtrees, and the leftmost box indicates an expected event.

The two remaining events in this row depend on outputs from the OR gates. An OR gate output occurs if any of its input events occur. The inputs to the OR gates are basic fault events and are enclosed in ellipses.

The presence of the second-level AND gate is favourable in preventing the top event, since all of the inputs to an AND gate must be present for a higher event to occur. In a simple tree such as Figure 20.1, the basic or undeveloped events that must occur for the top event to happen can be determined by inspection. Guided by the information in the fault tree, steps can be taken to make the product safer.

Events caused by users
Some of the basic fault events depend solely on the actions of the user and cannot be prevented by design changes. However, instructions for the proper use of the hair dryer and conspicuously placed warnings about the hazards of misuse, as discussed in the previous chapter, should decrease the probability of injury and lessen the potential liability of the engineer. In the previous example, inserting fingers or metal objects through the hair dryer openings can bring a person into contact with high-voltage components; this event can be made more difficult by placing the high-voltage components farther away from the openings. The insertion of objects through openings in electrical appliances is a well-recognized hazard, and Underwriters Laboratories [3] and others have standard probes to be used in the certification of products. Of course, this is only one of the requirements for certification.

The analysis described above is qualitative in nature; however, the next step in the analysis is usually quantitative. If failure probabilities for the basic fault events are known, the probability of the top event occurring can be calculated [4]. The failure probabilities for many standard components are available from reliability handbooks and other sources. Tests can be conducted to determine failure probabilities for components or systems for which data is not available. In a more complicated tree, Boolean algebra must be used to determine the sets of basic events that must all occur for the top event to happen. Computer programs exist for the construction and analysis of fault trees.

20.4   Safety in large systems

Systems containing many interconnected parts present distinguishing features of a different type than basic products, artifacts, or static structures. Examples for which safety analysis is difficult are space navigation, vehicle control, aircraft control, chemical refining, power generation, and almost any system with a computer embedded in it.

Complexity
The first such distinguishing feature is complexity. A system can be thought of as a graph, that is, a set of nodes representing system variables or parameters, joined by branches representing processes or basic subsystems. Then if the number of branches is of the same order as the number of nodes, the system is said to have low complexity, whereas high complexity corresponds to cases where the number of branches is large compared to the number of nodes.

Coupling   Another factor of importance is the coupling between parts of the system. In a tightly coupled system, a perturbation in one part may cause changes in many other parts. In a loosely coupled system, perturbations in one part have little effect on other parts.

Response rate   From a safety standpoint, the system response rate is very important, particularly as it affects the opportunity for operator intervention. Fast-response runaway processes require very careful safety measures, while a simple alarm may suffice for a process with very slow response time.

Stability   The concepts of feedback and stability are related to response time and complexity. Interdependence, through information and control feedback paths, can lead to unstable modes of operation. The avoidance of these modes is a prime consideration in safety design.

Robustness   Finally, a design is considered robust if it brings about fail-safe or fail-soft conditions in the event of trouble. A fail-safe system can suffer complete loss of function without any attending damage. A fail-soft system may suffer loss of functionality in the event of failure but retains a minimum level of performance and safety, such as the possibility of manual control.

In conclusion, a large system may be reasonably safe if it is not too complex and is loosely coupled, slow to respond, stable, and robust with respect to part failures. Safe operation will be difficult to achieve for a complex system that is tightly coupled, with fast response rate, many feedback paths, and sensitivity to part failures.

20.5   System risk

The engineering management of risk requires the quantification of the probability of hazardous events and their cost of occurrence. Because high-cost, high-probability events tend to be designed out of systems, higher probability will typically occur with the lower costs, and lower probability with the higher costs. Intense debates sometimes arise where the probability is very small but the cost is very severe, particularly when the cost applies only to very few beneficiaries of the process. Consider, for example, safety improvements on highways with low traffic flows: accidents may have a low probability but may involve fatalities when they occur.

One definition of risk r is the product of the probability p of an event and the event cost c. Thus,

$$p\,c = r. \tag{20.1}$$

Now take the logarithm of both sides to obtain

$$\log(p) + \log(c) = \log(r), \tag{20.2}$$

so that the constant-risk loci are straight lines when the above relation is plotted using log-log scales, as illustrated in Figure 20.2. The figure gives a simple interpretation of the tradeoff between event cost and probability; the difficulty in using such definitions is in obtaining precise estimates of these quantities.

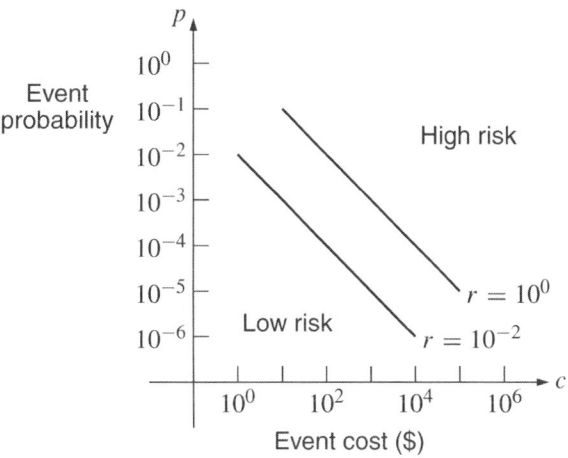

Figure 20.2 Illustrating lines of constant risk. Events in the upper-right part of the chart have high risk; low-risk events are in the lower left.

Defining the exposed population The probability of a hazardous event must be determined and defined relative to the population exposed to it. Some hazards have very low probabilities when averaged over the total population but may be more probable if only the participating population is considered. For example, the mortality of astronauts and coal miners must be evaluated relative to the number of participants, not the population as a whole. Similarly, when accident rates are compared for different modes of travel, they are often expressed in accidents per passenger-kilometre to give a defensible comparison.

Event cost The cost of event consequences may be even more difficult to define. Some costs, such as the cost of warranty replacement or repair and the cost of maintenance and service, may be relatively easy to estimate. However, the costs borne by others, including the total cost to society, may be impossible to estimate.

20.6 Expressing the costs of a hazard

In product safety analyses, costs of consequences are normally measured in monetary terms, including legal costs, which can be very high. The costs to society of an event, however, are not as simple to predict, particularly when the public environment is affected.

The risk of a particular hazard to the human population is usually termed *mortality:* the probability of death per year for exposed members of the population, or as loss of life expectancy expressed in days. The latter has the advantage that safety measures such as smoke alarms and air bags can be given a negative loss figure; that is, the gain in life expectancy by their use can be evaluated.

The wealth of an industrialized society depends on the creativity of its engineering sector, and technical advances almost always involve some risk. For example, in the last 160 years, life expectancy at birth in Canada has increased by a factor of 2.

This gain results primarily from improvements in safety and sanitation, communication, transportation, water supply, and food processing, and from medical and health improvements. The loss in life expectancy from the hazards of industrialization are small relative to the gains. Nevertheless, it is human nature, and thus engineering nature, to seek improvement. Large-scale projects now must be given a life-cycle cost analysis so that the designers see the full cost of the plant, including decommissioning and disposal of wastes, and not merely the construction and start-up costs.

20.7 Further study

1. Choose the best answer for each of the following questions.

(a) A design engineer who adheres to good practice
    i. has met the minimum acceptable level of hazard care expected but may still be liable for insufficient risk management.
    ii. may not be meeting the acceptable level of care expected.
    iii. must ensure that all hazards are eliminated.
    iv. guarantees that an acceptable level of risk has been met.

(b) Hazard checklists
    i. are designed to be used by people who are not knowledgeable about the product or process.
    ii. predict hazards on the basis of past experience.
    iii. encourage innovation.
    iv. are not useful for predictive risk analysis.

(c) A hazard and operability (HAZOP) study
    i. uses a structured procedure to avoid the unreliability of brainstorming a list of hazards.
    ii. may be a reasonable substitute for a hazard checklist.
    iii. generates a hazard checklist.
    iv. surveys the judgment of a team of experts to identify hazards.

(d) Identifying who is responsible for taking action to eliminate a hazard is a standard component of
    i. a fault-tree analysis (FTA).
    ii. a failure modes and effects analysis (FMEA) and a failure modes and effects criticality analysis (FMECA).
    iii. a failure modes and effects criticality analysis (FMECA).
    iv. a hazard and operability (HAZOP) analysis.

(e) Fault-tree analysis (FTA)
    i. can only provide qualitative results.
    ii. is a bottom-up risk analysis tool.

    iii. is a top-down risk analysis tool.

    iv. ensures that all possible failures are considered.

(f) A failure modes and effects analysis (FMEA)

    i. ensures all possible failures are considered.

    ii. is well suited to identifying concurrent system failure hazards.

    iii. only provides qualitative results.

(g) Top-down risk analysis tools include the following (identify all that apply).

    i. failure modes and effects analysis (FMEA)

    ii. hazard and operability studies (HAZOP)

    iii. fault-tree analysis (FTA)

    iv. checklists

(h) When financial risk is measured as the product of hazard event probability and cost, then

    i. risk increases with the cost of hazard event consequences.

    ii. a log-log graph is required to visualize the risk.

    iii. it is expected that the hazard event cost will include all costs.

    iv. risk accuracy may still remain a major difficulty.

(i) Mortality risk

    i. can be measured in more than one way, leading to more than one possible risk-management decision.

    ii. is a better measure of risk than monetary risk.

    iii. is uniquely defined in terms of loss of life expectancy.

    iv. measures the financial risk if someone dies.

(j) A large, complex engineering system is usually reasonably safe when

    i. the system is not too complex to model, responds quickly, is stable, has well established safety procedures, and exhibits robustness.

    ii. the system is not too complex to model, responds slowly, is stable, involves loosely coupled processes, and exhibits robustness.

    iii. the system is not too complex to model, responds quickly, involves loosely coupled processes, and exhibits robustness.

    iv. the system is not too complex to model, responds slowly, is stable, has well established safety procedures, and exhibits robustness.

2.　Suppose that you have designed a portable compressed-air supply system, as shown in Figure 20.3, for filling scuba tanks and for driving pneumatic tools in garages and factories. The system consists of an electric motor, an air compressor, an air tank, a regulator, and a pressure-relief safety valve. The regulator contains a pressure sensor connected to a power switch. The motor is switched on when the tank pressure drops

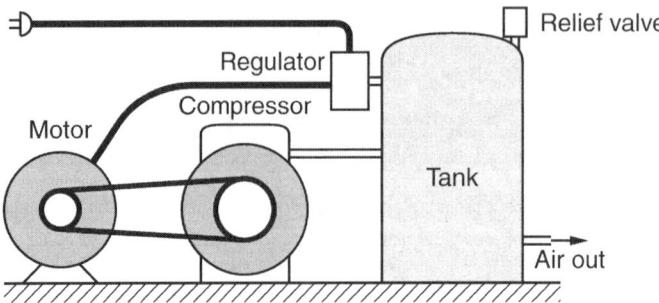

Figure 20.3 Compressed-air supply unit.

below a fixed value and off at a slightly higher pressure. The pressure-relief valve is set to open at a higher pressure than the regulator motor off-pressure. Explosion of the air tank is clearly a dangerous hazard. Three events are assumed to possibly lead to a tank explosion:

(a) an internal tank defect such as poor welding,

(b) an external cause, such as a plant vehicle colliding with the tank,

(c) excess pressure in the tank. The excess pressure can occur only if both the control unit and the pressure-relief valve fail:

- The pressure sensor might fail to shut off the motor because of switch failure or human error, such as setting the switch incorrectly or propping it open.

- The pressure-relief valve might fail to open because of mechanical valve failure or human error, such as incorrectly setting the valve or locking it shut.

Construct the fault tree for the system described above. Explosion of the air tank is the top event.

20.8 References

[1] B. Dodson and D. Nolan, *Reliability Engineering Handbook*. New York: Marcel Dekker, Inc., 1999.

[2] H. Kumamoto and E. J. Henley, *Probabilistic Risk Assessment and Management for Engineers and Scientists*. New York: IEEE Press, 1996.

[3] Underwriters Laboratories, *Household Electric Personal Grooming Appliances, Safety Standard 1727*. Northbrook, IL: Underwriters Laboratories, 2000.

[4] J. R. Thomson, *Engineering Safety Assessment: An Introduction*. New York: John Wiley & Sons, Inc., 1987.

Chapter

21

Environmental Sustainability

For centuries, human society regarded Earth's resources as inexhaustible. Previous generations concerned themselves only with finding ways to extract resources and discovering out-of-the way locations to dump waste. We assumed that the environment would be able to absorb and dissipate all the garbage, sewage, and other wastes we produced. In recent years we have come to realize that these assumptions are false. Our resource consumption and waste disposal are not sustainable. This simple fact has immense consequences for our society and for the next generations of engineers, who will be faced with solving many of the resulting problems. In this chapter we answer the following questions:

Figure 21.1 Billowing smokestacks were once considered signs of prosperity, but now are recognized as evidence of wasteful operations, health risks, and part of the cause of global warming.

- What is sustainable development?

- How is sustainability related to climate change and energy consumption?

- What do the licensing Associations expect professional engineers to know and do about sustainability?

- What can individuals and engineers do to make our future sustainable?

21.1   Defining sustainability

Sustainability is an old concept, summarized in the ancient proverb: "We do not inherit the earth from our ancestors; we borrow it from our children." The basic goal of sustainability is to ensure the continuing quality of the environment. In 1987, the Brundtland Commission of the United Nations defined sustainable development in its report, *Our Common Future:*

> Sustainable development is development that meets the needs of the present, without compromising the ability of future generations to meet their own needs [1].

Paul Hawken, in his book *The Ecology of Commerce*, defined sustainability more simply:

> Sustainability [...] can also be expressed in the simple terms of an economic golden rule for the restorative economy: leave the world better than you found it, take no more than you need, try not to harm life or the environment, make amends if you do [2].

To achieve sustainability, we must answer two fundamental questions: What kind of planet do we want for our children, and what kind of planet can we get? The first question has ethical implications, but the second question is both technical and societal. Engineers, scientists, and economists can solve the technical problems, but society must be willing to accept major changes in lifestyle. Also, society must act quickly, because decisions by previous generations have already limited what we can get.

Sustainable living and development are essential today, because the welfare of future generations is threatened by two vices of the past and the present: careless disposal of waste, and excessive consumption of resources. These are separate problems, but they are linked because emissions from burning fossil fuels contribute to global warming and climate change. Countries with high energy consumption generally have a large gross domestic product (GDP) that, in turn, is linked with large amounts of emissions and waste. For example, a large GDP typically involves manufacture of vehicles, machinery, and appliances that consume much energy when in service.

Figure 21.2 illustrates the links between GDP, energy consumption, and CO_2 emissions. First, GDP, energy consumption, and emissions are generally strongly related, as indicated by the nearly constant slopes before and after 1973. Second, massive societal change is possible with minimal effect on standard of living (GDP), as shown by the reduced slopes for energy consumption and CO_2 emissions, which indicate a massive shift to more efficient energy use following the 1973 Organization of Petroleum Exporting Countries (OPEC) oil embargo (also called the 1973 OPEC energy crisis).

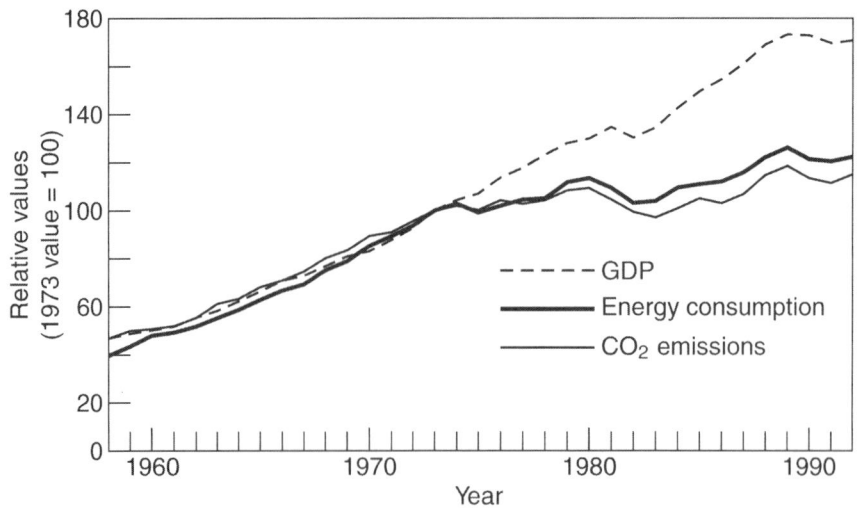

Figure 21.2 Canadian GDP, energy consumption, and CO_2 emissions relative to 1973. A deliberate reduction in oil deliveries known as the OPEC oil embargo occurred in 1973 (data from Chapter 11 of [3]).

Understanding the link between engineering achievements, such as increased efficiencies, and economics is critical as we strive for a sustainable future, and is part of the multidisciplinary engineering described later in this chapter.

| 21.2 | **An overview of sustainable thinking** |

When the Second World War ended in 1945, Canadians looked forward to an era of peace and prosperity. Industries, geared for war, converted their factories to produce appliances and vehicles. Petrochemical industries provided a selection of magical new materials—plastics—made from petroleum feedstock. Agro-chemical industries promised that new pesticides, herbicides and fertilizers would bring profitability to farms and the end of world hunger. The average person could fuel a car and heat a home effortlessly, using oil or natural gas drawn from massive reservoirs in the Canadian West.

A new era of electrical generation and distribution also started, as newly developed nuclear generating plants were built to augment electric power produced by hydroelectric and fossil-fuel plants. Visions for the future contained utopian predictions about nuclear energy, such as, "It is not too much to expect that our children will enjoy in their homes electrical energy too cheap to meter" [4]. In the 1950s, cheap electric power was made available to almost everyone in Canada. Jet engines, which were at an early experimental stage before the war, provided easy international travel, and vacations abroad became affordable. Air conditioning units and television sets, which were virtually non-existent in pre-war days, became common. At the same time, the world population continued to grow, from less than 1 billion people in 1800 to nearly 7 billion now. Many people aspired to a lifestyle that consumed more resources in a month than our pre-war ancestors consumed in a year. For decades, no one saw problems with such consumption-oriented lifestyles.

Silent Spring, 1962   In 1962, Rachel Carson described the dangers of pesticide use in her book *Silent Spring* [5]. Carson, a trained biologist, was investigating why songbirds did not return in the spring. She discovered that bird populations were dying because pesticides such as dichloro-diphenyl-trichlorothane (DDT) were being applied indiscriminately. Carson's book led to recognition that indiscriminate use of agricultural chemicals can be hazardous to bird, fish, animal, and human life.

The Population Bomb, 1968   Paul Ehrlich's book *The Population Bomb* predicted that population growth may eventually outpace agricultural food supply [6]. In a later book, Ehrlich linked population growth to environmental problems such as global warming, rain forest destruction, and air and water pollution [7]. Population growth is a basic challenge to sustainability, but it is an ethical and societal problem, beyond the control of engineers. However, as members of society, we all have a responsibility to monitor the problem and suggest solutions.

The Limits to Growth, 1972   In 1972, the Club of Rome, an organization concerned with the problems of humankind, published a report entitled *The Limits to Growth.* The report warned that uncontrolled human consumption had the potential to make our planet uninhabitable [8]. The report describes one of the first computer simulations of the behaviour of human populations. A simple "world model" simulated the creation or consumption of five basic elements over time: population, capital, food, non-renewable resources, and pollution. The "standard" run, which simulated behaviour from 1972 into the future, showed industrial output per capita peaking about the year 2000, with the production of non-renewable resources decreasing sharply thereafter. This computer analysis is naïve by current standards, but it is significant because it stimulated further research into global sustainability.

Gaia, 1979   James Lovelock's book, *Gaia, a New Look at Life on Earth*, likened our planet Earth to a self-regulating living being, with abilities to adapt and heal itself, like other organisms [9]. Many challenged the Gaia concept on the basis that the assumed healing phenomena did not exist very clearly. However, Gaia is a good metaphor. It emphasizes that we are the custodians of a living organism—Planet Earth—and we should manage it as we would a cherished family home that we want to pass on to future generations. If we damage the environment, we will regret our negligence.

The Brundtland Report, 1987   The Brundtland Commission, charged by the United Nations with the task of responding to growing concerns about the environment, issued a report: *Our Common Future* [1]. The Brundtland Report defined the concept of sustainable development, and proposed a compromise that accepted continued industrial development provided that it did not impair the ability of future generations to enjoy an equal level of prosperity. The report popularized the term "sustainable development." It is a well-accepted concept, but definitions still vary: to environmentalists who emphasize the word *sustainable,* it means that we must protect the environment from harmful change; whereas to many others who emphasize the word *development,* it means that business can continue as usual.

Montreal Protocol, 1987   The Montreal Protocol on Substances that Deplete the Ozone Layer was signed in 1987 to take effect in 1989 [10]. Certain useful fluorine and bromine-based compounds, called chlorofluorocarbons, have been shown to reduce the amount of ozone in the atmosphere. The purpose of the Montreal Protocol and its subsequent amendments is the eventual total elimination of these compounds. Although ozone is a pollutant at ground level, a layer of ozone in the stratosphere filters out harmful ultraviolet rays and is essential to life on earth. Fortunately, the Montreal Protocol has been very effective in stabilizing the ozone layer, and scientists predict that it will begin to recover in coming decades. Developing countries have, on average, a 10- to 15-year grace period to match Canada's commitments under the protocol. From acceptance and compliance perspectives, the Montreal Protocol is the most successful international environmental treaty.

IPCC, 1988   The World Meteorological Organization (WMO) and the United Nations Environment Programme (UNEP) established the Intergovernmental Panel on Climate Change (IPCC) in 1988. The IPCC members are scientists, experts from all over the world. The panel reports are comprehensive, scientific, and balanced. The IPCC does not conduct research, but monitors research around the world. It provides an objective opinion on climate change, its causes, its consequences, and how to reduce its effects or adapt to them. IPCC reports are essential documents for guiding discussions of global warming and climate change. The IPCC *Fourth Assessment Report* was released in 2007, and is discussed later in this chapter. The IPCC is a co-recipient of the 2007 Nobel Peace Prize.

The Earth Summit in Rio, 1992

An "Earth Summit" conference was held in Rio de Janeiro, where 165 nations, including Canada and the United States, voluntarily agreed to reduce greenhouse gas (GHG) emissions, because these are a known cause of global warming and climate change. This agreement, called the UN Framework Convention on Climate Change (UNFCCC), set a goal of reducing GHG emissions to 1990 levels by 2000. The goal was not achieved.

The Kyoto Protocol, 1997

More than 160 countries met in Kyoto, Japan, to negotiate new GHG emission targets. More than 80 countries agreed to reduce their emissions to an average level of 5.2 % *below* 1990 levels by the year 2010. Each country was allotted a different target. There was disagreement over issues such as credits for carbon dioxide "sinks" such as forests (which absorb carbon dioxide); whether countries could pay credits instead of reducing emissions; and what rules were fair for developing nations. In March 2001, the United States announced that it would no longer participate in the Kyoto Protocol. In December 2002, the Canadian Parliament voted to endorse the Protocol. Canada's target was to reduce its emissions of greenhouse gases to 570 Mt (megatonnes) of carbon dioxide by 2010. This amount is 6 % lower than total greenhouse gas emissions in 1990; however, emissions have increased since 1990. In fact, Canada's predicted emissions for 2010 are about 810 Mt, representing a 42 % increase over our 2010 Kyoto goal. In 2007, Prime Minister Harper announced that Canada could not meet the Kyoto target and would be negotiating a new international agreement, with more realistic "aspirational" goals [11].

Bali roadmap, 2007

In 2007, the United Nations Framework Convention on Climate Change (UNFCCC) adopted the Bali "roadmap," a process for negotiating an international agreement on climate change, to follow the Kyoto agreement when it ends in 2012 [12]. Key agreements were also reached on technology transfer and on reducing emissions from deforestation. Canada's position on future emissions targets in Bali was consistent with Canada's Clean Air Act. Tabled in 2006, this act sets intensity-based emissions targets, in sharp contrast to the total emissions targets of the Kyoto Protocol [13]. Intensity-based emissions targets require that the emissions per joule of energy consumed decrease, but permit total emissions to increase. In the Bali negotiations, Canada pushed for equal responsibility for developed and developing countries alike.

21.3   The process of global warming and climate change

The Industrial Revolution led to improved living conditions in the 1800s but, at the same time, it led to the beginning of human-induced global warming. Coal-fired steam engines replaced human and animal labour, making life easier. Trains, mills, and factories ran on steam power, and billowing smokestacks were a sign of industrial activity and prosperity. The use of coal grew so rapidly that it blackened many of the buildings in the industrial towns of central England—at the time this seemed to be a negligible effect compared to the benefits of steam power. However, the blackened buildings were omens of *global warming*, a general name for a four-step sequence that begins with burning fossil fuels and ends in climate change, as follows.

Gas emission   1. The combustion of coal, oil, natural gas, and other hydrocarbon fuels produces waste gas, mainly carbon dioxide, which is released into the atmosphere. Decaying foliage has always emitted carbon dioxide and methane, while living plants absorbed the carbon dioxide through photosynthesis, resulting in approximate equilibrium. However, after the Industrial Revolution, human activities began to produce more carbon dioxide than was absorbed.

Greenhouse effect   2. Components of the Sun's radiation pass through the atmosphere and warm Earth's surface. The absorbed energy is eventually re-emitted as thermal radiation, which is partially blocked by carbon dioxide (CO_2), methane (CH_4), and other greenhouse gases (GHGs) in the atmosphere. This transparency to some types of radiation and blocking of other types is called the "greenhouse effect." The products of combustion and other human activity contribute to this effect [15] by releasing greenhouse gases into the atmosphere. The average lifetime for carbon dioxide in the upper atmosphere (stratosphere) is many decades. Consequently, the impact of current greenhouse gas emissions will be with us for decades.

Global warming   3. The greenhouse effect is essential to human life, since it cushions Earth from the stark temperature extremes that exist on planets without an atmosphere. However, small deviations in Earth's solar energy balance can produce large effects leading either to global warming or to global cooling as occurs in ice ages. Over the past two centuries, greenhouse gas concentrations have increased dramatically in the atmosphere. For example, Figure 21.3 shows that atmospheric carbon dioxide remained nearly constant at 280 ppm (parts per million) for 2000 years until approximately the year 1800, when it began to rise to current levels in the neighbourhood of 380 ppm. This rise in CO_2 increases the greenhouse effect. It is therefore reasonable to expect Earth's surface temperature to increase. There is now widespread scientific agreement that global warming is occurring [12], although the severity of

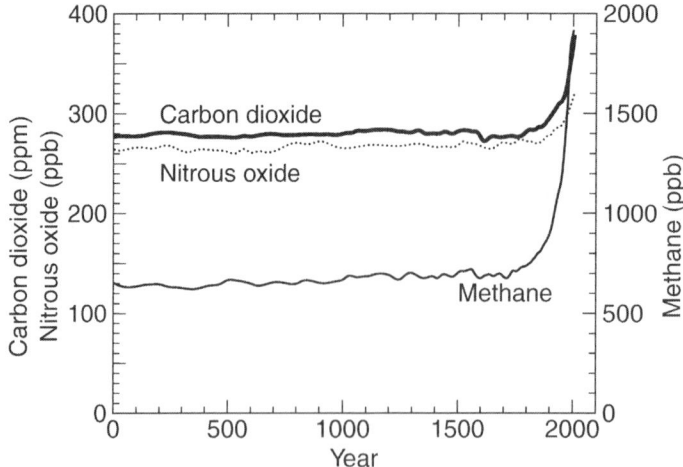

Figure 21.3   Concentrations of important long-lived greenhouse gases from year 0 to 2005. The units are parts per million (ppm) and parts per billion (ppb). (Data from Chapters 2 and 6 of [14])

the consequences are still under debate.

Climate change 4. Global warming is only a small deviation in Earth's solar energy balance, but it can produce dramatic climate changes such as severe storms, droughts, and floods.

21.4   The consequences of climate change

The research into global warming and climate change has had detractors and skeptics, of course. In 2001, Bjorn Lomborg's book *The Skeptical Environmentalist* challenged widely held conclusions from environment research, and predicted a rosy life for future generations [16]. The book continues to generate debate about climate change predictions [17].

In 2006, Al Gore, former U.S. vice-president, produced the award-winning documentary film *An Inconvenient Truth: A Global Warning*, based on his book of similar title. The film and book explain environmental research so that it is understandable by the average citizen, and the key message is that global warming is a real and present danger to our way of life [18]. Gore was a co-recipient of the 2007 Nobel Peace Prize.

21.4.1   Ending the debate about climate change

The debate about climate change effectively ended when the Intergovernmental Panel on Climate Change (IPCC) issued its comprehensive *Fourth Assessment Report* (AR4) in 2007. The AR4 took six years to prepare, and its conclusions are based on research by over 800 authors in 130 countries, reviewed by 2500 scientific reviewers. The four-part report is long and technical, but summaries are freely available on the Internet in reference [19] and companion documents.

The IPCC reports show great certainty, based on extensive evidence, that climate change is directly linked to global warming, caused by greenhouse gases that result mainly from human burning of fossil fuels. A few of the many IPCC conclusions are summarized as follows.

Human cause Worldwide atmospheric concentrations of carbon dioxide, methane, and nitrous oxide have increased markedly as a result of human activities since 1750. Current levels now greatly exceed pre-industrial values.

Temperature rise Global temperatures will increase by 1.8 °C to 4.0 °C in the next century, depending on future fossil fuel use, technological change, economic development, and population growth in the absence of climate-control initiatives such as the Kyoto Protocol. This expected temperature increase will not be uniform; the greatest temperature increases will be observed in the Arctic and Antarctic. Even if greenhouse gas concentrations are kept constant at year 2000 levels, a warming of about 1 °C will occur in the next century as a result of GHGs already emitted. The temperature rise may not appear to be great but the effects are surprising.

Sea level rise Observations of the decline in ice volume around the world lead to the prediction that sea levels will rise by 18 cm to 59 cm or more, a change that would produce the risk of floods in vulnerable coastal cities. Sea level predictions are less precise than those for temperature rise because of incomplete data on the melting of the Greenland ice sheet. If the entire sheet were to melt the sea level would rise 7 m, but that is not expected, although recent observations suggest faster melting than previously expected.

Present observations   Scientists and meteorologists have observed numerous long-term climate-related changes. These include changes in arctic temperatures and ice, ocean salinity, wind patterns, and extreme weather conditions including drought, floods, heat waves, and intense tropical cyclones. Eleven of the twelve years from 1995 to 2006 rank among the 12 warmest years in the instrumental record of global surface temperature since 1850.

Other consequences   The IPCC lists many possible consequences of climate change, including severe inland flooding and drought, melting of sea ice, insect plagues, extinction of species, bleaching of corals, and possibly even major changes in ocean currents.

Feedback   An important factor in the science of global warming and climate change is the effect of feedback [14, 20]. Feedback is studied in mathematical detail in advanced engineering courses, but it can be described roughly as the two-way interaction between different objects or quantities. The result of the interaction can be stabilizing or destabilizing, desirable or undesirable. Two examples of destabilizing, undesirable environmental feedback are the following. First, forest fires release carbon dioxide and other GHGs into the atmosphere, leading to global warming and drier conditions in which more forest fires occur. As a second example, sea ice is very effective in reflecting sunlight back into space, but as the ice melts, the darker water absorbs more energy from the Sun, thereby causing more ice to melt.

Irreversibility   A book by George Monbiot claims that if average global temperatures increase by $2\,°C$, which the book states to be probable, then vast peat bogs in the sub-arctic, presently under permafrost, will begin to decay and release greenhouse gases, leading to even greater warming that is irreversible [21].

Summary   The emission of greenhouse gases is changing our environment, and unless serious efforts are made to reduce these emissions, climate change will lower the quality of life of future generations. The burning of fossil fuels (coal, oil, gas) is the main source of increasing GHGs.

21.4.2 Ethical implications of climate change

The Brundtland Commission was concerned about environmental degradation because it leads to poverty and economic disparities between societies. The 2007 IPCC report showed that global warming exacerbates disparities, because the economic loss caused by climate change will fall hardest on the poorest nations. This disparity raises some very basic questions of fairness.

- Is it ethical or fair for richer countries to burn fossil fuels indiscriminately, thereby creating GHG emissions and indirectly imposing climate change on poorer countries? Is it ethical or fair to require poor or undeveloped third-world countries to meet the same emissions standards as Western countries, which have emitted about 20 times as much GHG per capita for the past century?

- How will the industrialized nations respond to the millions of "climate" refugees created in Africa when droughts reduce crops, or in the Netherlands, Bangladesh, and the Pacific islands when sea levels rise and flood these low-lying countries?

Ethically, sustainability is simple fairness. It is unfair and unethical to harm others through negligence, inefficiency, greed, or abuse. In this case, the "others" are future generations, including our own children. If the present generation fails to combat global warming, it will cause a serious reduction in the quality of life for future generations, particularly those living in the poorer countries.

21.5   Excessive consumption and the depletion of oil and gas

Climate change is not the only serious problem affecting sustainability. We have finally realized that the planet's resources are large but finite. We are consuming some resources so quickly that, within the lifespan of the present generation, these resources will become scarce and very expensive. If we cannot reduce consumption or find substitutes, our society will face serious disruption. Examples of such resources are fossil fuels (such as oil and gas but excluding coal), fresh water, fish stocks such as the Canadian East Coast cod fishery, and various species of wildlife. How can we maintain our standard of living without depriving future generations of these resources?

Peak oil and gas   Natural gas and oil are the most important resources that are approaching their predicted peak production rate. These resources are fundamental to our lifestyle: they provide energy to refine, manufacture, and distribute the commodities required by developed societies. They are also needed to generate the electric power that lights our homes, cooks our food, and even entertains us via computers and television. However, oil and natural gas production rates are now approaching the critical point at which they will no longer keep pace with demand. When demand exceeds supply, shortages occur and prices rise, as happened in 1973 during the OPEC oil embargo.

In order to predict future oil and gas supply and demand, we must know the current production and consumption rates, make educated guesses on how they may change, and estimate the amount of oil and gas remaining under Earth's surface.

Growth rates   Between 1850 and 2006, world population increased 520 % to 6.555 billion, corresponding to a growth rate of 1.06 % per year [22]. Over the same time interval, world energy consumption increased 850 % at an average growth rate of 1.38 % per year [23, 24]. Multiplying these two factors together (population × annual energy consumption per capita) yields the world's total annual energy consumption. Between 1850 and 2006, the total annual energy consumption increased by a factor of 43; that is, an astounding 4300 %. This total growth corresponds to a constant annual growth rate of about 2.4 %. Figure 21.4 shows this astonishing growth, categorized by energy source.

In 1850, total energy consumption per person was nearly all from wood, water, and wind power, consumed at a rate equal to 12 % of the total rate in 2006. In 2006, *renewable* per capita consumption was 11 % of the total, which means that the renewable energy consumed per person has remained essentially unchanged since 1850, and the substantial increase in standard of living since then has been fuelled entirely by non-renewable energy sources.

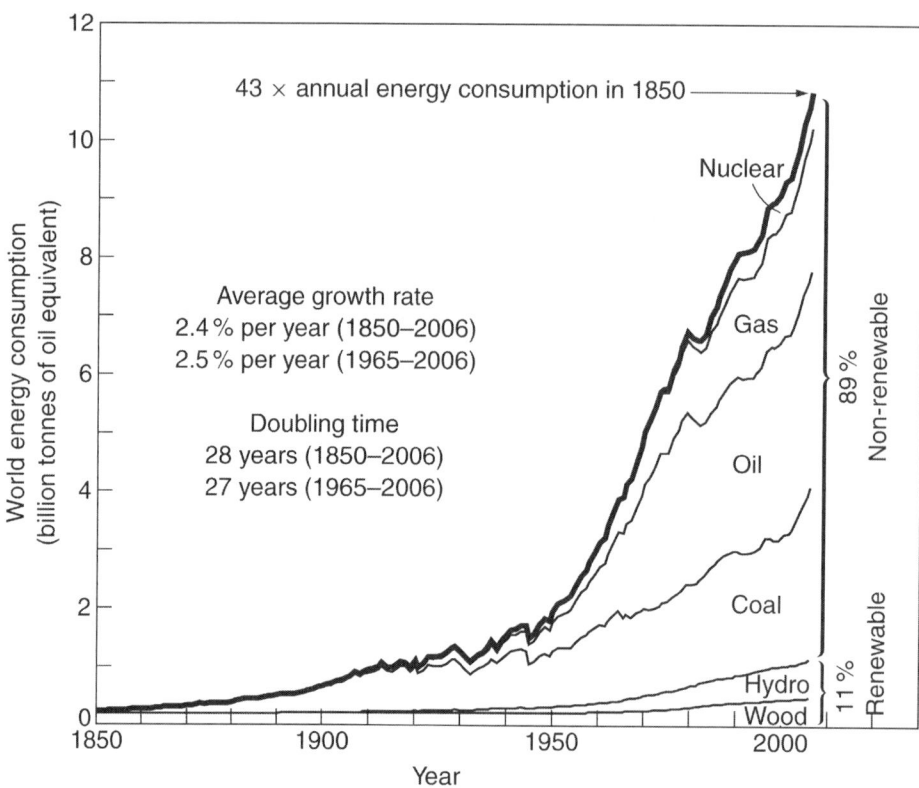

Figure 21.4 Primary annual world energy consumption 1850 to 2006 [22–24]. A tonne of oil equivalent equals 42 gigajoules (GJ) or 12 megawatt-hours (MWh) of thermal energy.

Doubling time The consumption curves in Figure 21.4 are approximately exponential. Growing exponential functions double in height in a fixed time, and the doubling times for two intervals with slightly different growth rates are noted on the figure. For an exponential curve with a 2.5 % relative annual rate of increase, the doubling time is approximately 27 years. Therefore, if total world energy consumption continues to increase at a constant rate equal to the current 2.5 %, the demand for energy will double in a mere 27 years (in the year 2035 approximately), or about one generation. More importantly, during this 27-year doubling interval, the total amount of fossil fuels we will consume will equal all the fossil fuels consumed by humans since the dawn of history! The global warming caused by the greenhouse gas emissions from this fossil fuel consumption should be obvious from the previous discussions—it will trigger irreversible climate change. Moreover, to double our consumption, we must ask whether fossil fuel production can meet this demand. This question requires us to discuss oil and gas production rates.

Peak production Oil and gas production (flow) rates have a known physical limit because, in the life of an oil or gas well, the discovery, exploitation, and depletion follow known patterns. When we add the production rates of all the wells in known reserves, we obtain the total oil and gas production rate. Taking account of well-production patterns, trustee Colin J. Campbell of the U.K. Oil Depletion Analysis Centre produced forecasts of total oil and gas peak production shown in Figure 21.5. The figure shows regional oil and natural gas peaks that have already occurred, as well as those forecast to 2050.

The best-known example of a predicted and verified peak in hydrocarbon production is the oil peak of 1970, known as Hubbert's peak after the author who predicted it in 1956 from a simple model for oil-well production in the 48 contiguous U.S. states [25].

Although some researchers argue that Campbell's forecasts in Figure 21.5 are overly pessimistic, all agree that a peak in oil production is coming soon, perhaps in a few years, but within two decades at most [24]. Natural gas is expected to peak soon afterwards. After the peak, production rates will drop because the oil and gas must come from less-accessible reserves, from stimulating older wells, or from mining oil sands, all of which are slower and more expensive methods of extraction.

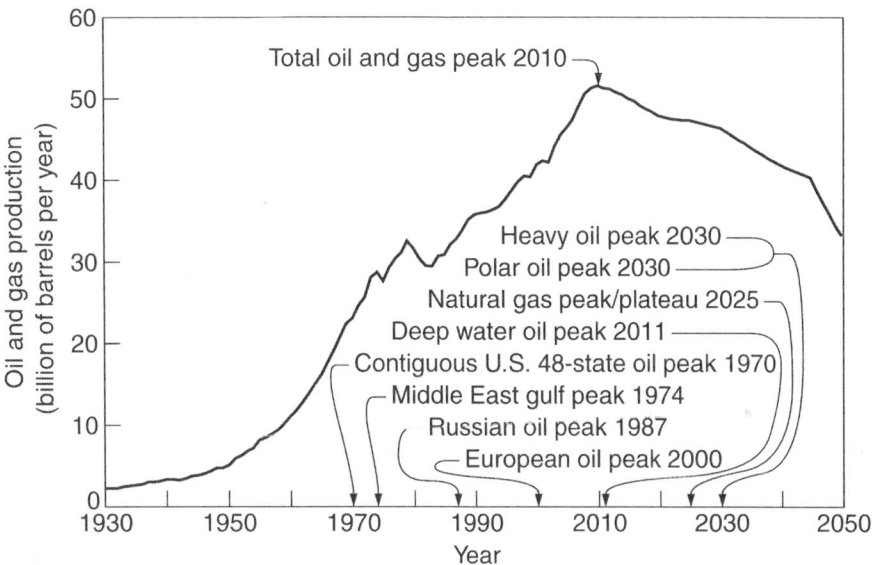

Figure 21.5   Campbell's 2006 total oil and gas production forecast [26, 27].

Gap between production and discovery   A production peak inherently lags a discovery peak. Figure 21.6 summarizes Campbell's evaluation of oil production and oil discoveries. Since 1984, world oil production has exceeded oil discoveries. This growing deficit between production and discoveries is the most important evidence confirming that a world oil-production peak is approaching.

Reducing fossil fuel dependence   We have only two choices if we want to reduce society's dependence on fossil fuels. Both will take a long time. We can

- decrease our energy consumption to a level that can be sustained with renewable sources such as solar, wind, and geothermal energy, or

- exploit another non-renewable energy resource such as nuclear energy which, being non-renewable, will itself eventually peak.

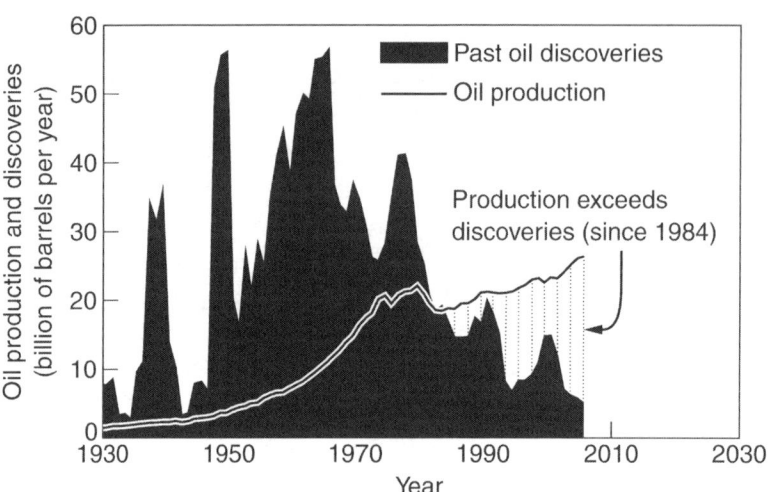

Figure 21.6   The growing gap between world oil production and discoveries [27].

Both of these choices involve great engineering challenges and opportunities, but we must realize a critical truth: *we consume enormous quantities of energy, so there is no quick fix.*

Nuclear energy is the classic example of a rapidly developed energy technology. Political and military interests sped the development of nuclear energy, but it still took about 40 years to develop the technology from scientific feasibility to engineering feasibility, then to commercial feasibility, utility integration, and finally, significant use. This example is evidence that all other major energy sources have needed or will need as much or more development time, perhaps much more. This implies that the full impact of global warming will be upon us before we can implement significant sustainable energy sources.

What next?   Unless there are large, rapid lifestyle changes to reduce energy use, or remarkable advances in renewable energy production, we will see increased dependence on nuclear energy and coal for the next century. For example, Ontario is planning more nuclear plants, and China is building hundreds of coal-fired generating plants to meet growing energy demands.

Coal is a dirty fuel; burning coal emits greenhouse gases and other serious pollutants. However, fossil fuels will be burned extensively for years to come in order to avoid economic collapse while we search for alternative energy sources. In reference [28], Jaccard suggests that carbon capture and sequestration (CCS) methods can allow the burning of fossil fuels without releasing greenhouse gases or other pollutants to the atmosphere. He further suggests that fossil fuels represent the cheapest source of clean energy for at least the next century. However, developing and applying the required techniques will take many years.

21.5.1   The ethical implications of a peak in oil and gas production

Developing countries are striving to reach the standard of living enjoyed by developed countries, and world population is increasing, so world energy consumption levels are almost certain to increase. If we cannot move away from fossil fuels or restrain the rising consumption, some very basic questions of fairness and personal sacrifice arise.

- Is it ethical or fair for the current generation to consume non-renewable resources inefficiently and thoughtlessly, thus depriving future generations, who may need these resources for survival in the midst of climate change and sharply rising prices?

- Is it ethical or fair for a government to restrict population growth? China, for example, introduced a "one child" policy many decades ago. Is it fairer to restrict population in an orderly way or to let wars, disease, and famine sort out the winners and losers?

The ethical issues raised by reaching the peak of oil and gas production are virtually identical to the ethics of climate change. As mentioned previously, sustainability is simple fairness. It is unfair and unethical to harm others through negligence, inefficiency, greed, or abuse. Obviously, we cannot accept the extreme options of wars, disease, and famine, so we must seek orderly methods to reduce our dependence on fossil fuels, to adopt a less energy-intensive lifestyle, and to increase the availability of renewable energy.

21.6   Guidelines for environmental practice

The most important guide for engineers is the same as the Hippocratic oath for medical doctors: "First, do no harm." Engineers must be able to say "no" when environmental degradation is proposed. In particular, engineers have a specific obligation to follow environmental regulations and guidelines, so that engineering projects do not make the sustainability problem worse. Environmental guidelines are published by many organizations, but those published by the engineering licensing Associations are among the most relevant, since the codes of ethics of the Associations require engineers to follow them.

In Alberta, for example, the first clause in the APEGGA code of ethics is that "Professional engineers, geologists, and geophysicists shall, in their areas of practice, hold paramount the health, safety and welfare of the public, and have regard for the environment." The APEGGA *Guideline for Environmental Practice* further explains this duty. The *Guideline* encourages licensed professionals to be involved actively with environmental issues, and to anticipate and prevent rather than react to environmental problems. Professionals are urged to seek a golden mean between "the two extremes of absolute preservation and unfettered development." One interpretation of this golden mean is sustainable development. Professionals should help to formulate environmental laws, to enforce them, and to be true stewards of the environment [29]. Briefly, the *Guideline* states that professional engineers, geologists, and geophysicists should:

1. Understand and monitor environmental and sustainability issues;

2. Use specialists in environmental and sustainability when needed;

3. Apply professional and responsible judgment;

4. Ensure that environmental planning and management is integrated into all activities that are likely to have any adverse effects;

5. Include the costs of environmental protection when evaluating the economic viability of projects;

6. Recognize the value of environmental efficiency and sustainability; consider full life-cycle assessment to determine the benefits and costs of environmental stewardship, and endeavour to implement efficient, sustainable solutions;

7. Solicit input from stakeholders in an open manner, and strive to respond to environmental concerns in a timely fashion;

8. Comply with regulatory requirements and endeavour to exceed or better them; disclose information necessary to protect public safety to appropriate authorities;

9. Work actively to improve environmental understanding and sustainability practices.

The Alberta *Guideline for Environmental Practice* was adapted by Engineers Canada, and is now proposed as a national standard [30]. Newfoundland and Labrador, Nova Scotia, Ontario, Saskatchewan, and British Columbia have adopted the Engineers Canada guideline, or have developed similar environmental guides [31, 32].

A British engineer, long involved in the oil and gas industry, made the following comment on sustainability:

> It is important to distinguish sustainability from environmental compliance. Compliance with environmental regulations is an essential, daily, operational requirement, but sustainability is a long-range strategy to combat a slow-moving, planet-wide terminal condition. Sustainability requires innovative thinking, persistence and possibly some sacrifice. Decades may pass before the social acceptance, legislation and treaties are devised to deal effectively with sustainability. Engineers must, of course, ensure that regulations are followed, but in planning projects and activities, engineers must aim their conceptual thinking at the higher hurdle of sustainability [33].

In summary, engineers have a duty to strive for efficiency, to gather information before making environmental decisions, and to get the best possible advice in order to assess projects and activities for their sustainability. Although the guidelines do not clearly say so, if we are to achieve a sustainable society, engineers must also decline to participate in projects or activities that are clearly unsustainable.

21.7 What we can do: personal lifestyles

Pursuing sustainability is important because anything less is equal to treating other species and future generations unfairly. It is unfair and unethical to harm others through negligence, inefficiency, greed, or abuse. In addition, every code of engineering ethics requires us to make the health, safety and welfare of the public our first concern (or "paramount" duty). This duty should not fall solely on engineers although they are normally part of the solution.

Examples of unsustainable inefficiencies abound. Our consumer society, even with recycling regulations in place, permits the following.

- Aluminum ore (bauxite) is extracted from Earth's crust, refined using large amounts of electrical energy, sold as aluminum foil, and discarded after a single use in thousands of dispersed landfill sites from which it cannot be recovered economically.

- Computers, television sets, radios, and other electronic products, manufactured using exotic minerals and materials, are discarded to landfill, even though they are still functional, as newer models come onto the market.

- Jet aircraft are a serious obstacle to sustainability. No technical substitutes exist for the jet engines that propel airplanes. The carbon dioxide emitted per passenger in a transatlantic flight is approximately equal to the annual use of gasoline in an automobile. Jet engines also emit water vapour, which has an effect similar to GHGs when emitted at high altitude. Travel of otherwise well-meaning people, for purposes as benign as visiting relatives or tourism, is a threat to sustainability [21].

- Plastic bags are used to carry groceries, plastic blister packs encase retail products from electronics to toys, and even chewing gum comes doubly wrapped. Excessive packaging for convenience or advertising, rather than protection, is the norm.

- Automobiles consume large quantities of fossil fuels, are a major source of greenhouse gases, and emit other unhealthy chemicals that pollute the air. Massive suburbs have grown outside cities, situating adults far from work and children far from schools. Coffee shops build drive-through lanes that accustom people to wasting energy while waiting. Automobiles are built and advertised for speeds that are illegal, for accelerations that are unnecessary, and for capacities that are never filled.

Reduce, reuse, recycle The above examples of wasteful consumption are the exact opposite of the sustainability strategy in the slogan "reduce, reuse, recycle." This practice, if followed conscientiously, is the first step to a sustainable future. For example, if we reduce our driving, heating, or use of electricity, we immediately reduce the need for fossil fuels.

To begin envisioning the next sustainability step beyond "reduce, reuse, recycle," consider the thinking of McDonough and Braungart in their book, *Cradle to Cradle: Remaking the Way We Make Things* [34]. They interpret "reduce, reuse, recycle" as an approach that "perpetuates our current one-way, 'cradle to grave' manufacturing model, dating from the Industrial Revolution, that creates such fantastic amounts of waste and pollution in the first place." In contrast, McDonough and Braungart propose a "cradle-to-cradle" principle, in which all materials used to create a product are returned to their original state, recycled to a state of readiness for reuse, or continually recycled. In effect, we discard nothing as waste in our landfills, waterways, or atmosphere. Cradle-to-cradle thinking is a challenge to designers and manufacturers, but it would be a key step toward sustainability.

The efficiency paradox

Consider next the history of the automobile, particularly the evolution of engine fuel economy. Starting with the Model T Ford in 1908, design improvements led to increases in the mass, room, comfort, power, and fuel economy of automobiles. However, by the late 1960s, room, comfort, and power were given higher priority than fuel economy. Manufacturers were responding to customer demand which, in an era of inexpensive gasoline and negligible government regulation, placed a low priority on fuel economy. This all changed with the 1973 OPEC oil embargo. Fuel prices skyrocketed. Smaller, more fuel-efficient foreign vehicles took a noticeable foothold in the North American market, and a rush to design smaller cars ensued. Fuel efficiency sold cars for a while but, by the late 1970s, smog was of greater concern to the public. This resulted in the introduction of catalytic converters. Catalytic converters reduce engine emissions, but they also reduce fuel economy, so for a time, engine designers focused on improving fuel efficiency to regain the losses from catalytic converters. Eventually however, priorities again focused on room, comfort, and power over fuel economy, as highlighted by the 2001 milestone when the sale of light trucks—SUVs, minivans, and pickups—surpassed car sales in North America for the first time. In 2001, the Union of Concerned Scientists observed, "With light trucks now accounting for more than half of all vehicles sold, the average new vehicle travels less [distance] on a gallon of gas than it did in 1980" [35].

Contrary to popular belief, efficiency improvements do not necessarily lead to a more sustainable future, a fact sometimes called the *efficiency paradox*. Although technology improvements allow for better energy efficiency, those savings are often lost to greater consumption. For example, the efficiency of air conditioning units has risen 17 % since 1990, but the number of units in use increased by 36 % over that time. Similarly, sales of light trucks, SUVs, and vans rose by 45 % between 1995 and 2005—nine times faster than passenger cars—but on average, they have 25 % lower fuel efficiency than cars [36]. Efficiency improvements are an important part of sustainability, so the work of engineers is critical, but without major societal changes in consumer thinking and expectations, achieving sustainability is unlikely in our lifetimes. Fortunately, there is a ray of hope. New U.S. government efficiency standards have begun to be applied to light trucks, the first "made in Canada" fuel economy standards for cars and light trucks were announced in 2008, and recent drastic fluctuations of fuel prices are having an effect on both travel volume and the types of vehicles purchased.

Rethink

The above examples should make it clear that the key to bringing sustainability changes to our consumer-oriented society is a re-assessment of our priorities and consumption. In short, we must rethink expectations and lifestyles. The sustainability mantra of "reduce, reuse, recycle" should be updated to read

Reduce, Reuse, Recycle, **RETHINK**

with emphasis on *rethink*. As a society, we must develop new ways to avoid inefficient or excessive consumption of energy.

| 21.8 | **What we can do: engineering for sustainability** |

Every code of ethics requires engineers to make the health, safety and welfare of the public our first concern: our "paramount" duty. However, this duty should not fall solely on engineers, because they are not the problem: they are part of the solution. Wasteful practices, accepted by society for centuries, must change. Fortunately, engineers can assist society to make these changes by promoting energy efficiency, encouraging research into alternative energy sources, modelling alternatives, and showing the environmental consequences of design decisions.

Recall the two fundamental questions of sustainability: What type of planet do we want? What type of planet can we get? Engineers have a key role to play in sustainability, because engineers have the knowledge and skills to determine what is technically possible. However, society must decide what it wants. Sustainability requires that our huge consumption-driven society must change its attitudes, laws, and way of life. This is a major disruption, and it will require debate, discussion, and courage. Engineers must provide feedback from our knowledge of the question "What type of planet can we get?" to help society answer "What type of planet do we want?"

Engineers rarely speak out in political and ethical debates—even in the debates about sustainability—but society needs engineers in these debates. The discussion below suggests how engineers can help society move toward sustainability.

Modelling   Global climate and general circulation models (GCMs) have steadily increased in sophistication as new knowledge and detail have been included. A 40-year review of climate modelling is found in reference [37].

GCMs are sophisticated tools for modelling the atmosphere and the possible effects of climate change. These models are applicable for the sustainable design of land use, coastal structures, erosion control, buildings, and towns. All models have some uncertainty, of course, but uncertain predictions of temperature and climate are more useful than complete ignorance. The design engineer must know the capabilities and limitations of any model and its result. From the pyramids to the space shuttle, engineers have used models with varying degrees of uncertainty to yield great designs. It is now time to use GCMs and other models to design for sustainability.

Life-cycle analysis   Life-cycle analysis (LCA) evaluates the total environmental impact of a product from cradle to grave or, in other words, from source material to disposal. The effects considered by some LCAs, for example, are those associated with emissions such as CO_2, SO_x, NO_x, phosphates, or other chemicals, or with the amount of water and energy consumed. Specific toxic emissions such as arsenic and lead are also sometimes considered, and efforts have been made to include mass and economic considerations [38].

One application of LCA is for sustainability decision-making. Consider, for example, the choice between buying either cotton-cloth diapers or biodegradable disposable diapers for a child. Environmental impact should be a factor in the decision. An LCA for the two alternatives may estimate the CO_2 emissions and effects on water and landfills as follows.

- Cloth: The LCA would likely start with the CO_2 emissions from the fertilizer and the tractor fuel needed to grow the cotton for the cloth diapers. The analysis would also include the water and energy consumed to wash and dry the diapers repeatedly, and the small mass eventually discarded to a landfill.

- Disposable: Correspondingly for the disposable diapers, the LCA would include the CO_2 emissions from manufacturing a much larger mass of absorbent material, the water and energy used in their manufacture and distribution, and the larger mass of diapers sent to a landfill over the length of the child's infancy.

Before an LCA comparison can be used to decide on the best environmental alternative, a case-specific analysis is generally required, along with an understanding of the effects of alternatives. For example, consider the diaper choice, above. If the cloth diapers are washed in cold water and dried in the open air, they will emit less CO_2, consume less energy, and contribute less mass to landfills, but will consume more water than the disposable diapers. The decision maker must then weigh the environmental-impact factors; for example, water consumption may be more critical than CO_2 emissions in a drought-stricken country.

The movement to create an international LCA standard by the International Organization for Standardization (ISO) [39] is presently hampered by serious restrictions on data and uncertainties about the correct theoretical scope. For confidentiality, cost, and image reasons, manufacturers are not willing to make the necessary LCA data publicly available. In addition, defining the scope is often difficult. For example, in the cotton diaper analysis above, does one include the energy needed to produce the farmer's food, or does one simply stop at the tractor and fertilizer energy consumed? In spite of these limitations, LCA is evolving into a vital tool for sustainability decision-making.

Multidisciplinary engineering and systems thinking

Economic concerns and outdated customs are the basic reasons why society is resisting the attack on global warming, climate change, and the depletion of our non-renewable resources. Engineers must bridge the gap between science and economics to arrive at good solutions. Scientists attempt to understand the technical issues and economists develop economic models, but it is the engineer who works in the world between the two, seeking technical solutions that are economically feasible. The technical–economic link is but one example of the types of multidisciplinary links that must be exploited in order to realize a sustainable future.

Multidisciplinary engineering requires multidisciplinary knowledge, for which experts in different fields collaborate. It also requires *systems thinking,* in which it is recognized that apparently unrelated activities often interact significantly.

The need for multidisciplinary collaboration arises because no individual can be an expert in all fields.

The need for systems thinking is evident in engineering design optimization: optimizing individual subcomponents of a system generally will not optimize the larger system. For example, typical automotive design methods minimize fuel consumption and emissions for standard highway and city driving, for typical cycles of acceleration, idling, and cruising. However, standard conditions exist only for a portion of real drivers. The larger system is not optimized. More importantly, optimizing the efficiency of an automobile that will likely spend its working life transporting only one or two people does not optimize the efficiency of our transportation system as a whole. We need to work toward optimizing the larger system. In brief, we need systems thinking.

The Gaia hypothesis (discussed previously) describes the world as a living organism, and is a good metaphor for systems thinking [9]. However, we also need numerical evaluations to bring systems thinking into practice. Life-cycle analysis is an example of a systems-thinking tool that helps us to understand the environment and the effect of humanity on it.

Multidisciplinary engineering helps us to understand how apparently independent systems affect one another. For example, although manufacturing, town planning, and earth science may be independent disciplines, no industry can build a factory without considering the proximity to employees, the distance and the quality of roadways linking the factory to raw materials and markets, the availability of water for manufacturing, and many similar factors. Engineers must seek expert advice from non-engineering disciplines such as environmental studies, political studies, economics, cultural studies, and psychology. New "linked" disciplines, such as industrial ecology, are beginning to meet the need for multidisciplinary engineering knowledge and for approaches better suited to tackling the problem of sustainability. These linked disciplines will continue to grow.

Next steps   As mentioned earlier, Jaccard [28] proposed that we use fossil fuels for many decades to come in order to avoid economic collapse while we search for alternative energy sources. If the required processes can be made to work as well as Jaccard suggests, then we can reduce greenhouse gases and other pollutants, but there is still an urgent need to develop alternative energy sources. Jaccard's book won the prestigious 2006 Donner Prize for the top Canadian book in public policy, and represents a Canadian voice describing how humankind can reach a sustainable future. The announcement by the Canadian federal government in 2008 that carbon capture and sequestration (CCS) will be compulsory by 2012 for new oilsands plants is encouraging [40]. Unfortunately, many other relatively modest and timely proposals provided by Jaccard and others continue to be ignored by governments [41].

There is an old saying: "Where there is a will, there is a way." Engineers can lead the way to a sustainable future, but at the same time, society must have the will and resolve to seek a sustainable future.

21.9   Conclusion

Ethics and justice require humanity to share equally the burdens of climate change and the decline of oil production. Although they are different problems, their main cause is the same: excessive consumption of fossil fuels. To maintain our standard of living, we must drastically reduce our dependence on fossil fuels and rapidly move to other energy sources. Unfortunately, few people are prepared to deny themselves what they see as their entitled standard of living. Society must develop the strength to change these outdated customs and habits, and to pass the legislation needed to achieve sustainability.

As engineers, we have a key role in the move toward sustainability. We must ensure that our projects and activities achieve maximum efficiency in terms of the energy, materials, and labour consumed. However, efficiency improvements alone will not be enough to avoid the crises of climate change and the peak of oil production. We must apply sustainability concepts to our everyday work, and we must try to lead within our spheres of influence. We need not be martyrs in the fight for sustainability, but for ethical reasons, we should avoid projects or activities that clearly lead to an unsustainable future.

Research, innovation, and new ideas are essential. We must work together to imagine, innovate, invent, and create a sustainable future. Fortunately, today's engineering students live in a world of unparalleled opportunity. The information revolution is still in its early stages, and it will continue to multiply our strength and intelligence.

Sustainability is the greatest challenge ever presented to engineers. If engineers can help society evolve to a high but sustainable standard of living without the predicted problems of unrest, deprivation, and suffering, it will be our greatest accomplishment.

21.10   Further study

1.   Choose the best answer for each of the following questions.

(a) In 1987, the Brundtland Commission wrote a report that

    i.   was made into a very popular motion picture.

    ii.   defined sustainable development in a meaningful way.

    iii.   was skeptical about the need for sustainable development.

    iv.   defined sustainability as an economic golden rule.

(b) Sustainable development is defined as development that

    i.   is achieved by voluntary effort.

    ii.   maximizes progress, growth, expansion, and consumption.

    iii.   does not reduce the ability of future generations to meet their needs.

    iv.   makes a continuously increasing profit.

(c) Global warming is caused by greenhouse gases (GHGs), which trap heat radiated away from the earth's surface. The GHGs are emitted by the

    i.   release of chlorofluorocarbon refrigeration gases.

    ii.   burning of coal, oil, and wood for heating homes.

    iii.   combustion of gasoline in automobiles.

    iv.   all of the above.

(d) Global warming is important because it will

    i.   lead to climate change, causing storms and floods.

    ii.   raise global temperatures and cause the extinction of many species.

    iii.   lead to climate change, causing droughts, agricultural losses and starvation.

    iv.   all of the above

(e) The debate over global warming effectively ended

    i. in 2001, when Bjorn Lomborg wrote a book about global warming.

    ii. in 2006, when Al Gore made a movie film about global warming.

    iii. in 2007, when the IPCC linked global warming to wasteful energy use.

    iv. in 2007, when Canada set intensity-based emission standards.

(f) The efficiency paradox refers to the observation, in the last few decades, that

    i. improved efficiency always reduces total energy consumption.

    ii. increased consumer demand always exceeds efficiency improvements.

    iii. bigger vehicles are always more efficient.

    iv. designers will not improve engine efficiency without government regulation.

(g) A constant, positive, energy-consumption growth rate

    i. is necessary for sustainability.

    ii. increases energy consumption linearly.

    iii. increases energy consumption exponentially.

    iv. is necessary to maintain a healthy economy.

(h) To become fully developed, a new renewable-energy technology will require at least

    i. 20 years.

    ii. 40 years.

    iii. 50 years.

    iv. 60 years.

(i) The best way for an engineer to work towards a sustainable future is to

    i. refuse to work for an oil company.

    ii. assist clients and employers to achieve whatever they want.

    iii. satisfy environmental guidelines and government regulations.

    iv. assist clients, employers, and society to understand the full impact of their engineering projects on the environment.

(j) Life cycle analysis (LCA)

    i. extends a product's useful life.

    ii. calculates a product's life expectancy using fatigue analysis.

    iii. evaluates the total environmental impact of a product from cradle to grave.

    iv. evaluates the total impact of the environment on the product.

2. Develop a life-cycle analysis (LCA) for a car. First, decide on the metrics you will consider, such as CO_2 emissions and energy consumption. Second, consider the LCA boundary. For example, will you or will you not consider the CO_2 emissions from the mine that extracted the iron ore used to produce the steel in the car? Third, draw a diagram showing the connections among all the elements that go into producing a car and contributing to your LCA metrics calculations. Next, attempt to quantify

an LCA-determined environmental impact from a car by using the *Greenhouse Gases, Regulated Emissions, and Energy Use in Transportation* (GREET) software available from Argonne National Laboratories at <http://www.transportation.anl.gov/modeling_simulation/GREET/>. Finally, discuss the environment impacts associated with a car.

3. Translate the energy consumption of Canadians into practical terms that would be better appreciated by the public. Today, Canadians consume energy at a rate of 13.5 kW per person. Convert this rate into units of 60 W light bulbs per person or 4.5-horsepower lawnmowers per person. Document how an average Canadian consumes at this rate. What can be done to reduce Canadian energy consumption?

21.11   References

[1] World Commission On Environment and Development, *Our Common Future*. New York: Oxford, 1987. Brundtland Report, <http://www.un-documents.net/wced-ocf.htm> (March 9, 2008).

[2] P. Hawken, *The Ecology of Commerce: A Declaration of Sustainability*. New York: HarperBusiness, 1993.

[3] Ministry of the Environment, *The State of Canada's Environment*. Ottawa: Environment Canada, 1996. <http://www.ec.gc.ca/soer-ree/English/SOER/1996report/Doc/1-1.cfm> (March 9, 2008).

[4] L. L. Strauss, "Speech to the National Association of Science Writers, New York City," *New York Times, September 17*, 1954.

[5] R. Carson, *Silent Spring*. Boston: Houghton Mifflin, 1962.

[6] P. R. Ehrlich, *The Population Bomb*. New York: Ballantine Books, 1968.

[7] P. R. Ehrlich and A. Ehrlich, *The Population Explosion*. Toronto: Simon and Schuster, 1990.

[8] D. H. Meadows, D. E. Meadows, J. Randers, and W. W. Behrens, *Limits to Growth: A Report for the Club of Rome's Project on the Predicament of Mankind*. New York: Universe Books, 1972.

[9] J. Lovelock, *Gaia: A New Look at Life on Earth*. New York: Oxford, 1987.

[10] United Nations Environment Programme (UNEP), *Handbook for the Montreal Protocol on Substances that Deplete the Ozone Layer*. Nairobi, Kenya: UNEP, seventh ed., 2006.

[11] A. Freeman, "Canada gets its way on climate change," *The Globe and Mail, November 24*, 2007.

[12] *United Nations Framework Convention on Climate Change (UNFCCC)*. Bonn, Germany, 2008. <http://unfccc.int> (March 9, 2008).

[13] Environment Canada, *Clean Air Act*. Ottawa, ON: Government of Canada, 2006. <http://www.ec.gc.ca/cleanair-airpur/Clean_Air_Act-WS1CA709C8-1_En.htm> (March 9, 2008).

[14] International Panel for Climate Change, "Summary for policymakers," in *Climate Change 2007: The Physical Science Basis. Contribution of Working Group I to the Fourth Assessment Report of the Intergovernmental Panel on Climate Change* (S. Solomon, D. Qin, M. Manning, Z. Chen, M. Marquis, K. B. Averyt, M. Tignor, and H. L. Miller, eds.), (New York), Cambridge University Press, 2007. <http://www.ipcc.ch/ipccreports/ar4-wg1.htm> (March 9, 2008).

[15] R. W. Jackson and J. M. Jackson, *Environmental Science: The Natural Environment and Human Impact.* Harlow, U.K.: Longman, 1996. page 317.

[16] B. Lomborg, *The Skeptical Environmentalist: Measuring the Real State of the World.* New York: Cambridge University Press, 2001.

[17] S. Schneider, J. Holdren, J. Bongaarts, and T. Lovejoy, *Misleading Math about the Earth.* New York: Scientific American, 2002.

[18] A. Gore, *Inconvenient Truth: The Planetary Emergency of Global Warming and What We Can Do About It.* Emmaus, Pa.: Rodale Press, 2006.

[19] International Panel for Climate Change, "Summary for policymakers," in *Climate Change 2007: Synthesis Report* (B. P. Jallow, L. Kajfež-Bogataj, R. Bojariu, D. Hawkins, S. Diaz, H. Lee, A. Allali, I. Elgizouli, D. Wratt, O. Hohmeyer, D. Griggs, and N. Leary, eds.), New York: Cambridge University Press, 2007. <http://www.ipcc.ch/ipccreports/ar4-syr.htm> (March 9, 2008).

[20] K. Knauer, ed., *Global Warming.* New York: Time Books, 2007.

[21] G. Monbiot, *Heat: How to Stop the Planet from Burning.* Toronto: Random House Anchor Canada, 2007.

[22] International Programs Center (IPC), Peter Johnson, *Historical Estimates of World Population.* Washington, D.C.: United States Census Bureau (USCB), 2008. <http://www.census.gov/ipc/www/worldhis.html> (March 9, 2008).

[23] BP International, *Statistical Review of World Energy 2007.* London: BP p.l.c., 2007. <http://www.bp.com/productlanding.do?categoryId=6848&contentId=7033471> (March 9, 2008).

[24] J. D. Hughes, "The energy sustainability dilemma: Powering the future in a finite world," in *Proceedings of Plug-in Hybrid Electric Vehicle (PHEV) Conference: Where the Grid Meets the Road*, (Winnipeg, Man), 2007. Keynote address, <http://www.pluginhighway.ca/proceedings.php> (March 9, 2008).

[25] M. K. Hubbert, *Nuclear Energy and Fossil Fuels.* San Antonio, TX: American Petroleum Institute Spring Meeting, 1956.

[26] C. Campbell, "Oil depletion–update through 2001," 2001. <http://www.hubbertpeak.com/campbell/update2002.htm>, (March 9, 2008).

[27] C. Campbell, 2006. Personal communication reported in reference [24].

[28] M. Jaccard, *Sustainable Fossil Fuels: The Unusual Suspect in the Quest for Clean and Enduring Energy.* New York: Cambridge University Press, 2006.

[29] The Association of Professional Engineers, Geologists and Geophysicists of Alberta, Edmonton, AB, *Guideline for Environmental Practice*, 2004. <http://www.apegga.org/pdf/Guidelines/18.pdf> (March 9, 2008).

[30] Canadian Engineering Qualifications Board, *National Guideline on Environment and Sustainability*. Ottawa: Canadian Council of Professional Engineers, 2006. <http://engineerscanada.ca/e/files/guideline_enviro_with.pdf> (March 9, 2008).

[31] L. Thorstad, C. Gale, H. Harris, J. Haythorne, and P. Jones, *Guidelines for Sustainablility*. Vancouver B.C.: The Associaton of Professional Engineers and Geoscientists of British Columbia, 1995. <http://www.apeg.bc.ca/ppractice/documents/ppguidelines/sustainabilityguidelines.pdf> (March 9, 2008).

[32] Professional Engineers Ontario, "Environmental guidelines for the practice of professional engineering in Ontario," in *Guideline of Professional Practice*, Toronto: Professional Engineers Ontario, 1988. Revised 1998, <http://www.peo.on.ca/Guidelines/Professional_practice_rev.pdf> (March 9, 2008).

[33] D. Burningham, 2007. Personal communication to G. C. Andrews.

[34] W. McDonough and M. Braungart, *Cradle to Cradle: Remaking the Way We Make Things*. New York: North Point Press, 2002.

[35] Union of Concerned Scientists, "Sales of SUVs, minivans, and pickups surpass cars for first time," 2008. <http://www.ucsusa.org/news/press_release/sales-of-suvs-minivans-and-pickups-surpass-cars-for-first-time.html> (March 9, 2008).

[36] S. McCarthy, "Dim prospects that 'energy efficient' will pay off: CIBC," *The Globe and Mail*, November 27, 2007.

[37] K. McGuffie and A. Henderson-Sellers, "Forty years of numerical climate modelling," *International Journal of Climatology*, vol. 21, pp. 1067–1109, 2001.

[38] M. Raynolds, M. D. Checkel, and R. A. Fraser, "The relative mass-energy-economic (RMEE) method for system boundary selection—a means to systematically and quantitatively select LCA boundaries," *Int. J. of Life Cycle Assessment*, vol. 5, no. 1, pp. 37–46, 2000.

[39] S. L. Jackson, *The ISO 14001 Implementation Guide: Creating an Integrated Management System*. Toronto, ON: John Wiley & Sons, Inc., 1997.

[40] D. Ebner and N. Scott, "Awash in cash, oil patch braces for changes," *The Globe and Mail*, March 11, 2008.

[41] J. Simpson, "Who will rid us of this troublesome GST cut?," *The Globe and Mail*, December 27, 2007.

Appendix: Answers to Quick Quiz and selected Further Study questions

Chapter 1

1. (a) iv, (b) iv, (c) i, (d) ii, (e) i, (f) ii, (g) ii, (h) ii, (i) iv, (j) i.

Chapter 2

1. (a) iv, (b) i, (c) iii, (d) iv, (e) ii, (f) i, (g) iv, (h) i, (i) ii. (j) iii.

Chapter 3

1. (a) iii, (b) i, (c) iii, (d) ii, (e) iv, (f) iii, (g) i, (h) v, (i) iv, (j) iii.

Chapter 4

1. (a) ii, (b) v, (c) i, (d) ii, (e) iv, (f) ii, (g) ii, (h) ii, (i) iv, (j) iv.

Chapter 5

1. (a) iii, (b) ii, (c) i, (d) iii, (e) iii, (f) iii, (g) iv, (h) iii, (i) ii, (j) iv.

Chapter 6

1. (a) i, (b) ii, (c) i, (d) iv, (e) i, (f) ii, (g) ii, (h) i, (i) iv, (j) iii.

Chapter 7

1. (a) ii, (b) i, (c) i, (d) iv, (e) v, (f) i, (g) ii, (h) iv, (i) i, (j) iv.

2. **A. Vocabulary:** (a) all right; (b) escaped; (c) censure; (d) composed; (e) continuous; (f) could not; (g) implied; (h) should have; (i) fewer; (j) likely; (k) himself; (l) lend; (m) memento; (n) could scarcely; (o) non-flammable.

B. Grammar and usage: (a) Americans speak differently. (b) Correct. (c) ... really big wrench ... (d) My sister is always bugging me. (e) The four students were quarrelling among themselves. (f) What do you think about those Blue Jays? (g) Driving down the road, we saw the poles go past us quickly. (h) Check your spelling thoroughly. (i) Correct. (j) Torquing a nut is the act of tightening it with a wrench.

C. Punctuation: (a) Correct. (b) ... lost its oil ... (c) Correct. (d) The crowd was mad; it strung him up. (e) Correct.

Chapter 8

1. (a) i, (b) iii, (c) i, (d) i, (e) i, (f) ii, (g) i, (h) ii, (i) ii, (j) i.

Chapter 9

1. (a) i, (b) iii, (c) iii, (d) ii, (e) ii, (f) ii, (g) iii, (h) i, (i) i, (j) iii.

2. (a) By inspection of the graph, $b = -2$, $m = 1.05$. (b) $a = -b/m$. (c) *Hint:* Draw a line tangent to the curve at the left end, which is the point (x_0, y_0). Let the vertical difference between the curve and the tangent line be given by $m_2(x - x_0)^2$. The curve then has the formula $y = y_0 + m_1(x - x_0) + m_2(x - x_o)^2$, in which all constants are known. Expand the formula to obtain b.

5. Taking the logarithm as for Equation (9.10) gives

$$\log_{10} n(t) = \log_{10}(n(1971) \times 2^{(t-1971)/T})$$
$$= \log_{10}(n(1971) \times 2^{-1971/T} \times 2^{t/T})$$
$$= \log_{10}(n(1971) \times 2^{-1971/T} \times 10^{(\log_{10} 2)\,t/T})$$
$$= \log_{10}(n(1971)2^{-1971/T}) + (\log_{10} 2)\,t/T,$$

which is a straight line with slope $(\log_{10} 2)/T$ when graphed using log-linear scales as in Figure 9.5. The coordinates for the 4004 processor are approximately (1971, 2300), and the coordinates for the Itanium 2 are approximately (2004, 4×10^8). Therefore, the slope is approximately

$$\frac{\log_{10} 2}{T} = \frac{\log_{10}(4 \times 10^8) - \log_{10}(2300)}{2004 - 1971}.$$

Solve for T.

7. Provided $f(\cdot)$ has an inverse function $f^{-1}(\cdot)$, then, as done for the exponential, let the vertical scale marks be drawn at positions proportional to this inverse function. Then with this choice of scales, points x, $f(x)$ on linear coordinates become x, $f^{-1}(f(x))$, producing a straight line with slope equal to 1.

Chapter 10

1. (a) i, (b) i, (c) v, (d) ii, (e) i, (f) iv, (g) iii, (h) i, (i) iv, (j) iv.

2. From the given formula, $I = My/\sigma$; therefore, the dimensions of I are

$$\frac{[F][L] \times [L]}{[F]/[L]^2} = [L]^4,$$

that is, the dimensions of I are length to the fourth power. This parameter is a shape factor that determines how beams of different cross-section resist bending. It is called the area moment of inertia of the beam section and is calculated as $\int y^2 \, dA$, where the integral is over the area of the beam cross-section.

3. From Newton's second law, $f = ma$, the mass is $m = (100\,\text{lb})/(32.17\,\text{ft/sec}^2) = 3.11$ slug. Mass is unaffected by gravity and is identical on the Moon, where its weight will be

$$3.11\,\text{slug} \times 1.62\frac{\text{m}}{s^2} \times \frac{1\,\text{ft}}{0.3048\,\text{m}} = 16.5\,\text{lb}.$$

4. The units of miles per gallon are length/volume $=$ length^{-2}, whereas the units of litres per 100 km are volume/length $=$ length^2, so to convert a figure given in miles per gallon, we invert it and multiply by

$$1\frac{\text{gallon}}{\text{mile}} \times \frac{1\,\text{mile}}{1.609\,\text{km}} \times \frac{3.785 \times 10^{-3}\,\text{m}^3}{1\,\text{gallon}}$$
$$\times \frac{1\,\text{litre}}{10^{-3}\,\text{m}^3} \times \frac{100\,\text{km}}{100\,\text{km}} = \frac{378.5\,\text{litre}}{1.609 \times 100\,\text{km}}$$
$$= 235.2\frac{\text{litre}}{100\,\text{km}}.$$

5. Energy is power $\times$ time, so the energy converted to heat is

$$(10^2 \times 200)\,\text{W} \times 3\,\text{min} \times \frac{60\,\text{s}}{1\,\text{min}} = 3.6\,\text{MJ}$$
$$= 3.6 \times 10^6\,\text{N} \cdot \text{m} \times \frac{1\,\text{lb}}{4.448\,\text{N}} \times \frac{1\,\text{ft}}{0.3048\,\text{m}}$$
$$= 2.66 \times 10^6\,\text{ft} \cdot \text{lb}.$$

6. The dimensional equation is

$$(\text{dimensions of } T) = (\text{dimensions of } \ell)^a$$
$$\times (\text{dimensions of } g)^b (\text{dimensions of } m)^c,$$

or, substituting,

$$[T] = [L]^a \times ([L]/[T]^2)^b \times [M]^c$$

from which we see that $c = 0$ to eliminate the dimension of mass [M], $b = -0.5$ to make [T] a factor, and $a = -b$ to eliminate [L]. Thus the period T is proportional to $\ell^{0.5} g^{-0.5} m^0 = \sqrt{\ell/g}$. In fact, from a detailed analysis, it turns out that $T = 2\pi\sqrt{\ell/g}$.

Chapter 11

1. (a) iv, (b) ii, (c) i, (d) iii, (e) iii, (f) iv, (g) iii, (h) ii, (i) iii, (j) i.

6. With the operands expressed to the same power of 10, the computation is as follows, where the trailing zeros in 3100 ($= 0.3100 \times 10^4$) are shown as possibly not significant.

2.840 2	$\times 10^4$
1.30	$\times 10^4$
$-0.31(00?)$	$\times 10^4$
0.003 289 734	$\times 10^4$
3.833 489 734	$\times 10^4$ (computed exactly)
3.83	$\times 10^4$ (rounded)

The trailing zeros in 0.3100×10^4 are less significant than the trailing zero of 1.30×10^4 and are dropped. The fourth operand has seven significant digits and is the most precise but contributes nothing to the rounded result.

Chapter 12

1. (a) i, (b) i, (c) ii, (d) iv, (e) iii, (f) i, (g) i, (h) ii, (i) i, (j) iv.

4. The solutions are as follows for the given measured quantities:

(a) Using the nominal measurement values, the formulas evaluate to $d = 73.7$, $e = 3.57$, $f = 1.48 \times 10^3$, $g = 28.1$.

(b) The functions are all monotonic with respect to their arguments, so the extreme values are found at extreme values of the arguments as

$$72.1 \le d \le 75.2, \quad 3.44 \le e \le 3.71$$
$$1.44 \times 10^3 \le f \le 1.53 \times 10^3$$
$$28.0 \le g \le 28.1$$

(c) The approximate error of each computed quantity is given by the formula $|S_x||\Delta x| + |S_y||\Delta y| + |S_z||\Delta z|$ where, for each function, the formulas for the partial derivatives with respect to x, y, and z are evaluated at the measured values to give the sensitivity constants S_x, S_y, S_z. The symbol $|_0$ associated with a formula indicates that the formula is to be evaluated at the measured values of its arguments.

$$d: S_x = \left.\frac{\partial d}{\partial x}\right|_0 = 1, \quad S_y = \left.\frac{\partial d}{\partial y}\right|_0 = 2,$$

$$S_z = \left.\frac{\partial d}{\partial z}\right|_0 = 3, \quad |\Delta d| \simeq 1.56$$

$$e: S_x = \left. yz^{-1}\right|_0 = 0.566, \quad S_y = \left. xz^{-1}\right|_0 = 0.387,$$

$$S_z = \left. xy(-1)z^{-2}\right|_0 = -0.219, \quad |\Delta e| \simeq 0.136$$

$$f: S_x = \left. 2xyz^{1/2}\right|_0 = 470, \quad S_y = \left. x^2 z^{1/2}\right|_0 = 161,$$

$$S_z = \left. x^2 yz^{-1/2}/2\right|_0 = 45.5, \quad |\Delta f| \simeq 43.2$$

$$g: S_x = \left. z^{-1}\right|_0 = 0.0613, \quad S_y = 3,$$

$$S_z = \left. x(-1)z^{-2}\right|_0 = -0.0237, \quad |\Delta g| \simeq 0.0443$$

(d) The approximate error of each computed quantity is given by $\sqrt{S_x^2 \Delta x^2 + S_y^2 \Delta y^2 + S_z^2 \Delta z^2}$, where the sensitivities for each function were calculated in the previous part: $|\Delta d| \simeq 1.50$, $|\Delta e| \simeq 0.112$, $|\Delta f| \simeq 29.6$, $|\Delta g| \simeq 0.0324$.

5. The problem can be solved as follows.

(a) The acceleration due to gravity $g = 9.72\ \text{m/s}^2$.

(b) The relative errors are as follows: (i) exact range: $-3.56\,\%$, $+3.76\,\%$; (ii) linear approximation: $\pm 3.66\,\%$; (iii) estimated: $\pm 3.51\,\%$.

6. The diagram below shows the measured area xy and the area $(x+\Delta x) \times (y+\Delta y)$ implied by the positive deviations Δx and Δy. The difference between the change in area and the linear estimate of the change in area is the area $\Delta x\,\Delta y$, since the linear estimate is the quantity $x\,\Delta y + y\,\Delta x$.

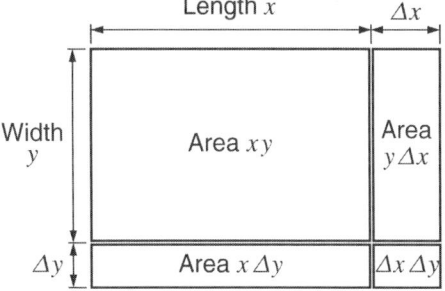

Chapter 13

1. (a) ii, (b) vi, (c) i, (d) i, (e) ii, (f) i, (g) iii, (h) ii, (i) ii, (j) i.

2. Mean $\bar{x} = 69.6$, median $= 73$, mode $= 74$, range $= 65$, $s^2 = 213$, $s = 14.6$, SIQR $= 6.5$

3. *Hint:* The sum of the observation values in a given group (bin) is equal to the number of observations in that group multiplied by the observation value of that group.

Chapter 14

1. (a) i, (b) iii, (c) i, (d) iii, (e) iv, (f) iii, (g) i, (h) iii, (i) iii, (j) ii.

2. $(a, r) = (0, 0), (1, 1), (-1, -1), (0, 1)$

3. A total of approximately 40 measurements are needed, assuming that only independent random errors affect the pressure measurements. Therefore, approximately 30 additional measurements are needed. The number of additional measurements needed is approximate because the uncertainty has been estimated using the sample standard deviation, which will vary from sample data set to sample data set.

Chapter 15

1. (a) ii, (b) iii, (c) iv, (d) i, (e) iii, (f) iv, (g) ii, (h) iv, (i) i, (j) iv.

Chapter 16

1. (a) iii, (b) i, (c) i, (d) i, (e) ii, (f) ii, (g) ii, (h) ii, (i) i, (j) iii.

Chapter 17

1. (a) iv, (b) iv, (c) iii, (d) ii, (e) i, (f) i, (g) iv, (h) ii, (i) i, (j) i.

2. (a) The function of a new pneumatic pump can be protected by patent. (b) Shoe design can be registered as an industrial design, if it is original. (c) A logo that identifies a product (shoes) may be registered as a trademark. (d) A slogan cannot be protected; however, a videotaped commercial containing the slogan is protected under copyright. (e) A new process for curing rubber can be protected under patent; it might also be possible to keep it a trade secret. (f) The circuit layout of the microcomputer chip can be protected as integrated circuit topography.

4. (a) 1173; (b) yes; (c) 1933; (d) yes.

6. (a) Non-disclosure agreements for proprietary information, fair remuneration, reference checks, computer access controls. (b) Trail of poor references; bad reputation on ethics and confidentiality.

7. (a) trademark; (b) copyright; (c) cannot be protected; (d) industrial design; (e) patent.

Chapter 18

1. (a) iv, (b) ii, (c) ii, (d) ii, (e) iii, (f) ii, (g) iii, (h) iv, (i) iv, (j) ii.

Chapter 19

1. (a) iii, (b) iv, (c) i, (d) ii, (e) ii, (f) i, (g) v, (h) iii, (i) ii, (j) i.

Chapter 20

1. (a) i, (b) ii, (c) iv, (d) iii, (e) iii, (f) iii, (g) iii, (h) iv, (i) i, (j) ii.

2. Part of the fault tree is shown below.

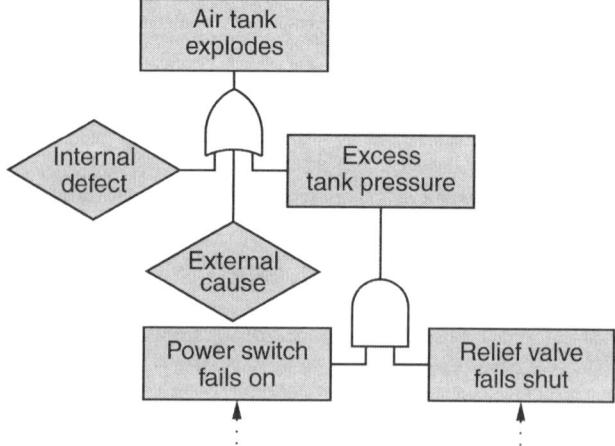

Chapter 21

1. (a) ii, (b) iii, (c) iv, (d) iv, (e) iii, (f) ii, (g) iii, (h) ii, (i) iv, (j) iii.

Index